Mineralogical and Geochemical Characterization of Geological Materials

Mineralogical and Geochemical Characterization of Geological Materials

Guest Editors

Teresa Pereira da Silva

Daniel P. S. De Oliveira

João Pedro Veiga

MDPI

Basel • Beijing • Wuhan • Barcelona • Belgrade • Novi Sad • Cluj • Manchester

Guest Editors

Teresa Pereira da Silva
Mineral Resources and Geophysics Research Unit
LNEG (National Laboratory for Energy and Geology)
Amadora
Portugal

Daniel P. S. De Oliveira
Mineral Resources and Geophysics Research Unit
LNEG (National Laboratory for Energy and Geology)
Amadora
Portugal

João Pedro Veiga
Department of Conservation and Restoration &
CENIMAT/i3N
NOVA School of Science and Technology
NOVA University of Lisbon
Campus de Caparica
Caparica
Portugal

Editorial Office

MDPI AG
Grosspeteranlage 5
4052 Basel, Switzerland

This is a reprint of the Special Issue, published open access by the journal *Minerals* (ISSN 2075-163X), freely accessible at: https://www.mdpi.com/journal/minerals/special_issues/MGCGM.

For citation purposes, cite each article independently as indicated on the article page online and as indicated below:

Lastname, A.A.; Lastname, B.B. Article Title. *Journal Name* **Year**, *Volume Number*, Page Range.

ISBN 978-3-7258-4155-4 (Hbk)
ISBN 978-3-7258-4156-1 (PDF)
https://doi.org/10.3390/books978-3-7258-4156-1

Contents

minerals

MDPI

Article

Artificial Thermal Quenching and Salt Crystallization Weathering Processes for the Assessment of Long-Term Degradation Characteristics of Some Sedimentary Rocks, Egypt

Marzouk Mohamed Aly Abdelhamid [1,†], B. G. Mousa [1,†], Hassan Waqas [2], Mohamed Abdelghany Elkotb [3,4], Sayed M. Eldin [5], Iqra Munir [6], Rashid Ali [7,*] and Ahmed M. Galal [8,9]

1 Department of Mining and Petroleum Engineering, Faculty of Engineering, Al-Azhar University, Cairo 11884, Egypt
2 National Engineering Research Center for Geographic Information System, School of Geography and Information Engineering, China University of Geosciences, Wuhan 430074, China
3 Mechanical Engineering Department, College of Engineering, King Khalid University, Abha 61421, Saudi Arabia
4 Mechanical Engineering Department, College of Engineering, Kafrelsheikh University, Kafr El-Sheikh 33516, Egypt
5 Center of Research, Faculty of Engineering and Technology, Future University in Egypt, New Cairo 11835, Egypt
6 State Key Laboratory of Information Engineering in Surveying, Mapping, and Remote Sensing (LIESMARS), Wuhan University, Wuhan 430079, China
7 School of Mathematics and Statistics, Central South University, Changsha 410083, China
8 Department of Mechanical Engineering, College of Engineering in Wadi Alddawasir, Prince Sattam bin Abdulaziz University, Wadi Addawaser 11991, Saudi Arabia
9 Production Engineering and Mechanical Design Department, Faculty of Engineering, Mansoura University, P.O. Box 35516, Mansoura 35516, Egypt
* Correspondence: rashidali@csu.edu.cn or rashidali0887@gmail.com
† These authors contributed equally to this manuscript.

Abstract: This research aims at investigating the deterioration of limestone rocks due to the influences of thermal quenching and salt crystallization weathering tests and predicting their long-term durability. Therefore, six types of limestones were quarried from different provinces of Egypt and subjected to 50 cycles of thermal quenching and 25 cycles of salt crystallization weathering processes. The porosity, Schmidt hammer rebound hardness, ultrasound pulse velocity, Brazilian tensile strength, and uniaxial compression strength were determined before and after weathering processes. In addition, the mathematical decay function model was developed to evaluate the degradation rate of samples against weathering processes. Results proved that the cyclic salt crystallization deteriorates the physico-mechanical characteristics of the studied limestone more strongly than the thermal quenching cycles do. The decay constant and half-life indexes obtained here indicate that the degradation rate differs for various limestone specimens under thermal and salt weathering processes. This model also showed that the deterioration rate of the studied rocks was higher during cyclic salt crystallization in comparison with thermal quenching. Therefore, the rock degradation rate and or long-term durability under cyclic thermal and salt processes can be estimated accurately. These results show that the studied limestones can be used as building stones in regions exposed to frequent cyclic thermal and salty weathering conditions for long periods without degradation. However, partial attention should be given to LSG limestone rocks characterized by increased porosity and water absorption characteristics.

Keywords: limestone rocks; thermal quenching; salt crystallization; decay function approach; weathering; long-term degradation

1. Introduction

Natural rocks have been widely used as building and construction stones, especially in historical structures and new engineering applications from the past up to the present times [1–4]. The long-term durability of stones is defined as its resistance to environmental conditions over an extended period of time [5]. Due to the long-term exposure to the atmosphere, rocks are always suffering different recurrent types of weathering processes, resulting in partial or full deterioration [6–9]. Temperature changes are among the major reasons for rock damage in ancient and new engineering structures [10]. All building materials expand when heated and shrink as they cool down. This process is named thermal shock weathering. Thermal quenching is created by sudden changes in temperature and leads to 'catastrophic' damage in the form of cracks, so the strength of rock material is suddenly reduced [5].

Many studies have been carried out for investigating the effect of thermal quenching process on the physico-mechanical characteristics of building stones [11–17]. Yavuz et al. [18] performed thermal quenching cycles on carbonate rocks in the laboratory and established a model for predicting the index properties of damaged rocks depending on the initial property and number of thermal quenching cycles. Hale and Shakoor [19] analyzed the variation in the compression strength for sandstone rocks against heating and cooling cycles. Sousa et al. [20] studied the influence of thermal quenching action on granite rocks and found that the compressional wave velocity of granite decreased after thermal quenching test. Wang et al. [21] experimentally studied the influences of thermal shock and freezing–thawing cycles on red sandstone rocks before and after accelerated tests. They indicated an important decrease in the physico-mechanical properties of red sandstone rocks after the thermal shock and freezing–thawing processes. Wang et al. [22] indicated induced change of physico-mechanical characteristics of red sandstone after 10, 20, 30, and 40 thermal quenching cycles. Gokceoglu et al. [23] developed models for estimating the weathering degree of granitic rocks from some parameters, including porosity, ultrasound wave velocity, and unconfined compression strength. Mutluturk et al. [24] established a decay function model for predicting the decrease in rock integrity against heating and cooling. Similarly, salt crystallization is also the most powerful weathering process, especially in marine and mild environmental conditions [25]. Salt crystallization is also one of the significant reasons for the degradation of rocks in nature and in stones used in construction [26]. Crystallization pressure is the most critical damage mechanism during the salt weathering action, which depends mainly on the pore size and super saturation degree [27]. When the crystallization pressure reaches the tensile strength of the rock, new microcracks are developed and the present ones are deepened and widened; thus, this process leads to rock decay [28].

Different studies have been carried out to determine the rock degradation induced by salt crystallization weathering [29–42]. Angeli et al. [43] investigated the impact of temperature and salt concentration on the salt weathering process of a sedimentary rock with sodium sulfate. Cultrone et al. [44] investigated the petro-physical and durability tests on sedimentary rocks to assess their quality as building materials. Vazquez et al. [45] measured the changes in the weight of sedimentary rocks used as building stones against the salt weathering process. Yavuz and Topal [46] studied the impacts of thermal shock and salt crystallization on different types of marble specimens from Western Anatolia, Turkey. Benavente et al. [47] performed laboratory tests on five different kinds of porous stones, concluding that the mechanical characteristics of rocks have a statistically significant weight in the prediction of salt crystallization process, with a small contribution to water transport and pore structure parameters. Yavuz et al. [48] investigated the durability of green andesite rocks through aging weathering processes such as salt crystallization and freezing–thawing cycles. Ruedrich and Siegesmund [49] studied the physical weathering processes under the crystallization of salt and ice in the pores of sandstone rocks. They proved that these pores have a crucial influence on the behavior of rock damage.

Despite the numerous studies conducted on this topic, few lead to the modeling of the alteration phenomenon. Experimental investigations of rocks are essential for the thermal and salt weathering processes, whilst it is also significant to develop damage models and to estimate the degradation rate. Few damage models have been established for evaluating the freezing–thawing and thermal quenching weathering processes [13,22,24,28,50,51], whilst estimating deterioration models against salt crystallization process have not yet reported and there are almost no damage models that take into consideration the mechanical strength properties to evaluate the rock integrity loss due to the cyclic salt crystallization action.

The aims of this research are (1) to better understand the degradation of the selected limestone rock specimens that occurred by the artificial weathering processes, including thermal quenching and salt crystallization, and correlate it with their durability characteristics; (2) to determine their response to the artificial weathering processes through the assessment of their physico-mechanical characteristics described as porosity, Schmidt hammer rebound hardness, ultrasound pulse velocity, and Brazilian tensile and uniaxial compression strengths; and (3) to statistically assess and estimate the long-term durability (integrity decrease rate) of the mechanical strength parameters for limestones against cyclic thermal quenching and salt crystallization by use of a decay function model. In general, this paper contributes to preliminary design stage, engineering structures, replacement works, and safety or stability evolution of rock applications in mild or humid and salty environments.

2. Materials and Methods

2.1. Materials

In this research, six types of limestone rocks were taken out from different localities in Egypt, including South Sinai, Suez, El-Minia, Qena, Sohag, and Aswan (Figure 1). These rocks are used in construction industries. Large blocks were extracted for each rock type ranging in size from 0.3 m × 0.3 m × 0.35 m to 0.35 m × 0.4 m × 0.4 m. Cylindrical and cubical rock samples were obtained in the laboratory for various experiments tests. The tested rock samples were free of any discontinuities or joints to avoid the impact of anisotropy on the measurements. The studied limestone samples were firstly examined under the polarized optical microscope for petrographical analysis using the Eclipse E600POL, based on Nikon's popular Eclipse E600 microscope, which features a 12-volt, 100-watt internal tungsten halogen illumination system. The microscopic features of the studied limestones are shown in Table 1. The main mineral constituent of all tested specimens is calcite associated with some rare amounts of accessory and or opaque minerals. The chemical analysis was carried out at the Central Laboratories Sector at the Egyptian Mineral Resources Authority, Cairo, Egypt using a ZSX PRIMUS X-ray fluorescence spectrometer (XRF). The ZSX Primus is a 3 or 4 kW WDXRF spectrometer with the thinnest end-window tube (30 microns) and 6 auto-selectable diameters: 35, 30, 20, 10, 1, and 0.5 mm (sample size). The chemical analysis of the studied limestones is presented in Table 2.

Table 1. Petrographic features of the tested limestone rocks.

Rock Type	Code	Locality	Microstructure
Limestone	LSS	South Sinai	Medium-grained calcite texture, oolitic, porous, packstone, cemented, pores filled with rare minerals
Limestone	LSZ	Suez	Fine-grained calcite texture, calcareous massive, slightly porous
Limestone	LEM	El-Minia	Fine-grained calcite texture, slightly hard stone, clastic
Limestone	LQ	Qena	Very fine-grained calcite texture, clastic, compact, marly, hard stone
Limestone	LSG	Sohag	Coarse-grained calcite texture, porous, weak stone
Limestone	LA	Aswan	Medium- to fine-grained calcite texture, semiporous, pack stone

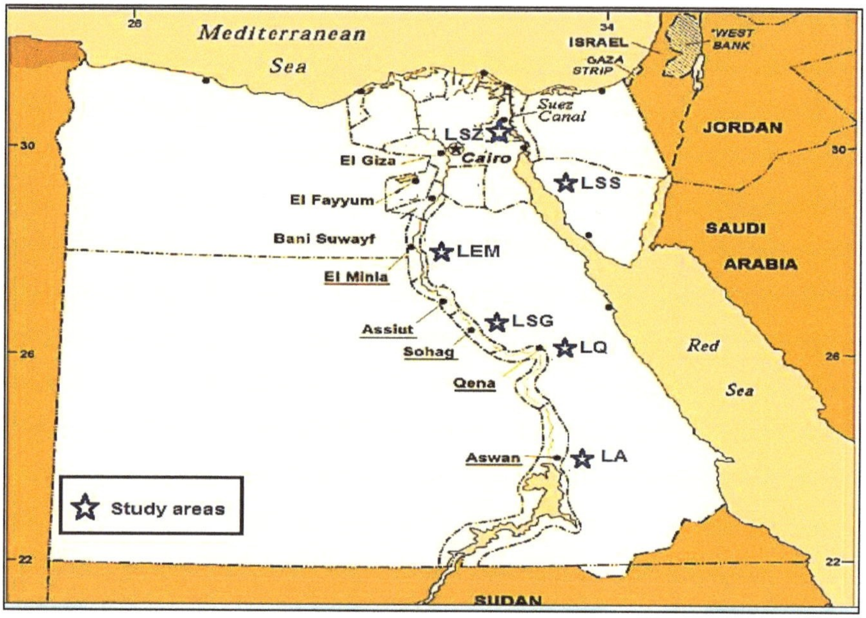

Figure 1. Location map of the studied limestone specimens.

Table 2. XRF analysis of the studied limestones.

Rock type	Locality	CaO	K_2O	MgO	Al_2O_3	Na_2O	Fe_2O_3	SiO_2	LOI	Total
LSS	South-Sinai	55.06	0.05	0.30	0.11	0.09	0.15	2.07	42.09	99.92
LSZ	Suez	53.13	0.07	0.60	0.16	0.05	0.17	2.77	43.20	100.15
LEM	El-Minia	51.87	0.04	0.43	0.09	0.04	0.11	3.28	44.25	100.11
LQ	Qena	50.61	0.02	0.33	0.07	0.01	0.10	4.97	43.88	99.99
LSG	Sohag	56.47	0.09	0.66	0.19	0.06	0.23	1.80	40.54	100.04
LA	Aswan	53.39	0.07	0.61	0.15	0.05	0.20	3.78	41.68	99.93
Absolute error (%)		0.030	0.001	0.006	0.002	0.001	0.004	0.013	0.00	

2.2. Methods

Laboratory experiments were carried out for determining the Physico-mechanical properties including, the bulk density, absorption of water, apparent porosity, the compressional wave velocity, the Schmidt hammer rebound hardness, the unconfined compression strength, and the tensile strength (Brazilian). The physical properties, including the density, absorption by weight, and porosity, were determined for rock samples of size (side 5 cm) according to the test methods carried out by TS EN 1936 [52] and TS EN 13755 [53]. Five specimens were tested for each limestone type and their mean values were obtained.

The mechanical properties including the unconfined compression strength and the Brazilian tensile strength were determined. The unconfined compression strength test was carried out on specimens of cubic form (side 10 cm) in accordance with the test methods proposed by ISRM [54] and TS EN 1926 [55]. Five samples were tested for each limestone type and their average values were obtained.

The tensile strength test (Brazilian) was applied in according with the test methods proposed by ISRM [56]. This test was carried out on diametrically shaped samples of 5.0 cm, and the ratio of height to diameter was 0.5. The number of rock samples used in this test was five, and the mean values were taken.

The Schmidt hammer rebound hardness was performed according to the test procedures suggested by ISRM [56] and Aydin [57]. The Schmidt hammer (N-Type) test was carried out on samples of cubic shape (side 10 cm). Twenty individual impacts were performed on each rock specimen. The highest and lowest rebound values were excluded and then the arithmetic mean was taken.

The compressional wave velocity was determined using the ultrasonic system. This test was carried out at the Housing & Building National Research Center (HBRC), Cairo, Egypt. Cylindrical rock specimens with a diameter of 5.0 cm and the ratio of height to diameter of 2.5 were prepared according to the procedures outlined by Martinez-Martinez et al. [58]. The number of tested rock specimens was five, and mean values were determined.

2.3. *Artificial Weathering Processes*

The thermal quenching weathering test was applied according to EN-14066 [59] recommendations, and salt crystallization was performed in accordance with the general procedures suggested by Rothert et al. [60] for studying the resistance of rock samples to determine the physico-mechanical characteristics and change rates.

2.3.1. Thermal Quenching

A thermal shock weathering test was carried out on specimens of cylindrical shape (5.0 cm). All studied limestone samples were subjected to fifty cycles of thermal quenching weathering. The thermal quenching process consisted of two stages (Figure 2a). Firstly, all limestone specimens were heated in an air-ventilated oven at 70 °C for 18 h. Finally, all specimens were taken out from the oven, and rapidly saturated in distilled water at +20 °C for 6 h. Each thermal shock cycle requires 24 h to be completed. All studied limestone samples were dried to constant weight in an oven at 70 °C before and after cyclic thermal quenching. In this research, two series of limestone rock specimens were prepared; one set of these specimens was used to determine the initial physico-mechanical characteristics before the thermal shock process (unweathered specimens). The second set of specimens was subjected to 10, 20, 30, 40, and 50 cycles, and then the mechanical characteristics, including unconfined compression strength and Brazilian tensile strength, were recorded. Meanwhile, the porosity, the Schmidt hammer rebound hardness, and the ultrasound compressional wave velocity were determined at the end of the 50th cycle.

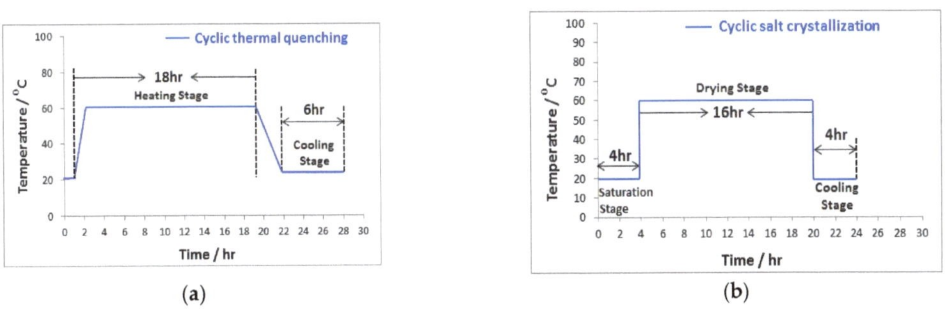

Figure 2. The temperature–time schematic diagram: (**a**) thermal quenching; (**b**) salt crystallization.

2.3.2. Salt Crystallization

The salt crystallization weathering process was performed on cubic samples of 5.0 cm by total saturation in a sodium chloride solution (10%). All rock samples were subjected to twenty five cycles of salt crystallization. The salt weathering process consisted of three stages (Figure 2b). Firstly (saturation), the clean and dried rock specimens were placed in a container and saturated in the solution of sodium chloride for a period of 4 h. Secondly (drying), the specimens were taken out from the solution and dried in an air-ventilated oven at 60 °C for 16 h. Thirdly (cooling), the specimens were cooled at room conditions of

20 °C for a period of 4 h. The cycle of the salt crystallization requires 24 h to be completed. In this process, the uniaxial compression strength and the Brazilian tensile strength were carried out at 5, 10, 15, 20 and 25 cycles, whereas the porosity, the Schmidt hammer rebound hardness, and the ultrasound compressional wave velocity were obtained at 25 cycles.

3. Results and Discussions

As studied by many researchers, the long-term deterioration of rocks due to the weathering processes is a function of their physico-mechanical characteristic. Consequently, determination of these properties is significant for the assessment of the stone durability. The mean values of the physico-mechanical properties and standard deviations of the studied limestone rocks before the accelerated weathering processes (fresh samples) are presented in Table 3. This table shows that the average values of dry densities for studied rocks range between 2.44 for the LQ limestone samples and 2.63 for the LSG specimens. The mean values of water absorption of these limestones range between 1.56% (LQ specimens) and 5.5% (LSS samples). The average values of porosity range between 6.89% (LQ samples) and 9.82% (LSS samples). The mechanical strength tests, including uniaxial compression strength and Brazilian tensile strength, were investigated. Table 3 displays the mean values of mechanical strength characteristics of studied rocks. This table indicates the highest uniaxial compression strength of 73.77 MPa for LQ limestone samples in comparison with the lowest (38.08 MPa) for LSS specimens. Similarly, the highest tensile strength of 5.83 MPa was recorded by the LQ limestone samples and the lowest of 3.19 MPa regarding the LSS specimens. A nondestructive test including Schmidt hammer rebound hardness and ultrasonic wave velocity was obtained. The average values of Schmidt rebound hardness range between 30.5 for the LSS limestone samples and 40.7 for the LQ specimens. Moreover, the mean values of ultrasound compressional wave velocity of these rocks range between 3.33 Km/s (LSS limestone samples), and 5.02 Km/s for LQ samples (Table 3). Rock strength is influenced by its pores and voids [61]; hence, permeability, which simplifies the presence of water in the stone, is strongly affected by the value of the porosity [62]. As rock specimens in the current research have porosity <10%, their permeability behavior is determined by their porosity. Consequently, the porosity value is a critical physical parameter that influences the rock characteristics and controls its durability [63]. It is necessary to better understand the correlation between the porosity and the physico-mechanical characteristics of studied rocks. Many studies investigate the relationships between porosity and rock properties. Tuğrul [64] and Diamantis et al. [65] determined relationships between porosity and unconfined compressive strength of rocks. Figure 3 shows the correlation between the porosity and physico-mechanical properties of the studied limestones. Figure 3b–e show an inverse strong linear relationship between porosity and physical and mechanical properties. Therefore, the physico-mechanical characteristics of the tested limestone specimens were influenced by their porosity. However, no important correlation was found between porosity and dry density (R^2 = 0.0316) (Figure 3a). Based on these results, the physico-mechanical properties of the studied limestone rocks decrease with increased porosity. Figure 4 displays a three-dimensional view of the effect of porosity (n) on the Schmidt rebound hardness (SCH), ultrasound pulse velocity (UPV), tensile strength (TS), and uniaxial compression strength (UCS) for the tested limestone rocks.

Table 3. Average values of the physico-mechanical properties and standard deviations of fresh specimens.

Rock Type	ρ_d (g/cm^3)		W (%)		n (%)		SCH		UPV (Km/s)		Ts (MPa)		UCS (MPa)	
	Mean	SD	Mean	SD	Mean	SD	Mean	SD	Mean	SD	Mean	SD	Mean	SD
LSS	2.46	0.08	5.50	0.64	9.82	0.95	30.5	1.65	3.33	0.39	3.19	0.37	38.08	1.74
LSZ	2.61	0.11	2.26	0.41	7.99	1.17	36.9	0.99	4.67	0.65	5.15	0.62	54.95	1.71
LEM	2.49	0.07	1.61	0.44	7.47	0.83	38.3	1.06	4.81	0.44	5.42	0.72	66.90	1.75

Table 3. *Cont.*

Rock Type	ρ_d (g/cm^3)		W (%)		n (%)		SCH		UPV (Km/s)		Ts (MPa)		UCS (MPa)	
	Mean	SD	Mean	SD	Mean	SD	Mean	SD	Mean	SD	Mean	SD	Mean	SD
LQ	2.44	0.09	1.56	0.38	6.89	0.90	40.7	1.41	5.02	0.53	5.83	0.99	73.77	1.65
LSG	2.63	0.10	4.80	0.40	9.33	0.72	33.8	1.18	3.78	0.88	3.99	0.65	41.05	2.36
LA	2.52	0.06	4.30	0.52	8.80	0.91	35.4	1.13	4.02	0.75	4.19	0.60	45.97	2.02
No of tested specimens	5		5		5		5		5		5		5	

(ρ_b) Dry density, (W) water absorption, (n) porosity, (SCH) Schmidt hammer rebound hardness, (UPV) ultrasound compressional pulse velocity, (Ts) Tensile strength, and (UCS) uniaxial compression strength. LSS (Limestone—South-Sinai), LSZ (Limestone—Suez), LEM (Limestone—El-Minia), LQ (Limestone—Qena), LSG (Limestone—Sohag), and LA (Limestone—Aswan).

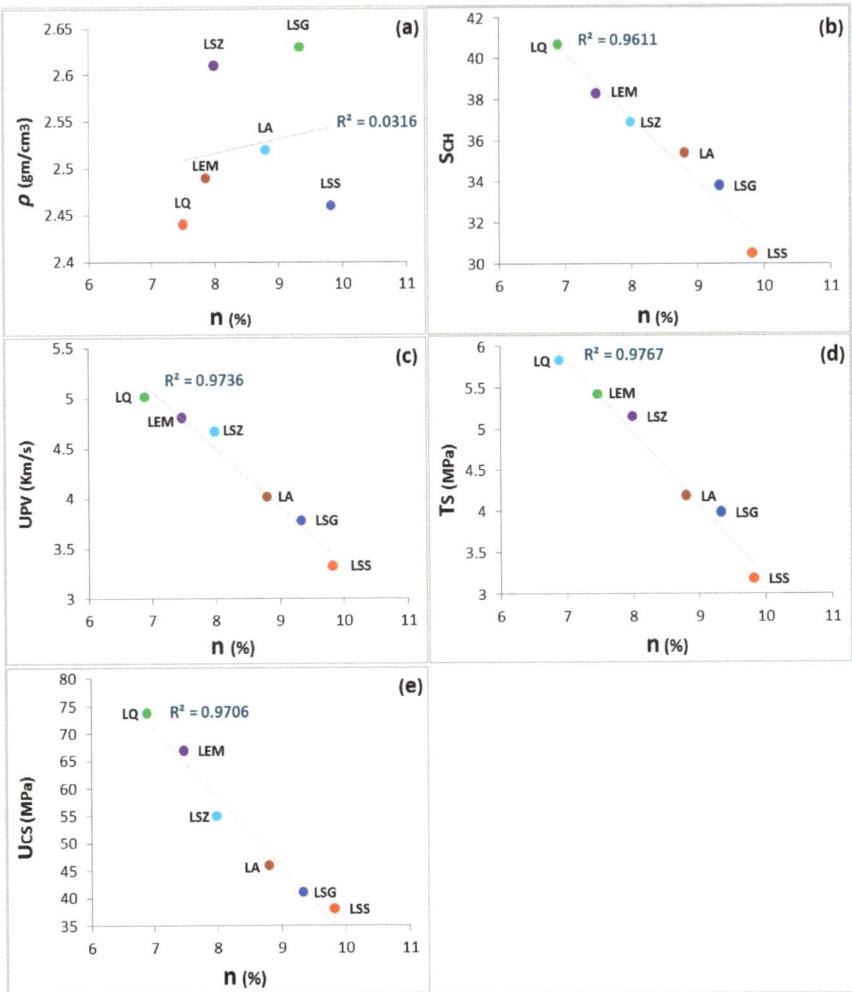

Figure 3. Relationship between porosity, n (%) and (**a**) desity, ρ (gm/cm^3), (**b**) Shmidt hammer rebound, SCH, (**c**) ultra sound pulse velocity, UPV (Km/s), (**d**) tensile strength, Ts (MPa), and (**e**) uniaxial compression strength, UCS (MPa).

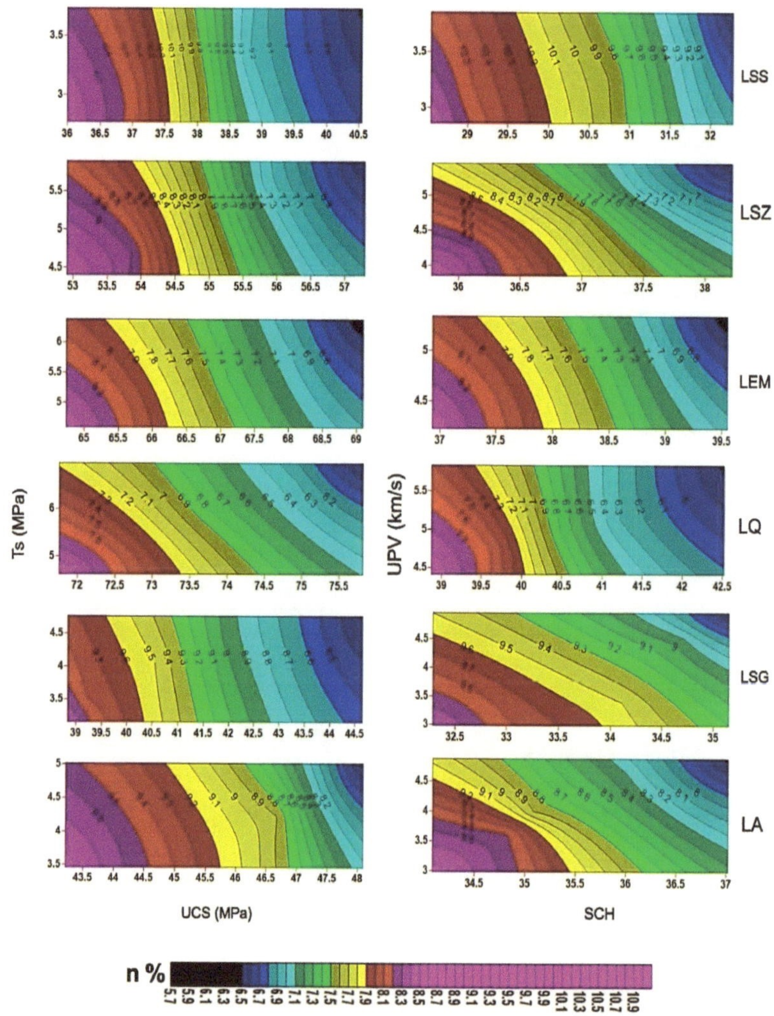

Figure 4. Influence of porosity (n) on TS, UCS, UPV, and SCH for the tested limestone specimens.

3.1. Influence of Cyclic Thermal Quenching Weathering on Physicomechanical Characteristics

Table 4 shows the assessment of the physico-mechanical characteristics before and after cycles of thermal quenching weathering. The porosity, Schmidt hammer rebound hardness, and ultrasound pulse velocity are determined at the end of 50 cycles. Meanwhile, the tensile strength and uniaxial compression strength are determined at 10, 20, 30, 40 and 50 cycles, and then their mean values and standard deviations are presented in Table 4. Generally, the tested limestone specimens did not indicate an important visual degradation after the thermal quenching weathering process. There were no visible granular disintegrations and edge fissures in the tested rock specimens after thermal quenching cycles. As can be seen in Table 4, the results show that the porosity values increase for the tested rock specimens with the number of thermal quenching cycles, whereas the Schmidt hammer rebound hardness, ultrasound pulse velocity, Brazilian tensile strength, and uniaxial compression strength decrease. As seen in this table, after the end of the 50 cycles, the LSG limestone samples exhibit the highest increase in porosity value, while the lowest was for LQ specimens.

Table 4. Mean values and standard deviations of the physico-mechanical properties under cyclic thermal quenching.

Cyclic Number	LSS		LSZ		LEM		LQ		LSG		LA	
	Mean	SD	Mean	SD	Mean	SD	Mean	SD	Mean	SD	Mean	SD
n (%)												
0	9.82	0.95	7.99	1.17	7.47	0.83	6.89	0.90	9.33	0.72	8.80	0.91
50	9.94	0.99	8.14	1.12	7.55	0.95	6.95	0.99	9.51	0.66	8.97	1.07
SCH												
0	30.51	1.65	36.90	0.99	38.44	1.06	40.71	1.41	33.8	1.18	35.4	1.13
50	29.11	1.09	35.05	1.11	37.10	1.27	39.70	1.33	31.9	1.29	33.7	1.28
UPV (Km/s)												
0	3.23	0.39	4.67	0.65	4.81	0.44	5.02	0.53	3.78	0.88	4.02	0.75
50	3.10	0.44	4.53	0.77	4.70	0.62	4.95	0.50	3.56	1.04	3.82	0.80
Ts (MPa)												
0	3.19	0.37	5.15	0.72	5.42	0.62	5.83	0.99	3.99	0.65	4.19	0.60
10	2.95	0.39	5.03	0.77	5.26	0.78	5.75	1.10	3.65	0.66	3.98	0.50
20	2.83	0.28	4.83	0.80	5.10	0.80	5.69	1.04	3.40	0.70	3.57	0.67
30	2.56	0.44	4.55	0.96	4.99	0.77	5.55	0.90	3.12	0.98	3.33	0.89
40	2.35	0.50	4.28	0.91	4.72	0.84	5.33	0.95	2.75	0.91	2.88	0.70
50	2.30	0.45	3.80	0.76	4.41	0.93	5.07	0.87	2.25	0.84	2.59	0.99
UCS (MPa)												
0	38.08	1.74	54.95	1.71	66.90	1.75	73.77	1.65	41.05	2.32	45.97	2.02
10	36.55	2.10	51.15	1.66	65.11	2.12	73.02	1.59	37.36	2.55	41.57	2.22
20	33.82	2.04	47.10	1.77	63.50	1.80	71.89	1.87	34.56	2.23	36.54	2.43
30	31.75	1.99	45.15	2.05	60.14	1.98	69.18	2.14	31.23	2.67	33.70	2.30
40	28.65	1.67	42.98	1.99	56.84	1.95	66.65	2.23	26.66	2.11	30.54	2.25
50	27.83	1.89	41.29	1.85	55.9	2.02	65.99	1.77	25.00	1.99	29.32	2.61

The main reason for degradation in the thermal shock process is the difference in the thermal shrinkage and contraction of the inside and outside area of the mineralogical composition of the samples [15]. Heating water in the micro-fractures during the thermal shock test reduces the rock structure through moving the soluble minerals toward the stone surface [19]. Therefore, changes in the porosity value related to variations in mineralogical and chemical composition, especially the percentage of calcite, which has a higher tendency to shrinkage and contract than other minerals. It can also be seen that after 50 cycles, the LQ limestone specimens show the lowest decrease in the Schmidt hammer rebound, while the highest was for LSG samples. Table 4 also shows that after 50 cycles, the LSG and LA limestone samples exhibit the highest reduction in ultrasound pulse velocity, while the lowest was for the LQ and LEM specimens. This is due to the thermal quenching process producing new internal cracks that lead to a decrease in ultrasound wave velocity. Therefore, it can be said that the studied limestone specimens, namely LQ and LEM, are less sensitive to the devastating influence of thermal quenching.

Generally, it can be concluded that the change in porosity values for the studied limestones is less sensitive to the influence of cyclic thermal quenching compared to the Schmidt hammer rebound and ultrasound wave velocity values. Figure 5 shows a three-dimensional view of the influence of the number of thermal quenching cycles on the tensile strength and uniaxial compression strength for the tested limestones. This figure shows that an increase in the thermal quenching cycles decreased Brazilian tensile strength and uniaxial compression strength values.

Table 4 and Figure 5 show that after 50 cycles, the LSG limestone samples have the lowest values of Brazilian tensile strength and uniaxial compression strength, while the highest values were for LQ specimens. Due to the cyclic thermal quenching, the loss of tensile strength and unconfined compression strength are related to porosity and/or micro-fissures. Therefore, these specimens were more affected than those LQ specimens due to having the highest increase in porosity. Overall, it is concluded that the limestone specimens, namely LQ and LEM, exhibited the highest durability under the influence of thermal quenching cycles, whilst the lowest durability was for the LSG and LA rock specimens.

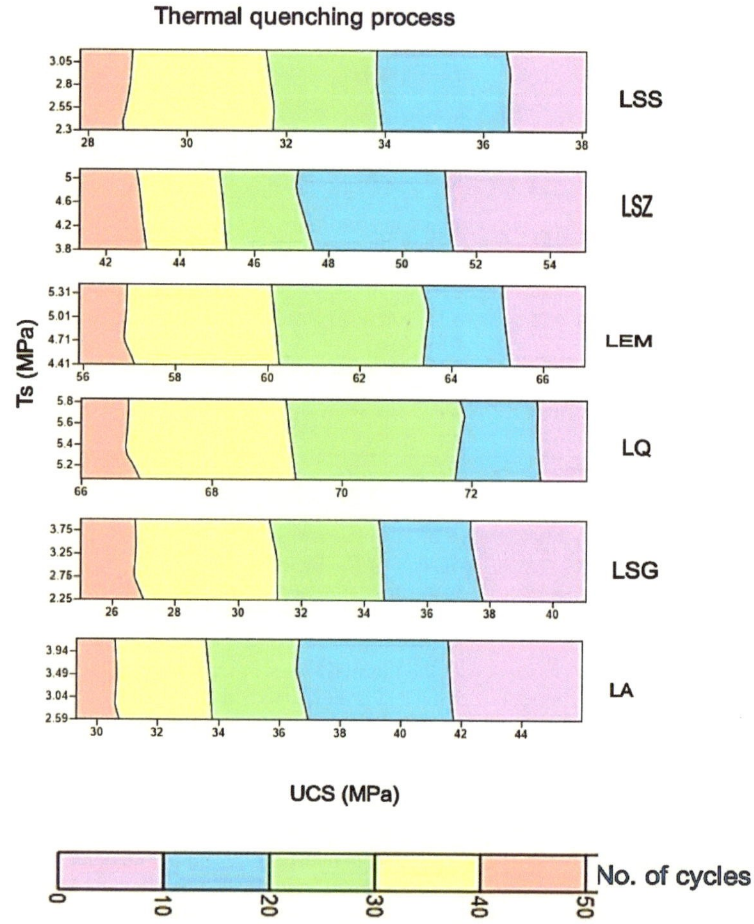

Figure 5. Relationship between TS and UCS values and number of thermal quenching cycles.

3.2. *Influence of Cyclic Salt Crystallization Weathering on Physicomechanical Properties*

Table 5 presents the mean values of the physico-mechanical characteristics before and after the salt crystallization cycles. The porosity, Schmidt hammer rebound hardness, and ultrasound pulse velocity were determined after the end of 25 cycles. The tensile strength and uniaxial compression strength were determined at 5, 10, 15, 20 and 25 cycles, and then their average values and standard deviations are displayed in Table 5. As seen in this table, the results show that the porosity values increased with the number of salt crystallization cycles, whilst the Schmidt rebound hardness, ultrasound pulse velocity, Brazilian tensile strength, and uniaxial compressive strength decreased. As observed in this table, at the end of 25 cycles, the LSG limestone specimens showed the highest increase in porosity, while the lowest was for LQ samples. The changes in porosity values under salt crystallization are linked to the variations in the size of the pores size within the rock specimens. Therefore, different degradations may occur.

Table 5. Average values and standard deviations of physico-mechanical properties due to salt crystallization cycles.

Cyclic Number	LSS		LSZ		LEM		LQ		LSG		LA	
	Mean	SD	Mean	SD	Mean	SD	Mean	SD	Mean	SD	Mean	SD
n (%)												
0	9.82	0.95	7.99	1.17	7.47	0.83	6.89	0.90	9.33	0.72	8.80	0.91
25	10.03	1.11	8.24	1.30	7.67	0.90	7.05	1.04	9.64	1.0	9.08	1.20
SCH												
0	30.51	1.65	36.90	0.99	38.44	1.06	40.71	1.41	33.80	1.18	35.43	1.13
25	28.00	1.77	33.80	1.09	36.60	1.13	39.34	1.47	30.20	1.11	32.10	1.29
UPV (Km/s)												
0	3.23	0.39	4.67	0.65	4.81	0.44	5.02	0.53	3.78	0.88	4.02	0.75
25	3.02	0.49	4.44	0.73	4.62	0.55	4.86	0.76	3.48	0.97	3.76	0.66
Ts (MPa)												
0	3.19	0.37	5.15	0.72	5.42	0.62	5.83	0.99	3.99	0.65	4.19	0.60
5	2.94	0.44	4.90	0.88	5.18	0.85	5.74	1.02	3.55	0.87	3.86	0.65
10	2.77	0.50	4.65	0.90	5.01	0.90	5.63	1.11	3.25	0.73	3.29	0.75
15	2.41	0.41	4.26	0.88	4.88	0.80	5.40	0.92	2.95	0.90	2.99	0.78
20	42.17	0.70	3.90	1.05	4.56	0.66	5.29	0.96	2.45	0.88	2.62	0.59
25	2.03	0.66	3.35	1.11	4.30	0.78	4.95	0.77	1.86	1.10	2.25	0.81
UCS (MPa)												
0	38.08	1.74	54.95	1.71	66.90	1.75	73.77	1.65	41.05	2.32	45.97	2.02
5	35.90	2.27	49.85	1.89	64.55	2.14	72.12	1.70	35.16	2.40	39.88	2.16
10	32.99	2.14	46.12	1.75	61.88	1.99	70.79	1.94	30.66	2.15	34.12	2.22
15	30.15	1.90	43.55	2.22	59.24	1.86	68.22	2.01	27.67	2.60	31.07	2.07
20	27.12	1.99	39.58	1.80	56.77	1.90	65.25	2.10	23.02	2.43	27.89	2.28
25	24.65	2.18	37.12	1.99	54.02	2.22	63.98	1.99	21.22	2.70	26.64	2.35

Based on the degree of disintegration, the development of minute fissures at the corners and surfaces was observed in the LSG specimens after the salt crystallization cycles. Similarly, the LSG limestone samples had the highest decrease in Schmidt rebound hardness, while the lowest was for the LQ specimens. As also seen in Table 5, at the end of 25 cycles, the LSG limestone specimens had the highest decrease in ultrasound pulse velocity, while the lowest was for LQ samples. The compressional wave velocity depends on porosity, pore space, strength, and degree of disintegration [66]. The micro-fissures and pores are the most crucial parameters for decreasing the ultrasound wave velocity. This is evident in LSG limestone specimens. Therefore, it can be said that the loss of ultrasound wave velocity for limestone specimens after the salt crystallization cycles is related to the dilation and enlargement of pores and fractures. Figure 6 presents a three-dimensional view of the effect of the number of salt crystallization cycles on the tensile and uniaxial compression strengths for the studied limestones. As seen in this Figure, the progression of the number of salt crystallization cycles caused a reduction in the values of Brazilian tensile strength and uniaxial compression strength. Table 5 and Figure 6 show that after 25 cycles, the LSG limestone specimens had the lowest values of uniaxial compression and Brazilian tensile strengths, while the highest values were for LQ samples.

It is said that the variation of tensile strength and unconfined compressive strength values is due to the widening and dilation of pre-existing micropores during the salt crystallization process affecting the mechanical behavior of limestone specimens [15].

Previous studies proved that porosity and water absorption are the most significant parameters affecting the behavior of rocks and durability control against the salt weathering test [63]. Therefore, it can be believed that the limestone rocks with the highest increase in porosity will be subject to a high level of disintegration and result in lower durability compared to rocks with the lowest increase in porosity. Therefore, it is said that the limestone rock samples, namely LSG, exhibited the highest level of degradation against the salt crystallization weathering cycles, whereas the lowest level of deterioration was for LQ rock samples. The major reason for such disintegration of the physico-mechanical parameters against the cyclic salt crystallization process is the high pressure caused by precipitation of salt crystals within the pores and discontinuities of rock specimens. Hence, new microcracks were developed, and finally, the rock was damaged [67].

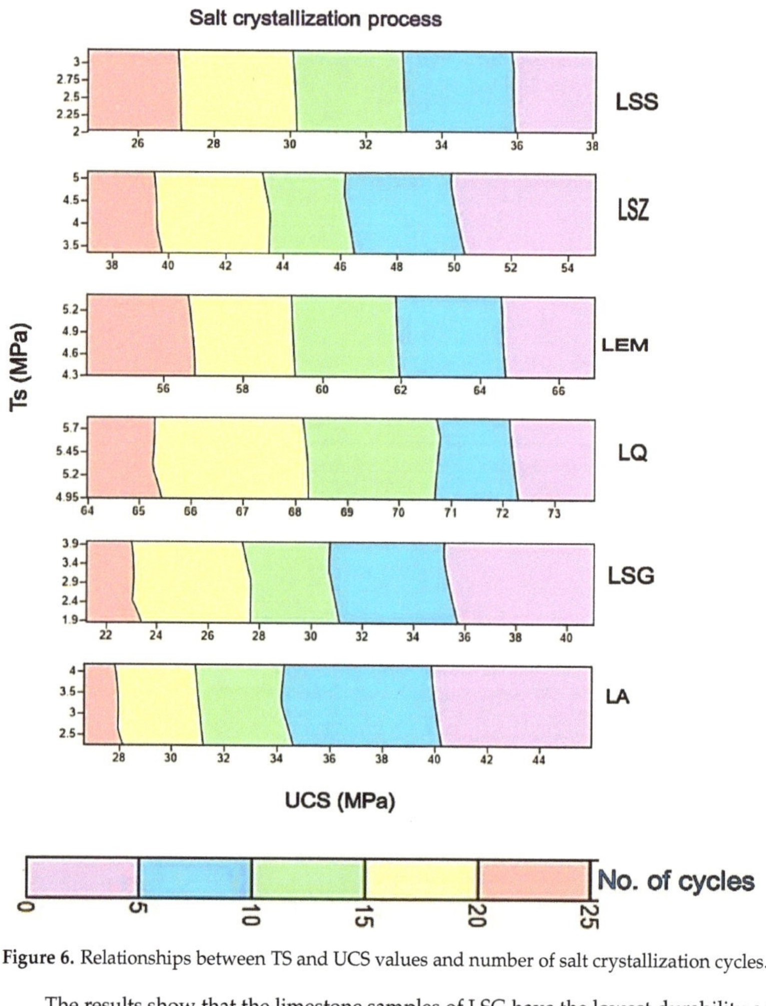

Figure 6. Relationships between TS and UCS values and number of salt crystallization cycles.

The results show that the limestone samples of LSG have the lowest durability against the salt crystallization cycles due to having the highest reduction in the values of tensile strength and uniaxial compression strength, whereas the LQ and LEM rocks have the lowest decrease concerning these properties (Figure 6).

Overall, it is concluded that the change of physico-mechanical properties of the studied limestone rocks is more sensitive to the destructive influences of cyclic salt crystallization weathering than the thermal quenching process.

Brazilian tensile strength and uniaxial compression strength are the most sensitive parameters against the influences of thermal quenching and salt crystallization weathering tests, so their rates of variation are obtained after the end of every ten cycles of the thermal process and after the end of every five cycles of the salt weathering process. The variation rate determines the degree of rock material deterioration. Figure 7 shows the relationship between the change rate of tensile and uniaxial compression strengths (ΔTS % and ΔUCS %) against the cyclic thermal quenching and salt crystallization weathering processes. As seen in Figure 7, after the end of the cyclic weathering processes, there was a high decrease in the change rate of tensile strength and uniaxial compression strength for limestone samples, namely LSG, while LQ specimens showed the lowest decrease compared to other rocks.

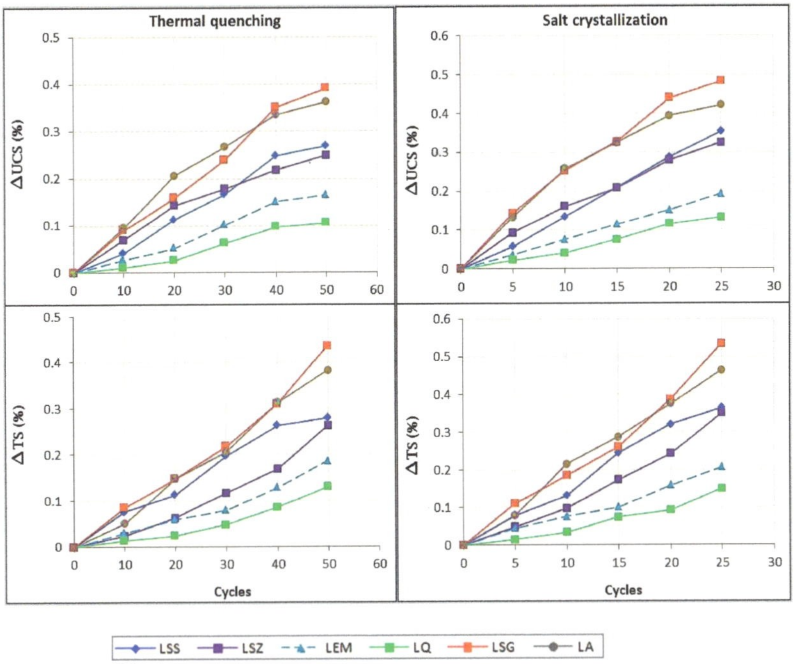

Figure 7. Change rate of TS and UCS values due to the cyclic thermal quenching and salt crystallization.

As this Figure shows, after the end of 50 cycles of thermal quenching, the highest percentage decreases in uniaxial compression strength and tensile strength were 39.1% and 43.6 %, respectively, for LSG limestone specimens, whilst the lowest percent reduction was for LQ samples (10.5% and 13%, respectively). Similarly, for the salt crystallization process, it is seen that after the end of 25 cycles, the highest percentage decreases in uniaxial compression strength and tensile strength were 48.3% and 53.4%, respectively, for LSG limestone specimens, whereas the lowest percent decrease was for LQ samples (13.3% and 15.1%, respectively). Therefore, it is observed that limestone rocks, namely LSG have low durability under the influence of accelerated weathering processes compared to other rocks, whereas limestone rocks, namely LQ and LEM, have the highest resistance and durability to the effect of weathering processes. Based on these results, it is concluded that the studied limestone rock specimens are more sensitive to the destructive effect of the salt crystallization process than the thermal quenching. In general, it is said that the change limits of the data obtained for the LSG and LA limestone samples are high, so these types cannot withstand cyclic weathering processes compared to other rocks. Therefore, these stones may not be suitable for construction purposes in humid and salty environments.

3.3. Degradation Model of the Mechanical Strength Parameters

Thermal quenching and salt crystallization weathering processes in the lab are significant for obtaining the necessary data on rock decay, yet insufficient to assess the long-term deterioration.

Mutlutürk et al. [24] postulated the decay function model by using a decay constant (λ) and half-life ($N_{1/2}$) parameters as Equation (1) to express the degradation rate of rock under cyclic freeze–thaw and heat–cool weathering tests, which could also be expressed in an exponential model as Equation (2) by integrating Equation (1) between the initial integrity (I_0) and the integrity after N cycles (I_N). To determine the long-term durability of rock, the decay constant (λ) indicates the mean relative decrease in rock integrity by the work of any single cycle, and the half-life ($N_{1/2}$) refers to the number of cycles required to decrease the rock integrity to its half-value, which is obtained as Equation (3) by replacing ($I_0/2$) with I_N in Equation (2):

$$-(dI/dN) = \lambda N \quad (1)$$

$$I_N = I_0\ e^{-\lambda N} \quad (2)$$

$$N_{1/2} = 0.6930/\lambda \quad (3)$$

where $e^{-\lambda N}$ is the decay factor, which shows the proportion of the remaining integrity after number of cycles, i.e., (I_N /I_0); I is the rock integrity; N is the number of cycles; (dI/dN) is the standard for the degradation rate. The model postulated by Mutlutürk et al. [24] considered only the physical parameters, and other mechanical properties were ignored. Therefore, in our study, for investigating the model validity and evaluating the model parameters, laboratory tests of Brazilian tensile strength and uniaxial compression strength of limestone rocks for predicting the integrity decrease rate due to cyclic thermal quenching and salt crystallization were performed. A simple regression analysis method was used to determine the decay constant (λ). According to the model, the data obtained from the laboratory investigations of tensile strength and uniaxial compression strength were exponentially fitted. The degree of fit to this model (curve) can be measured by the value of determination coefficients (R^2), which measure the proportion of variation in the dependent variable [68]. The high values of the determination coefficients indicate that the proposed model fits the data well and facilitates the prediction of the rock sample's properties after any cycle of artificial weathering processes. The fitted curves and experimental data results of tensile strength and uniaxial compression strength for the different limestone rocks against cyclic thermal quenching and salt crystallization weathering processes are illustrated in Figure 8. Based on the previously mentioned experimental results of the Brazilian tensile strength and uniaxial compression strength for the limestone rock types against the cyclic thermal quenching and salt crystallization (Tables 4 and 5), the values of the decay constant (λ) are determined graphically in Figure 8 and listed in Tables 6 and 7; in addition, the half-life ($N_{1/2}$) values obtained from Equation (3) are detailed in Tables 6 and 7.

3.3.1. Decay Constant (λ)

Figure 8 and Tables 6 and 7 show the decay constant values and the determination coefficients (R^2) obtained from the results of mechanical strength parameters, including Brazilian tensile strength (Ts, MPa) and uniaxial compression strength (UCS, MPa) of the studied limestone rocks due to the effect of thermal quenching and salt crystallization cycles. As seen in Figure 8, the integrity loss of the uniaxial compression strength and tensile strength show relatively high values of the decay constant and a rapid integrity decrease rate (rapid durability rate) of the studied rocks, especially in the salt crystallization process. It can also be seen that the high values of the determination coefficients indicate that the proposed model fits the data well and is reliable for predicting the loss of mechanical strength parameters and/or long-term durability of the studied limestone rock specimens after any thermal quenching and/or salt crystallization cycle.

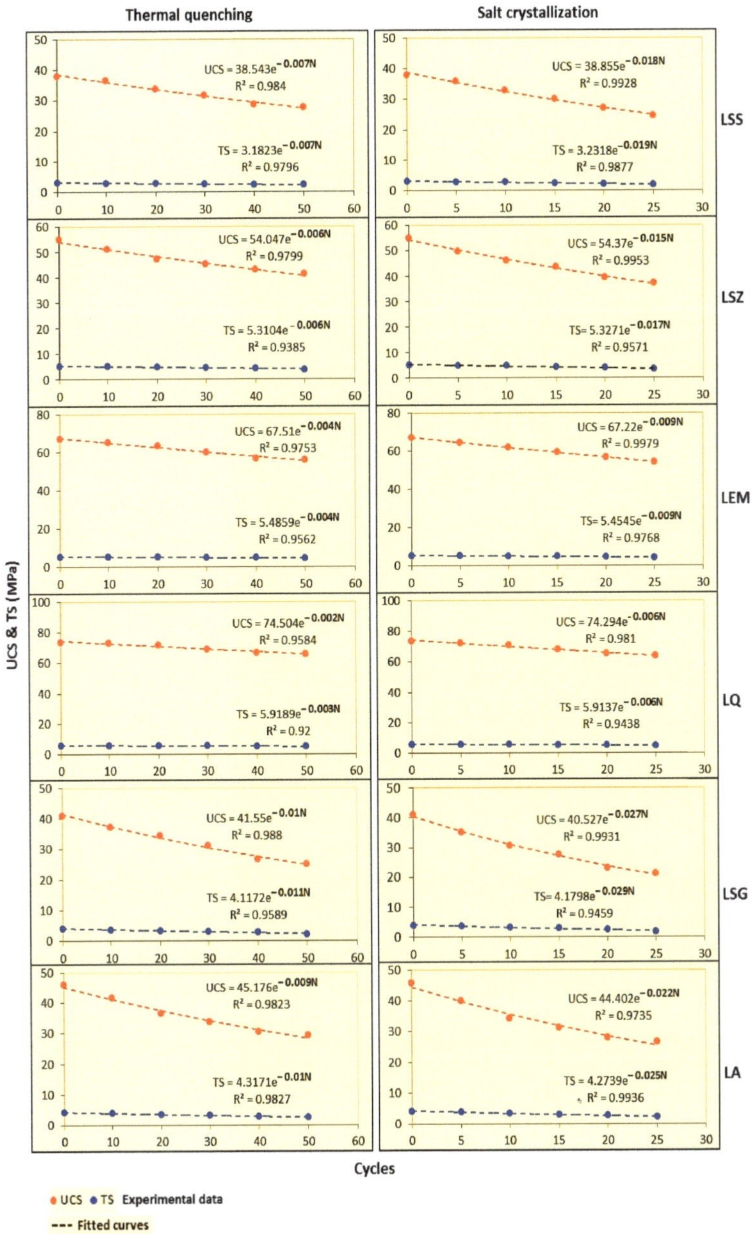

Figure 8. Determining decay constant (λ) based on TS and UCS data against cyclic thermal and salt weathering.

Table 6. Decay constant (λ), half-life ($N_{1/2}$), and coefficient of determination (R^2) of tensile strength parameter due to cyclic thermal shock and salt crystallization.

Tensile Strength TS (MPa)						
	Thermal Quenching			Salt Crystallization		
	λ	$N_{1/2}$	R^2	λ	$N_{1/2}$	R^2
LSS	−0.007	99.0	0.9796	−0.019	36.5	0.9877
LSZ	−0.006	115.5	0.9385	−0.017	40.8	0.9571
LEM	−0.004	173.3	0.9562	−0.009	77.0	0.9768
LQ	−0.003	231.0	0.9200	−0.006	115.5	0.9438
LSG	−0.011	63.0	0.9589	−0.029	23.9	0.9459
LA	−0.010	69.3	0.9827	−0.025	27.7	0.9936

Table 7. Decay constant (λ), half-life ($N_{1/2}$), and coefficient of determination (R^2) of uniaxial compression strength parameter due to cyclic thermal shock and salt crystallization.

Uniaxial Compression Strength UCS (MPa).						
	Thermal Quenching			Salt Crystallization		
	λ	$N_{1/2}$	R^2	λ	$N_{1/2}$	R^2
LSS	−0.007	99	0.9840	−0.018	38.5	0.9928
LSZ	−0.006	115.5	0.9799	−0.015	46.2	0.9953
LEM	−0.004	173.3	0.9753	−0.009	77	0.9979
LQ	−0.002	346.5	0.9584	−0.006	115.5	0.981
LSG	−0.010	69.3	0.9880	−0.027	25.7	0.9931
LA	−0.009	77.0	0.9823	−0.022	31.5	0.9735

It is concluded that the sedimentary rocks of the same kind provide good information about their long-term deterioration against the impact of cyclic weathering processes. For example, as seen in Table 6, one of the highest durable limestone samples (LQ) and also one of the least durable specimens (LSG), due to the influence of thermal quenching process (i.e., showing the lowest and highest decay constant values, (λ = −0.003 and −0.011, respectively), are of the same rock type. In fact, the tensile strength value of the LQ specimens decreased by 0.3% compared to their initial value, on average, after one thermal quenching cycle, whereas the LSG specimens decreased by 1.1%. Therefore, the integrity decrease rate (degradation rate) of LSG limestone specimens is 3.7 times more than that of the LQ specimens. Similarly, for the cyclic salt crystallization, some of the highest durable samples (LQ) and the least durable specimens (LSG) (i.e., exhibiting the lowest and highest decay constant, (λ = −0.006 and −0.029, respectively). Indeed, the LQ specimens lost only 0.6% of their main tensile strength value, on average, after one cycle of salt crystallization, while the LSG specimens decreased by 2.9%. As a result, the integrity loss rate of the LSG specimens is 4.8 times more than that of the LQ specimens (Table 6). Therefore, according to the presented results in Tables 6 and 7, it can be said that the studied limestone specimens exposed to the salt crystallization cycles exhibited the highest values of the decay constant compared to the thermal quenching process.

3.3.2. Half-Life ($N_{1/2}$)

The values of the half-life ($N_{1/2}$) of the loss of tensile and uniaxial compressive strengths of the studied limestones due to the cyclic thermal quenching and salt crystallization are also listed in Tables 6 and 7 and graphically illustrated in Figures 9 and 10, respectively. Based on the results of the tensile strength under the cyclic thermal quenching presented in Table 6 and Figure 9, the limestone specimens, namely LQ, have the longest half-life (231 cycles), whilst the LSG samples have the shortest half-life (63 cycles). For the cyclic salt crystallization, it can also be seen that the LQ samples exhibit the longest half-life (115.5 cycles), while the shortest half-life is for LSG specimens (23.9 cycles). Similarly, according to the data of the uniaxial compression strength against the cyclic thermal quenching listed in Table 7 and Figure 10, the LQ limestone samples show the longest half-life (346.5 cycles), whilst the LSG specimens have the shortest half-life (69.3 cycles). For the salt crystallization process, it can also be seen that the LQ samples exhibit the longest half-life (115.5 cycles), while the shortest half-life is for LSG specimens (25.9 cycles). As shown in Figures 9 and 10, the half-life values based on the tensile strength and uniaxial compression strength parameters, respectively, are similar in some of the studied limestone specimens. The half-life values have no meaningful pattern with respect to the rock kind. Actually, rocks linked to the same origin can indicate a clear variation in long-term deterioration due to the various weathering processes in some cases. Therefore, the type of rock alone cannot provide sufficiently useful information related to the durability of rock specimens under the artificial weathering processes. Eren and Bahali [69] measured the reduction in weight of two different types of limestone under cyclic freezing–thawing. They noticed that the rocks regarding the same origin could indicate clearly different behaviors under freezing–thawing action, and this is due to change in their porosity. Mutluturk et al. [24] measured the shore hardness changes of some various rock types against freezing–thawing cycles. They showed that the rock type alone does not give a clue for the half-life of rocks due to cyclic freezing–thawing.

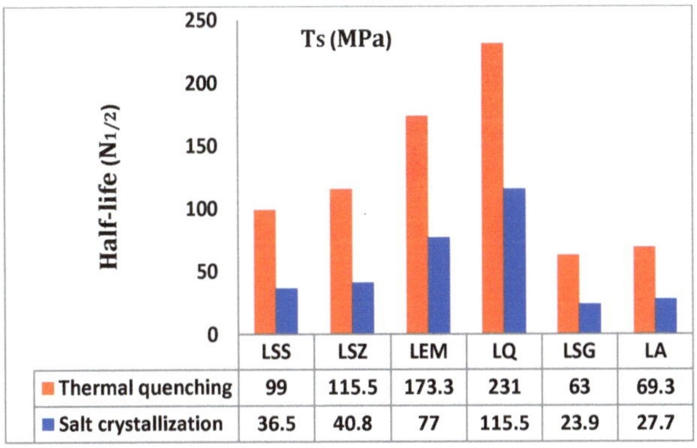

Figure 9. Half-life values of tensile strength of limestones under cyclic thermal and salt weathering.

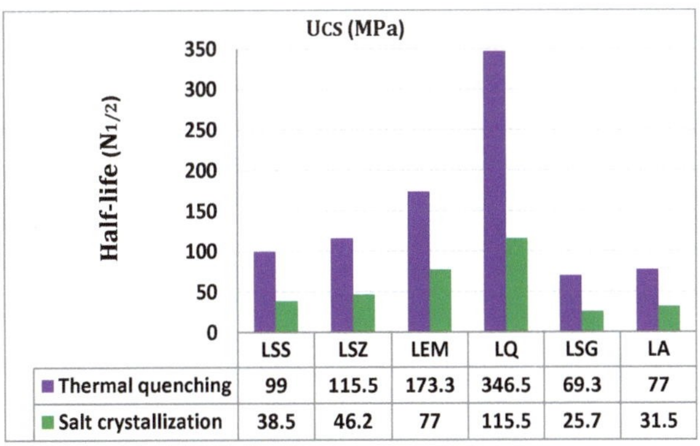

Figure 10. Half-life values of uniaxial compression strength of limestones under the thermal and salt weathering.

According to the current results, it is seen that the half-life values of the integrity loss of the strength properties (Ts and UCS) for the studied limestone rock specimens due to the impact of salt crystallization cycles are lower than those the thermal quenching process.

Overall, it can be concluded that studied limestones were more sensitive to deterioration under the salt crystallization process than thermal quenching.

4. Conclusions

Rock deterioration subject to various physical weathering conditions was assessed by performing the laboratory's cyclic thermal quenching and salt crystallization tests and using the decay function model parameters (decay constant and half-life) to provide helpful information for the assessment of their durability. In addition, the studied limestone rock specimens were also subjected to a series of laboratory experimental tests for determining their physico-mechanical characteristics. In this study, the cyclic thermal quenching and salt crystallization weathering processes were carried out on six different limestone types used for construction purposes in Egypt to evaluate their long-term degradation. In general, the present results show the strong impact of salt crystallization cycles on the rock physico-mechanical characteristics compared to thermal quenching. Therefore, the major following findings can be drawn from the current work:

1. The studied limestone rocks are more sensitive to the destructive effect of the salt crystallization action than the thermal quenching weathering process. The tested limestone rock samples, namely LSG, showed the most significant increase in porosity (3.30%) and the highest percentage decrease in uniaxial compression strength and tensile strength (48.3% and 53.4%), respectively, after the cyclic salt crystallization compared to the other limestone samples. Therefore, the thermal quenching process is not very destructive to the tested limestone samples with high strength and durability compared to salt crystallization. After the end of the cyclic salt crystallization, a slight disintegration was observed at the edges and corners of the LSG samples. Therefore, the studied limestone samples can be used as building stones in mild or humid and salty areas for a long time without degradation. Nevertheless, partial attention must be given to the LSG limestone type.
2. The lowest deterioration level was recorded for the LQ rock samples due to their low porosity and water absorption capacity. Therefore, these specimens were more resistant to weathering processes than other samples.
3. The tested rock specimens under cyclic thermal quenching showed low variations in their physico-mechanical properties. The major reason for these changes is the presence of calcite mineral that leads to anisotropic pressures. These pressures may reopen micro-fractures or pores and finally damage the rock. Overall, the cyclic salt weathering process deteriorated the physical and mechanical characteristics of the rock samples compared to the thermal quenching action.
4. An exponential decay model is applied for predicting the integrity loss characteristics (disintegration rate) of the mechanical strength properties for the tested limestones following cyclic thermal quenching and salt crystallization weathering. This model exhibited a higher sensitivity of limestone rock samples to the cyclic salt weathering process. High values of the decay constant were attained under salt crystallization weathering in comparison with the thermal quenching action, so this model is capable of predicting the degradation rate (integrity loss) of limestone rock specimens without performing any testing and is appropriate for rocks quarried from the same type.

After investigating the artificial weathering processes of the studied rocks, the information attained may be very useful in new engineering applications and restoration works. According to the current results, the tested limestone specimens may be appropriate for use in areas exposed to frequent thermal and salt weathering action. However, future work should investigate the impact of weathering processes on different types of rocks for obtaining a complete evolution of the rock deterioration characteristics. Furthermore, the decay function model used here to estimate the rock durability characteristics in projected future constructions might be usefully applied on various rocks against weathering conditions for use in other studies. Therefore, the subsequent studies must select more than one appropriate model and compare their durability changes. This might give a good impression of overall model performance and compare the results of these intermodal.

Author Contributions: Methodology, M.M.A.A. and B.G.M.; formal analysis, M.M.A.A., B.G.M. and H.W.; writing—original draft preparation, M.M.A.A., B.G.M., I.M., M.A.E. and R.A.; writing—review and editing, M.M.A.A., B.G.M. and A.M.G.; resources, S.M.E. Furthermore, R.A. supervised the whole work. All authors have read and agreed to the published version of the manuscript.

Funding: This research was funded by the Research Group Program under grant no. RGP. 2/19/43.

Institutional Review Board Statement: Not applicable.

Informed Consent Statement: Not applicable.

Data Availability Statement: Not applicable.

Acknowledgments: The authors extend their appreciation to the Deanship of Scientific Research at King Khalid University, Abha, Saudi Arabia, for funding this work.

Conflicts of Interest: The authors declare no conflict of interest.

References

1. Ozcelik, Y.; Careddu, N.; Yilmazkaya, E. The effects of freeze–thaw cycles on the gloss values of polished stone surfaces. *Cold Reg. Sci. Technol.* **2012**, *82*, 49–55. [CrossRef]
2. García-Del-Cura, M.A.; Benavente, D.; Martínez-Martínez, J.; Cueto, N. Sedimentary structures and physical properties of travertine and carbonate tufa building stone. *Constr. Build. Mater.* **2012**, *28*, 456–467. [CrossRef]
3. Momeni, A.; Abdilor, Y.; Khanlari, G.R.; Heidari, M.; Sepahi, A.A. The effect of freeze– thaw cycles on physical and mechanical properties of granitoid hard rocks. *Bull. Eng. Geol. Environ.* **2016**, *75*, 1649–1656. [CrossRef]
4. Ali, R.; Khan, I.; Ali, A.; Mohamed, A. Two new generalized iteration methods for solving absolute value equations using M-matrix. *AIMS Math.* **2022**, *7*, 8176–8187. [CrossRef]
5. Zhang, Z.P.; Shao, Y.F.; Song, F. Characteristics of crack patterns controlling the retained strength of ceramics after thermal shock. *Front. Mater. Sci. China* **2010**, *4*, 251–254. [CrossRef]
6. Abad, S.A.N.K.; Mohamad, E.T.; Komoo, I. Dominant weathering profiles of granite in southern Peninsular Malaysia. *Eng. Geol.* **2014**, *183*, 208–215. [CrossRef]
7. Ercoli, L.; Zimbardo, M.; Nocilla, A. Rock decay phenomena and collapse processes in the Latomiae del Paradiso in Syracuse (Sicily). *Eng. Geol.* **2014**, *178*, 155–165. [CrossRef]
8. Abad, S.V.A.N.K.; Mohamad, E.T.; Komoo, I.; Kalatehjari, R. A typical weathering profile of granitic rock in Johor, Malaysia based on joint characterization. *Arab. J. Geosci.* **2015**, *8*, 2191–2201. [CrossRef]
9. Abad, S.V.A.N.K.; Mohamad, E.T.; Komoo, I.; Kalatehjari, R. Assessment of weathering effects on rock mass structure. *J. Tekh.* **2015**, *72*, 71–75.
10. Khanlari, G.; Sahamieh, R.Z.; Abdilor, Y. The effect of freeze–thaw cycles on physical and mechanical properties of Upper Red Formation sandstones, central part of Iran. *Arab. J. Geosci.* **2015**, *8*, 5991–6001. [CrossRef]
11. Shushakova, V.; Edwin, R.; Fuller, E.R., Jr.; Heidelbach, F.; Mainprice, D.; Siegesmund, S. Marble decay induced by thermal strains: Simulations and experiments. *Environ. Earth. Sci.* **2013**, *69*, 1281–1297. [CrossRef]
12. Andriani, G.F.; Germinario, L. Thermal decay of carbonate dimension stones: Fabric, physical and mechanical changes. *Environ. Earth. Sci.* **2014**, *72*, 2523–2539. [CrossRef]
13. Ghobadi, M.H.; Babazadeh, R. Experimental studies on the effects of cyclic freezing– thawing, salt crystallization, and thermal shock on the physical and mechanical characteristics of selected sandstones. *Rock Mech. Rock Eng.* **2015**, *48*, 1001–1016. [CrossRef]
14. Vazquez, P.; Shushakova, V.; Gomez-Heras, M. Influence of mineralogy on granite decay induced by temperature increase: Experimental observations and stress simulation. *Eng. Geol.* **2015**, *189*, 58–67. [CrossRef]
15. Çelik, M.Y.; Sert, M. Accelerated aging laboratory tests for the evaluation of the durability of hydrophobic treated and untreated andesite with respect to salt crystallization, freezing–thawing, and thermal shock. *Bull. Eng. Geol. Environ.* **2020**, *79*, 3751–3770. [CrossRef]
16. Lisci, C.; Sitzia, F.; Pires, V.; Mirão, J. Building stones durability by UVA radiation, moisture and spray accelerated weathering. *J. Build. Rehabil.* **2022**, *7*, 60. [CrossRef]
17. Karakaş, A.; Morali, G.; Coruk, ö.; Bozkurtoğlu, E. Geomechanical, durability–hygrothermal and thermal shock properties of Kocaeli Kandira stone used as building stone in historical structures. *Environ Earth Sci.* **2021**, *80*, 130. [CrossRef]
18. Yavuz, A.B. Deterioration of the volcanic kerb and pavement stones in a humid environment in the city centre of Izmir, Turk. *Environ Geol.* **2006**, *51*, 211–227. [CrossRef]
19. Hale, P.A.; Shakoor, A. A laboratory investigation of the effects of cyclic heating and cooling, wetting and drying, and freezing and thawing on the compressive strength of selected sandstones. *Environ. Eng. Geosci.* **2003**, *9*, 117–130. [CrossRef]

20. Sousa, L.M.O.; Suarez, L.M.; Calleja, L.; de Argandoña, V.G.R.; Rey, A.R. Influence of micro fractures and porosity on the physico-mechanical properties and weathering of ornamental granites. *Eng. Geol.* **2005**, *77*, 153–168. [CrossRef]
21. Wang, P.; Xu, J.; Liu, S.; Wang, H.; Liu, S. Static and dynamic mechanical properties of sedimentary rock after freeze-thaw or thermal shock weathering. *Eng. Geol.* **2016**, *210*, 148–157. [CrossRef]
22. Ali, R.; Pan, K.; Ali, A. Two generalized successive overrelaxation methods for solving absolute value equations. *Math. Theory Appl.* **2020**, *40*, 44–55.
23. Gokceoglu, C.; Zorlu, K.; Ceryan, S.; Nefeslioglu, H. A comparative study on indirect determination of degree of weathering of granites from some physical and strength parameters by two soft computing techniques. *Mater. Charact.* **2009**, *60*, 1317–1327. [CrossRef]
24. Mutluturk, M.; Altindag, R.; Turk, G. A decay function model for the integrity loss of rock when subjected to recurrent cycles of freezing–thawing and heating–cooling. *Int. J. Rock. Mech. Min. Sci.* **2004**, *41*, 237–244. [CrossRef]
25. Jefferson, D.P. Building stone: The geological dimension. *Q. J. Eng. Geol. Hydrol.* **1993**, *26*, 305–319. [CrossRef]
26. Doehne, E.; Price, C.A. *Stone Conservation: An Overview of Current Research*; The Getty Conservation Institute: Second edition; J. Paul Getty Museum Publications: Los Ángeles, CA, USA, 2010; Volume 158.
27. Benavente, D.; del Cura, M.A.G.; Fort, R.; Ordóñez, S. Thermodynamic modeling of changes induced by salt pressure crystallization in porous media of stone. *J. Cryst. Growth* **1999**, *204*, 168–178. [CrossRef]
28. Jamshidi, A.; Nikudel, M.R.; Khamehchiyan, M. Predicting the long-term durability of building stones against freeze–thaw using a decay function model. *Cold Reg. Sci. Technol.* **2013**, *92*, 29–36. [CrossRef]
29. Pitzurra, L.; Moroni, B.; Nocentini, A.; Sbaraglia, G.; Poli, G.; Bistoni, F. Microbial growth and air pollution in carbonate rock weathering. *Int. Biodeterior.* **2003**, *52*, 63–68. [CrossRef]
30. Sandrolini, F.; Franzoni, E. An operative protocol for reliable measurements of moisture in porous materials of ancient buildings. *Build. Environ.* **2006**, *41*, 1372–1380. [CrossRef]
31. Bozdağ, A. The Effect of Salt (NaCl) Crystallization on Engineering Parameters of Rocks. Ph.D. Thesis, Selcuk University, Konya, Turkey, 2013.
32. Korkanç, M. Deterioration of different stones used in historical buildings within Nigde Province, Cappadocia. *Constr. Build. Mater.* **2013**, *48*, 789–803.
33. Barone, G.; Mazzoleni, P.; Pappalardo, G.; Raneri, S. Microtextural and microstructural influence on the changes of physical and mechanical proprieties related to salts crystallization weathering in natural building stones. The example of Sabucina stone (Sicily). *Constr. Build. Mater.* **2015**, *95*, 355–365. [CrossRef]
34. Erdogan, O.; Ozvan, A. Evaluation of strength parameters and quality assessment of different lithotype levels of Edremit (Van) Travertine (Eastern Turkey). *J. Afr. Earth. Sci.* **2016**, *106*, 108–118. [CrossRef]
35. Jamshidi, A.; Nikudel, M.R.; Khamehchiyan, M. Evaluation of the durability of Gerdoee travertine after freeze-thaw cycles in fresh water and sodium sulfate solution. *Eng. Geol.* **2016**, *202*, 36–43. [CrossRef]
36. Menendez, B.; Petranova, V. Effect of mixed vs single brine composition on salt weathering in porous carbonate building stones for different environmental conditions. *Eng. Geol.* **2016**, *210*, 124–139. [CrossRef]
37. Ruffolo, S.A.; La Russa, M.F.; Ricca, M.; Belfiore, C.M.; Macchia, A.; Comite, V.; Pezzino, A.; Crisci, G.M. New insights on the consolidation of salt weathered limestone: The case study of Modica stone. *Bull. Eng. Geol. Environ.* **2017**, *76*, 11–20. [CrossRef]
38. Çelik, M.Y.; Tığlı, R. The investigation of the water repellent chemical influence on salt crystallization in high porous building stones. *Gazi Univ. J. Eng. Arch.* **2018**, *34*, 535–552.
39. Khodabandeh, M.A.; Rozgonyi-Boissinot, N. The Effect of Salt Weathering and Water Absorption on the Ultrasonic Pulse Velocities of Highly Porous Limestone. *Period. Polytech. Civ. Eng.* **2022**, *66*, 627–639. [CrossRef]
40. Lezzerini, M.; Tomei, A.; Gallello, G.; Aquino, A.; Pagnotta, S. The Crystallization Effect of Sodium Sulfate on Some Italian Marbles, Calcarenites and Sandstones. *Heritage* **2022**, *5*, 1449–1461. [CrossRef]
41. Çelik, M.Y.; İbrahimoglu, A. Characterization of travertine stones from Turkey and assessment of their durability to salt crystallization. *J. Build. Eng.* **2021**, *43*, 102592. [CrossRef]
42. Oguchi, C.T.; Yu, S. A review of theoretical salt weathering studies for stone heritage. *Prog. Earth Planet. Sci.* **2021**, *8*, 32. [CrossRef]
43. Angeli, M.; Heber, R.; Menendez, B.; David, C.; Bigas, J.-P. Influence of temperature and salt concentration on the salt weathering of a sedimentary stone with sodium sulphate. *Eng. Geol.* **2010**, *115*, 193–199. [CrossRef]
44. Cultrone, G.; Luque, A.; Sebastián, E. Petrophysical and durability tests on sedimentary stones to evaluate their quality as building materials. *Q. J. Eng. Geol. Hydrog.* **2012**, *45*, 415–422. [CrossRef]
45. Vázquez, P.; Alonso, F.; Carrizo, L.; Molina, E.; Cultrone, G.; Blanco, M.; Zamora, I. Evaluation of the petrophysical properties of sedimentary building stones in order to establish quality criteria. *Constr. Build. Mater.* **2013**, *41*, 868–878. [CrossRef]
46. Yavuz, A.B.; Topal, T. Thermal and salt crystallization effects on marble deterioration: Examples from Western Anatolia, Turkey. *Eng. Geol.* **2007**, *90*, 30–40. [CrossRef]
47. Benavente, D.; Cueto, N.; Martínez-Martínez, J.; Del Cura, M.A.G.; Cañaveras, J.C. The influence of petrophysical properties on the salt weathering of porous building rocks. *Environ. Geol.* **2007**, *52*, 197–206. [CrossRef]
48. Yavuz, A.B.; Kaputoglu, S.A.; Çolak, M.; Tanyu, B.F. Durability assessments of rare green andesites widely used as building stones in Buca (Izmir), Turkey. *Environ. Earth Sci.* **2017**, *76*, 211. [CrossRef]
49. Ruedrich, J.; Siegesmund, S. Salt and ice crystallisation in porous sandstones. *Environ. Geol.* **2007**, *52*, 225–249. [CrossRef]

50. Yavuz, H. Effect of freeze–thaw and thermal shock weathering on the physical and mechanical properties of an andesite stone. *Bull. Eng. Geol. Environ.* **2011**, *70*, 187–192. [CrossRef]
51. Liu, Q.S.; Huang, S.B.; Kang, Y.S.; Liu, X.W. A prediction model for uniaxial compressive strength of deteriorated rocks due to freeze–thaw. *Cold Reg. Sci. Technol.* **2015**, *120*, 96–107. [CrossRef]
52. *TS EN 1936*; Natural Stone Test Methods-Determination of Real Density and Apparent Density and of Total and Open Porosity. Turkish Standards Institute: Ankara, Turkey, 2010; p. 10.
53. *TS EN 13755*; Natural Stone Test Methods-Determination of Water Absorption at Atmospheric Pressure. Turkish Standards Institute: Ankara, Turkey, 2009; p. 10.
54. ISRM. *Rock Characterization, Testing, and Monitoring*; ISRM Suggested Methods, Pergamon Press: Oxford, UK, 1981.
55. *TS EN 1926*; Natural Stone Test Methods-Determination of Uniaxial Compressive Strength. Turkish Standards Institute: Ankara, Turkey, 2007; p. 19.
56. ISRM. Suggested methods for determining tensile strength of rock materials. *Int. J. Rock. Mech. Min. Sci. Geomech. Abstr.* **1978**, *15*, 99–103. [CrossRef]
57. Aydin, A. ISRM suggested method for determination of the Schmidt hammer rebound hardness: Revised version. In *The ISRM Suggested Methods for Rock Characterization, Testing and Monitoring: 2007–2014 SE-2*; Ulusay, R., Ed.; Springer: Berlin/Heidelberg, Germany, 2015; pp. 25–33.
58. Martínez-Martínez, J.; Benavente, D.; García-del-Cura, M.A. Spatial attenuation: The most sensitive ultrasonic parameter for detecting petrographic features and decay processes in carbonate rocks. *Eng. Geol.* **2011**, *119*, 84–95. [CrossRef]
59. *Turkish Standards Institute*; Natural Stone Test Methods—Determination of Resistance to Ageing by Thermal Shock. Institute of Turkish Standards: Ankara, Turkey, 2004; p. 3.
60. Rothert, E.; Eggers, T.; Cassar, J.; Ruedrich, J.; Fitzner, B.; Siegesmund, S. Stone properties and weathering induced by salt crystallization of Maltese Globigerina Limestone. *Geol. Soc. Lond. Spec. Publ.* **2007**, *271*, 189. [CrossRef]
61. Ruedrich, J.; Kirchner, D.; Siegesmund, S. Physical weathering of building stones induced by freeze–thaw action: A laboratory long term study. *Environ. Earth. Sci.* **2011**, *63*, 1573–1586. [CrossRef]
62. Benavente, D.; Pla, C.; Cueto, N.; Galvañ, S.; Martínez-Martínez, J.; García-Del-Cura, M.; Ordóñez, S. Predicting water permeability in sedimentary rocks from capillary imbibition and pore structure. *Eng. Geol.* **2015**, *195*, 301–311. [CrossRef]
63. Akin, A.; Ozsan, A. Evaluation of the long-term durability of yellow travertine using accelerated weathering tests. *Bull. Eng. Geol. Environ.* **2011**, *70*, 101–114. [CrossRef]
64. Tuğrul, A. The effect of weathering on the pore geometry and compressive strength of selected rock types from Turkey. *Eng. Geol.* **2004**, *75*, 215–227. [CrossRef]
65. Diamantis, K.; Gartzos, E.; Migiros, G. Study on uniaxial compressive strength, point load strength index, dynamic and physical properties of serpentinites from Central Greece: Test results and empirical relations. *Eng. Geol.* **2009**, *108*, 199–207. [CrossRef]
66. Ahmad, A.; Pamplona, M.; Simon, S. Ultrasonic testing for the investigation and characterization of stone—A non-destructive and transportable tool. *Stud. Conserv.* **2009**, *54*, 43–53. [CrossRef]
67. La Iglesia, A.; González, V.; López-Acevedo, V.; Viedma, C. Salt crystallization in porous construction materials. I. Estimation of crystallization pressure. *J. Cryst. Growth* **1997**, *177*, 111–118. [CrossRef]
68. Johnson, R.A.; Wichern, D.W. *Applied Multivariate Statistical Analysis*, 5th ed.; Prentice-Hall: Englewood Cliffs, NJ, USA, 1999.
69. Eren, O.; Bahali, M. Some engineering properties of natural building cut stones of Cyprus. *Constr. Build. Mater.* **2005**, *19*, 213–222. [CrossRef]

minerals

MDPI

Article

Copper-Bearing Mineralisation in the Upper Devonian Limestones: A Case Study from the Historical Teresa Adit in the Świętokrzyskie Mountains, Poland

Agnieszka Ciurej [1,*], Monika Struska [2], Anna Wolska [1], Marek Szczerba [3] and Janusz Olszak [4]

1 Institute of Geography, Geology Department, Pedagogical University of Krakow, Podchorążych 2, 30-084 Kraków, Poland
2 Okręgowy Urząd Górniczy w Kielcach, Wrzosowa 44, 25-211 Kielce, Poland
3 Institute of Geological Sciences, Polish Academy of Sciences, 31-002 Kraków, Poland
4 Faculty of Geology, Geophysics and Environmental Protection, AGH University of Science and Technology, 30-059 Kraków, Poland
* Correspondence: agnieszka.ciurej@up.krakow.pl

Abstract: The studied copper ore deposit is located in Miedzianka Mountain (Świętokrzyskie Mountains, central Poland). This deposit was exploited from the 13th century to the 1950s; therefore numerous historical adits are currently present. One of these is Teresa adit (established in 1805), consisting of underground mining corridors and natural cave developed in the Upper Devonian limestones, partially transformed by mining works. Samples of copper- and rock-forming minerals in limestones collected at seven sites within the richest copper-bearing mineralisation in this adit were studied with petrographic investigation of thin sections, micro-area chemical analysis (EDS), and XRD. This study shows the presence of various minerals: (a) Cu-Fe sulphides (relics of chalcopyrite) and Cu sulphides (covellite, chalcocite), (b) Cu-Fe oxides (cuprite and hematite), (c) Ca and Cu carbonates (calcite, azurite, and malachite), (d) clay minerals (Fe-Mg illite), and (e) micro-crystalline silica (quartz). For the first time in the studied deposit, we described chalcopyrite relics in cuprite pseudomorphosis, hematite with admixture of vanadium in pinkish-creme veins in limestones, and the presence of an admixture of Fe-Mg illite and microcystaline silica within cracks of limestones. In addition, for the first time, unit cell parameters of malachite and azurite from Miedzianka Mountain were determined, indicating very low substitutions of atoms other than Cu in their structures. We suppose that the minerals studied were formed during three types of copper mineralisation processes: (a) hydrothermal (relics of chalcopyrite), (b) secondary weathering (chalcocite, covellite, cuprite, hematite), and (c) adsorptive mineralisation (azurite, malachite). The latter stage is related to residuum, which consists of a mixture of Fe-Mg illite and micro-crystalline quartz, which was formed during the dissolution of limestones in karst processes in some crevices. We proposed a model of the formation of copper carbonates in the adsorption stage of the copper-bearing mineralisation in Miedzianka Mountain deposits. Two generations of calcite veins (older—red calcite and younger—crème-pinkish calcite) were also detected. Mineralogical–petrographical studies of samples revealed a high scientific and educational value. Due to the fact that the Teresa adit is planned to be made available to geotourists, this work is worth presenting to the public either in the adit and/or in a local museum in Miedzianka village.

Keywords: Miedzianka Mountain deposit; mining heritage; copper minerals; azurite; malachite

1. Introduction

Miedzianka Mountain is located in the southeastern part of Poland, about 20 km west of Kielce (Figure 1). The massif lies in the southwestern part of the Świętokrzyskie Mountains, in the northwest part of the Chęciński range. It consists of two rocky peaks: the eastern one, the higher of the two, 354 m above sea level, and the lower, western one, 350 m.a.s.l., separated by a mountain pass (Figure 2a,b). At the foot of Miedzianka Mountain lies Miedzianka village.

Figure 1. Position of Miedzianka Mountain in Poland.

Figure 2. Field photos showing historical mining activities in Miedzianka Mountain: (**a**) the summit of the mountain with shallow morphological depressions, which are the remains of collapsed mining excavations (funnels); a view towards the northwest; (**b**) peak of Miedzianka Mountain, with visible excavation remnants; a view towards the east; (**c**) an exposition of rocks, mining instruments, and machines in the Museum Chamber of Ore Mining in Miedzianka; (**d**) abandoned Devonian limestone quarry on the southeast slope of Miedzianka Mountain, showing the entrance (yellow arrow).

Miedzianka Mountain was a thriving centre of ore mining (copper, silver) since the 13th century [1]. Currently the deposit has not been exploited since 1958, when the Miedzianka Mountain Reserve was formed in the mountain and adjacent area. The history, geology, tectonics, and mineralogy of Miedzianka Mountain deposit have been studied for almost 200 years, e.g., [1–7]. Forty minerals have been found in this deposit so far [8]. Two adits named Teresa and Zofia located on Miedzianka Mountain are planned to be opened for tourists [9]. Recent exploration of the Teresa adit has shown that it has unique geological, mining, mineralogical, and historical features [10]. This work is a comprehensive study of the minerals in the Teresa adit, which could constitute an important educational and scientific contribution, especially when this adit is made publicly available.

The aim of this study was mineralogical and petrographical investigation of copper and accompanying minerals at selected points in the Teresa adit. A comprehensive documentation of the minerals in the Teresa adit is presented in the context of making it available to tourists and constructing the mineralogical museum's collection. For the first time in this adit, in situ samples were collected from seven selected points with the richest mineralisation, and then studied using petrographic investigation of thin sections, micro-area chemical analysis (EDS), and XRD analysis. The results presented in this study are a significant supplement to the information for the proposed geotourist route in the Teresa adit [10,11].

2. Outline of the History of Mineral Investigation in the Miedzianka Mountain Deposit

2.1. Exploitation of the Deposit

Historical sources indicate that Miedzianka Mountain was an important source of copper and silver in Poland through the centuries [1]. For example, according to Czarnocki [5], from 1820 to 1840 almost 18 thousand (17.623) tons of Cu ores were excavated here. The copper content in this deposit was 1%–4% [5]. Later data showed that from 1915 to 1918 about 1360 tons of copper ore were extracted [5]. At present, numerous traces of old mining activity, including buried shafts, mine craters, and heaps, are observed on Miedzianka Mountain (Figure 2a,b). There are also three adits: Zofia, Teresa, and Łaszczyńscy, as well as the Austrian shaft. The Teresa and Zofia adits were actively operated from the beginning of the 19th century to the 1920s and are currently unavailable.

Due to the historical importance of this place, its legal protection was planned from the 1950s. Finally in 1958 Miedzianka Mountain was granted the status of an inanimate nature reserve. The area of the Miedzianka Mountain Reserve is approx. 0.25 km^2. There is also the Museum Chamber of Ore Mining in Miedzianka, where, among other exhibits, the old mining machines (Figure 2c) and the Piotr shaft tower are presented. This reserve is located in the Kielce-Chęciny Landscape Park and is one of the most beautiful points of the Świętokrzyski Archaeological and Geological Trail. Already in the 19th century, there were plans to open the adits for tourism. In 2017, preliminary research on making underground workings available for geotourism (geotouristic path) were published [9]. Miedzianka Mountain is also one of the geological sites located in the Świętokrzyski Geopark (Holy Cross Mountains Geopark) which in the 2021 was granted UNESCO Global Geopark status.

2.2. Minerals Described from the Deposit

The first general descriptions of the Miedzianka Mountain deposit and copper mineralisation were made by Stanisław Staszic [2], who is named the father of geology in Poland, due to his precursor geological and deposit research in Poland. In 1841, a monograph on the deposits and their exploitation was published [12]. From the mid-19th century, the copper ore deposit on Miedzianka Mountain was abandoned for economic and political reasons. At the beginning of the 20th century, operations were resumed by the mining company of the Łaszczyński brothers. In the first half part of the 20th century three new phases were described by Morozewicz: "staszicite" [3], "lubekite" [4], and "miedziankite" [13]. They were not accepted by IMA CNMNC (International Mineralogical Association, Commission on New Minerals, Nomenclature and Classification). Detailed studies in the 21st century showed that the "staszicite" is a zincian variety of conichalcite (fomula: $CaCu(AsO_4)(OH)$ [14], "lubeckite" is an admixture of a cuprian–cobaltoan wad (formula: $(Mn,Cu,Co)_xO_y$) [8], whereas "miedziankite" is Zn-rich tennantite-Cu (formula: $(Cu,Zn)_{12}(As,Sb)_4S_{13}$) [15].

Mineralogical studies on minerals in the Miedzianka Mountain deposit were later carried out by Łaszkiewicz [16]. In 1928, Czarnocki [17] described the relationship between tectonics and copper mineralisation. After the Second World War, the research was continued; however, in the 1950s it was decided that the Miedzianka Mountain ore was not suitable for mining for economic reasons. Despite the fact that the ore deposit is no longer exploited, mineralogical, geochemical, and geological studies are still being carried out e.g., [1,6–8,10,11,18–23].

Recently, for the purposes of mineralogical and geochemical research, samples are usually taken from heaps, but occasionally from underground adits, e.g., [8,18]. Forty minerals from this deposit have been described so far [8] and references therein. They are mainly copper ore minerals (sulphides, sulphosalts, oxides, and carbonates), but also very rare minerals (such arsenates, vanadates, and phosphates) [6,7,18,19,24–27].

2.3. The Origin of the Deposit

The origin and age of mineralisation in the Miedzianka Mountain deposit were discussed in many papers. It is assumed that the Miedzianka Mountain deposit has a hydrothermal origin, related to migration of low-temperature fluids (epithermal) [6,28] or high-temperature ones [7]. Kozłowski [8] suggested the primary mineral assemblages are of meso- to (mostly) epithermal origin; many of the mineral assemblages are of the tele-thermal type. The deposit was probably formed during the Variscan movements in the Lower Permian [29] and before the Upper Permian because there is no copper mineralisation in the Buntsandstein sandstones and Zeichstein conglomerates. The ore concentration in the oxidation and weathering zone of the primary copper deposit was described by Urban [30] as impregnations in tectonic zones and/or vein ores.

Ore mineralisation in the Miedzianka Mountain deposit is described as a complex process consisting of three stages [6]. According to Rubinowski [6], the first hydrothermal stage of the formation of primary copper ore in Devonian rocks is associated with hydrothermal solutions. The presence of such solutions was associated with the intrusion of diabase and lamprophyre glands into the Cambrian and Lower Devonian rocks in the area of the Świętokrzyskie Mountains [6,31–33]. Based on the geological position of these rocks in the exposures and drill cores, it was assumed that their intruded age and the formation of ore-bearing hydrothermal solutions are Upper Carboniferous/Lower Permian [31,33,34]. This primary mineralisation is related to the older longitudinal fault parallel to the axis of the Chęciny anticline [28] and forms a vein deposit located in the tectonic fractures. New data [35] using K-Ar and Ar-Ar methods for diabases and lamprophyres showed that these rocks are between 391.7 and 254.5 Ma (Lower Devonian–Permian).

The formation of secondary copper sulphides and the second stage of ore mineralisation were linked by Rubinowski [6] with the processes of primary deposit weathering and called the weathering stage. The third adsorption stage [6] was related to the "capture" of copper ions by the clay and carbonate residue resulting from the karstification process of Devonian limestones, which led to the formation of copper carbonates.

Minerals that occur in the primary ore zone belonging to the primary paragenesis are chalcopyrite ($CuFeS_2$), Zn-rich tennantite-Cu [$(Cu, Zn)_{12}$ $(As,Sb)_4S_{13}$], and galena (PbS) [7]. "Miedziankite" oxidises quickly, and the unoxidised mineral phase can only observed at the mining site on the sides of the adit tunnels [18]. The "Miedziankite" phase is also called grey copper ore (so-called "szaromiedniak") [36].

Kozłowski [8] also included native gold as a primary mineral of the Miedzianka Mountain deposits. It was probably also an admixture of sulphides (chalcopyrite) or Cu-sulphosalts (tennantite series). Chemical weathering of primary ores resulted in the release of gold from their crystal lattice and its secondary accumulation in hypergenic minerals—Fe oxides and hydroxides and malachite [7,8,37,38].

Secondary minerals related to the oxidation zone described in the Miedzianka Mountain deposit [6–8,39] are copper carbonates (malachite and azurite), chalcocite (Cu_2S), covellite (CuS), bornite (Cu_5FeS_4), cuprite (Cu_2O), tenorite (CuO), and iron oxides. Secondary minerals are found in veins/veinlets as well as in karst funnels into Devonian limestones.

In the primary zone of the deposit, there is also chalcocite (Cu_2S), which, according to Rubinowski [6], is the dominant mineral in the Miedzianka deposit. The origin of this mineral is still under discussion. Rubinowski [6] described that this chalcocite occurs in a rhombus-shaped modification (the so-called β-chalcocite) and its formation was associated with the weathering process of the deposit. On the other hand, Balcerzak et al. [7] suggested that chalcocite occurs in veins in the zone of primary ores, but it cannot be ruled out that these veins were formed as a result of chalcocite migration from the ore zone changed by distention solutions. These authors reported that chalcocite occurs in two overgrown forms that are crystallographically related and do not take up relics of bornite and chalcopyrite.

Gold found in some minerals, i.e., hematite and malachite, was also described from this copper deposit [3,7,8,38,40]. As suggested by Balcerzak et al. [7] gold was formed during the hypergenic processes and is located in the lower part of weathering ores. In the 16th century, gold and silver were extracted from malachite and other Cu-bearing minerals that were mined at Miedzianka Mountain. Extracted gold was used to make the golden dome of Sigismund's Chapel (a funerary chapel for the last kings of the Jagiellonian dynasty) in the Wawel Cathedral in Kraków, Poland [41].

3. Study Area

Miedzianka Mountain is located in the Świętokrzyskie Mountains (Kielce zone) in the south wing of the Chęciny anticline (Czarnocki vide Konon [42]). It is composed of Devonian massive organogenetic limestones (Upper Givetian, Frasnian) and plate limestones (Frasnian–Famennian [6]). On the southern slopes there are Triassic sediments that inconsistently lie on the Palaeozoic formations [21]. In the east part, clay-mudstone Cambrian rocks occur, and build the core of the Chęciny anticline (Figure 3). The contact between the Palaeozoic series is tectonic because the Devonian rocks are overlapped by the Cambrian rocks [6].

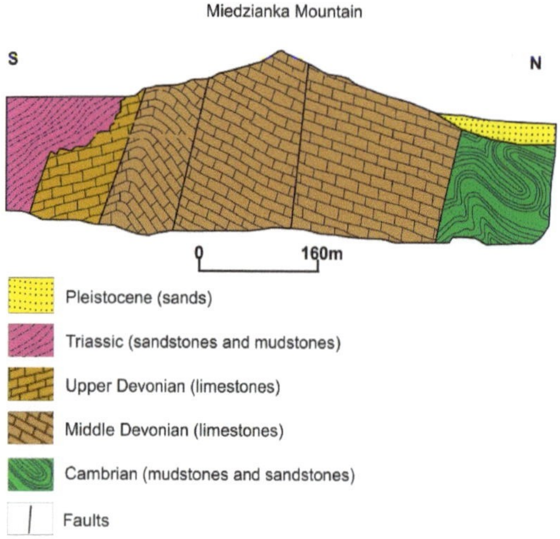

Figure 3. Geological cross-section of Miedzianka Mountain [28], simplified.

Our investigations were carried out in the Teresa adit, which is located on the eastern slope of Miedzianka Mountain. It is hollowed out within the Upper Devonian (Frasnian–

Famennian) limestones, which are cut by cherry, marly shales (Famennian) [6]. The entrance is located in the eastern part an abandoned quarry (Figure 2d). The Teresa adit is a 523 m system of underground corridors including 270 m of natural karst caves, altered by mining works (Figure 4). There are also two shafts that are inaccessible. Detailed descriptions of the Teresa adit are included in a paper published by Ciurej et al. [10].

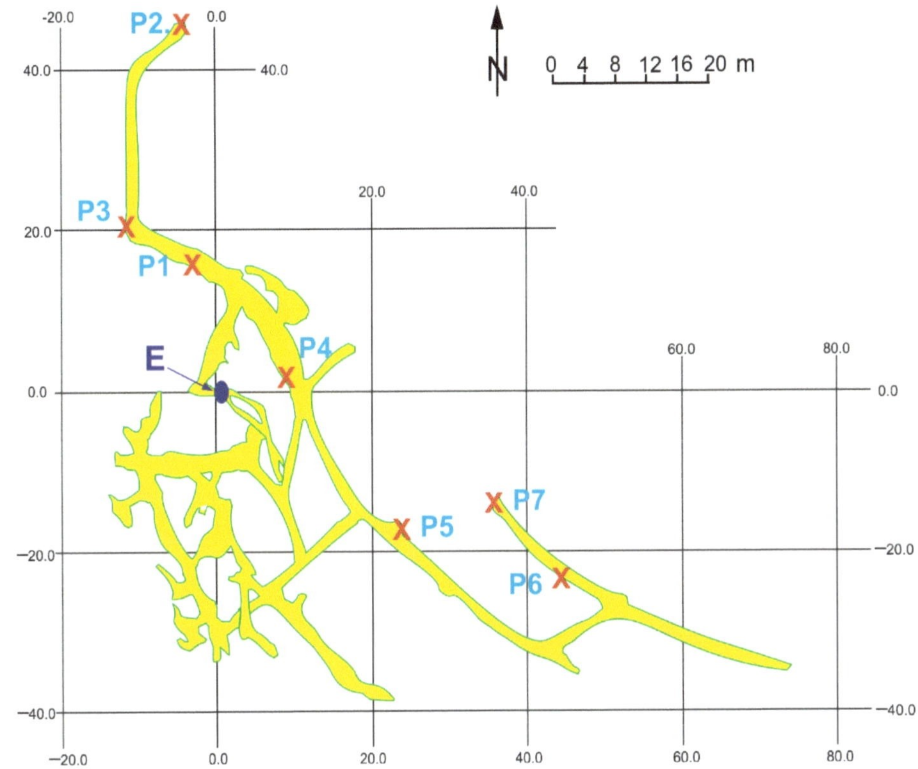

Figure 4. Simplified map of the cave in the Teresa adit inside Miedzianka Mountain [43], with the location of the studied sites and samples collected for mineralogical studies. X—location of studied sites and E—entrance.

4. Research Methods

The Teresa adit is not publicly available and it was necessary to obtain permission in order to explore and collect samples from the Regional Director for Environmental Protection in Kielce. The study of the Teresa adit took place during winter 2018/2019, by a team of specialists in the fields of geology and speleology. Exploration covered 150 m of the adit's natural caves, corridors, and underground galleries. Two shafts were not examined, because they are collapsed and their exploration could be associated with high risk. Seven representative sites with symbols P1–P7 were selected for detailed examination and laboratory studies (Figure 4). For detailed descriptions of the studied sites see Ciurej et al. [10]. In total, twenty-three rock samples were obtained. The samples were chipped (detached) with a hammer and also drilled using a core drill with a diameter of 2.5 cm and 5 cm (Figure 5). Twenty-four petrographic, uncovered thin sections of about 5 cm × 3 cm size were made. All thin sections were observed with a Nikon Eclipse LV100N POL (Japan) polarizing optical microscope with a photographic camera and NIS-Elements BR (ver. 4.30) software located at the Department of Geology, Pedagogical University in Krakow.

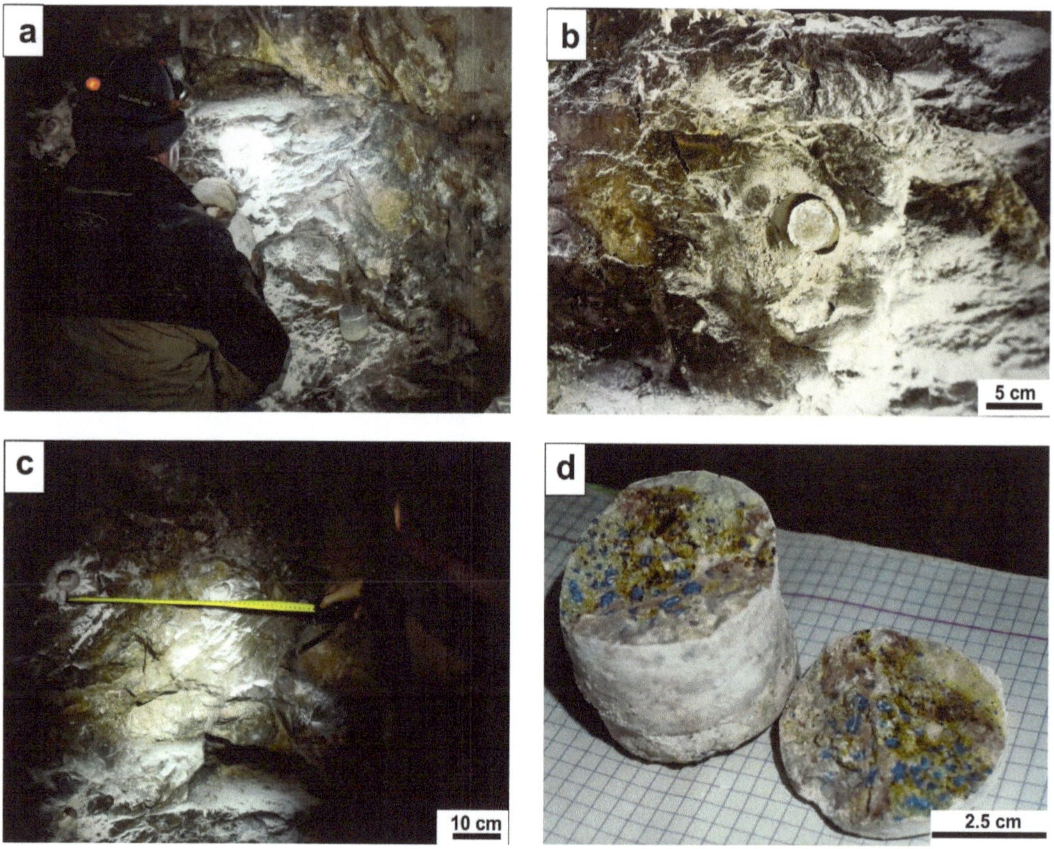

Figure 5. Sampling procedure: (**a**) drilling a sample with a core drill in dead end walkway, site P7; (**b**) sample P3A1-2 drilled in the sidewalk wall, site P3; (**c**) pulling out the core of sample P7A1-1 from the sidewalk wall, site P7; (**d**) core of sample P6A1-1 obtained by using a core drill, note distinct deep blue colour of azurite mineral dispersed in rock, site P6.

The chemical composition of ore minerals was determined using the SEM-EDS method. Ten selected samples, including thin sections and rock chips of about 5 cm × 3 cm size, were examined. Carbon-coated polished thin sections and rock chips of the studied samples were analysed using a HITACHI S 4700 (Japan) microscope with a Vantage (NORAN EDX) spectrometer. The time of analysis was 100 s for point at an acceleration voltage of 20 kV. The ZAF correction algorithm was used. The study was carried at the Laboratory with Field Scanning Emission Microscopy and Microanalysis at the Institute of Geological Sciences of the Jagiellonian University, Krakow, Poland.

The mineralogy of five selected samples was determined on randomly oriented powder specimens. The material was precisely extracted with a dental scalpel and the aim was to obtain information on minerals from selected crusts. The samples were then milled in an agate mortar. X-ray diffraction measurements were carried out using a Bruker D8 diffractometer (USA) located in the Clay Minerals Laboratory in the Institute of Geological Sciences, Polish Academy of Sciences, Kraków, Poland. Diffractograms were recorded with Co radiation in the range of 4–85° 2 Θ, with a step size of 0.014°, registered every 3 s. BGMN/Profex (ver. 4.3) software was used for quantitative analysis. The range below 8° or 10° 2 Θ was excluded from fitting due to high background at lower angles and artificial peaks caused by diffraction of X-rays that were not fully filtered by Fe foil. Peak positions were corrected by assuming sample displacement. All major phases were calculated with corrections for crystallite sizes, microstrain, as well as preferred orientation. A LaB_6 crystallite size standard (SRM 660a) was used as a test for correct instrumental function calculation. Unit cell parameters of the identified Cu-bearing phases were determined and compared with parameters of these minerals from the American Mineralogist Crystal Structure Database (AMCSD).

The samples are archived at the Institute of Geography, Geology Department, Pedagogical University of Krakow, Podchorążych 2, 30-084 Kraków, Poland; Collection reference: Góra Miedzianka, 2019/ P1-P7.

5. Results

Devonian limestones with calcite (Figure 6a) and copper (Figure 6b) mineralisation were collected for this study. The presence of secondary copper minerals—secondary copper sulphides, copper oxides— and carbonates—malachite (Figure 6c) and azurite (Figure 6d)—as well as iron minerals (Fe-oxides) and a mixture of Fe-Mg illite with microcrystalline quartz/silica (in karst fractures) were found.

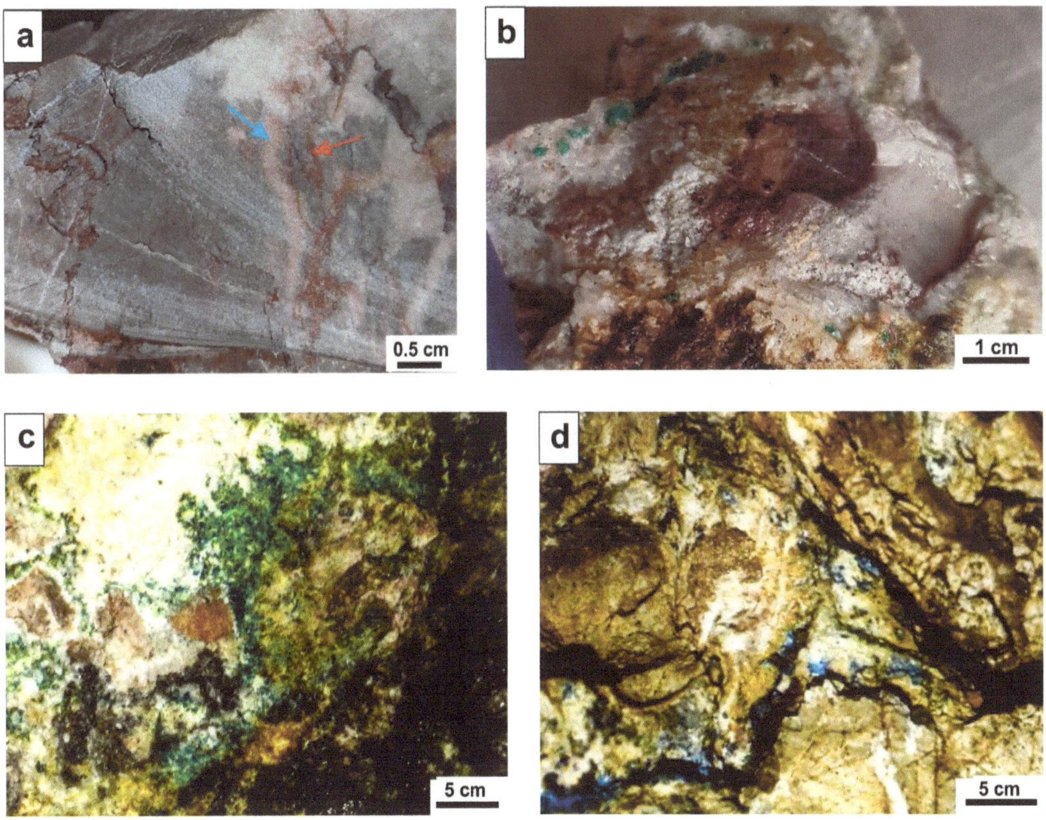

Figure 6. Devonian limestone with various styles of mineralisation in the Teresa adit: (**a**) fine-grained micrite limestones with different-colour calcite (rosa-like calcite type) vein: old—red (red arrow) and young—creme-pinkish (blue arrow) generations, sample P3A3-1; (**b**) breccia of Devonian limestones cut with veins of red calcite and malachite, sample P1A1; (**c**) underground photo showing the malachite (green colour) and black mineralisation in creme-pinkish calcite vein, site P7; (**d**) underground photo showing the nest azurite (blue colour) mineralisation in creme-pinkish calcite vein, site P6.

XRD tests showed the presence of the above mentioned minerals (Figure 7, Tables 1 and 2). The determined unit cell parameters show that azurite is within precision, practically identical to the one from the database, indicating a lack or small substitutions of Cu in the structure. For malachite the unit cell parameters are also very close to those from the database, but the a and b parameters are slightly lower, while c is slightly higher than in the AMCSD.

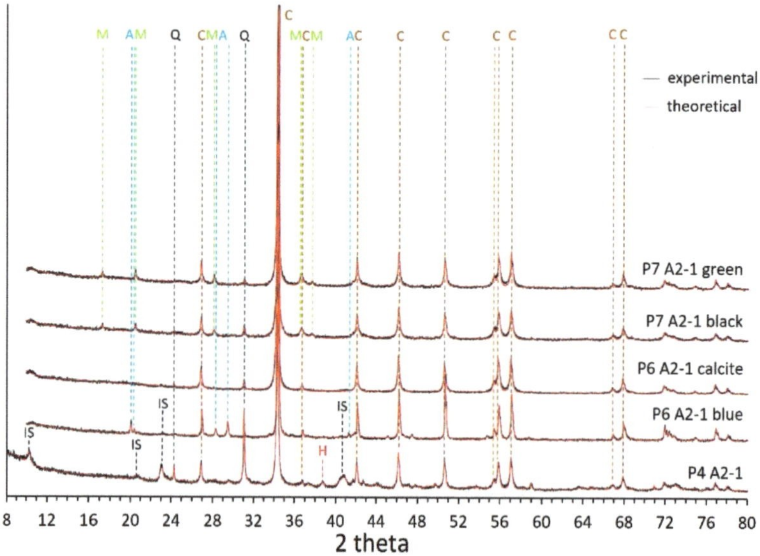

Figure 7. Comparison of experimental (black) and calculated (red) diffractograms of selected samples. C—calcite, Q—quartz, A—azurite, M—malachite, IS—mica/illite, H—hematite.

Table 1. Quantitative composition of precisely extracted crusts. Optimised unit cell parameters of copper-bearing phases are expressed in Å and degrees. For sample P4A2-1, unit cell parameters are not shown because of the low content of this phase, and thus its low precision. Unit cell values in brackets are from the AMCSD database [44,45].

Sample	Quartz	Calcite	Azurite	Malachite	Mica/Illite	Hematite	Siderite	SUM
P4A2-1	8.1	40.3	0.8		47.7	2.5	0.6	100.0
P6A2-1 calcite	2.5	97.5						100.0
P6A2-1 blue	0.4	85.5	8.8 a = 5.0150 ± 0.0063 (5.011) b = 5.8482 ± 0.0092 (5.850) c = 10.349 ± 0.012 (10.353) β = 92.48° ± 0.10 (92.41)		5.3			100.0
P7A1-1 green	1.3	87.9		10.8 a = 9.492 ± 0.027 (9.502) b = 11.946 ± 0.025 (11.974) c = 3.2441 ± 0.0082 (3.240) β = 98.56° ± 0.20 (98.75)				100.0
P7A1-1 black	3.0	89.0		8.0 a = 9.493 ± 0.038 (9.502) b = 11.943 ± 0.033 (11.974) c = 3.243 ± 0.012 (3.240) β = 98.55° ± 0.30 (98.75)				100.0

Table 2. Reflections used for qualitative analysis of copper-bearing phases.

Phase	2 Theta (Co)	Relative Intensity	D-Spacing	hkl
Malachite	17.18	48.11	5.9870	020
	20.41	67.53	5.0484	120
	28.02	68.15	3.6948	220
	36.51	100.00	2.8556	−201
	37.57	43.29	2.7777	−211
Azurite	19.92	61.54	5.1719	002
	20.23	29.12	5.0921	011
	28.17	55.14	3.6753	−102
	29.33	100.00	3.5336	111
	41.14	26.31	2.5460	022

Malachite was found in both samples: P7 green and P7 black, based on XRD analysis (Figure 7, Table 1). The black mineralisation in the sample P7 black was not detected by XRD. The difference in colour in both samples can be related to admixtures of other minerals; sample P7 black can contain cuprite, that is below the sensitivity of XRD.

5.1. Characteristics of Devonian Limestones

Ore mineralisation occurs in host Middle Devonian limestones. The limestones are light grey to grey, fine-grained (Figure 6a), and contain fauna (e.g., *Amphipora* sp.; *Stromatopora* sp., *Megalodon* sp., *Lexonema* sp., *Tetracorallia* sp.) [11,46]. The studied limestone was described as biolithite/boundstone and biostrome [47,48]. The calcite in the limestone is fine-grained and occurs in the micrite variety (Figure 8a). Micro-area chemical analysis of the micrite calcite showed that it is a calcite with an admixture of magnesium (MgO from 0.6 to 0.5 wt.%) (Figure 9a—sample P6A1).

The limestones are strongly tectonically fractured and cut by numerous veins and veinlets (0.5–0.1 cm). The veins and veinlets are of various colours (white, pink, black, brown, blue, and green—Figures 6a,b and 9b) because they are filled with different minerals—white and creme: pinkish calcite (Figures 8a and 9c); black: Cu oxides (tenorite?) (Figure 8b) and cuprite (Figure 8c,d); blue and green: Cu carbonates—azurite (Figure 8e) and malachite (Figure 8f); red-brown, red Fe hydroxides (Figure 8f) and oxides (hematite) (Figure 8g,h).

In thin sections, the studied limestones are micritic (Figure 8a) with stylolite seams (Figure 8b). The rocks are cut by numerous veins and veinlets. The carbonates from white and pink veins and veinlets contain sparitic, colourless calcite (Figure 8a). In transmitted light, this mineral has variable relief, low to high (n_ω = 1.640–1.660, n_ε = 1.86), high birefringence (δ = 0.154–0.174) [49], and high-order interference colour. In fine-grained aggregates (micritic aggregates), the rhombohedral cleavage is not observed, while in the veins, rhombohedral cleavage is visible in the sparitic calcite crystals (coarse grains) as well as polysynthetic lamellar twinning in two directions (Figure 8a). These twins form as a result of relatively low mechanical stress during the tectonic deformation of rocks.

In karst nests, which were formed due the weathering processes in the Permian/Early Triassic, the Fe-Mg illite blasts and micro-crystalline quartz grains occurred. The presence of these minerals in the studied limestones was confirmed by the XRD data (Figure 7—sample P4A2-1, Table 1) as well as the EDS investigations (Figure 9d–f—sample P6A1). The small broadening of ~10° 2Θ peak of illite indicates a low percentage of the smectitic component (10%–20%).

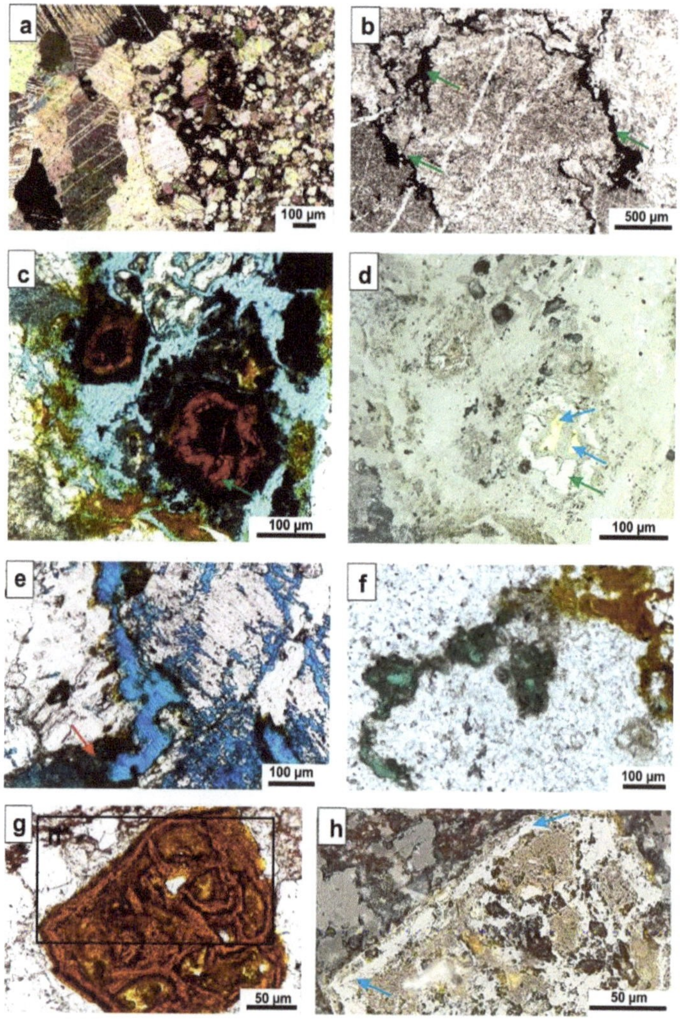

Figure 8. Minerals in optical microscope views: (**a**) host micritic Devonian limestone (right part) cut by veinlet of sparite, creme-pinkish calcite (left part), sample P1A1 (transmitted light XPL); (**b**) stylolite impregnated by the so-called "black/tarry copper ore" (green arrows), sample P3A1 (transmitted light PPL); (**c**) cuprite (cochineal red colour, green arrow, left) in azurite nests (blue colour) in the centre of the pseudomorphosis chalcopyrite relics (black), sample P6A1-1 (transmitted light PPL); (**d**) cuprite (bluish white colour, green arrow) in azurite nests (light gray). In the centrecan be observed yellow chalcopyrite relics (blue arrows), which are replaced by cuprite, sample P6A1-1 (reflected light, PPL); (**e**) fibrous azurite (deep blue) locally turns into malachite (red arrow), sample P6A1-1, (transmitted light PPL); (**f**) fibrous malachite (green colour) nests/veinlets (left) and secondary Fe-hydroxide (geothite) (brown colour), on the right side, sample P6A1-1 (transmitted light PPL); (**g**) hematite (red-brown colour) pseudomorphosis (called "kidney ore") after primary sulphides, sample P1A1 (transmitted light PPL); (**h**) greyish-white hematite (blue arrows) in pseudomorphosis after primary sulphides, sample P1A1, enlarged from (**g**) (reflected light, PPL).

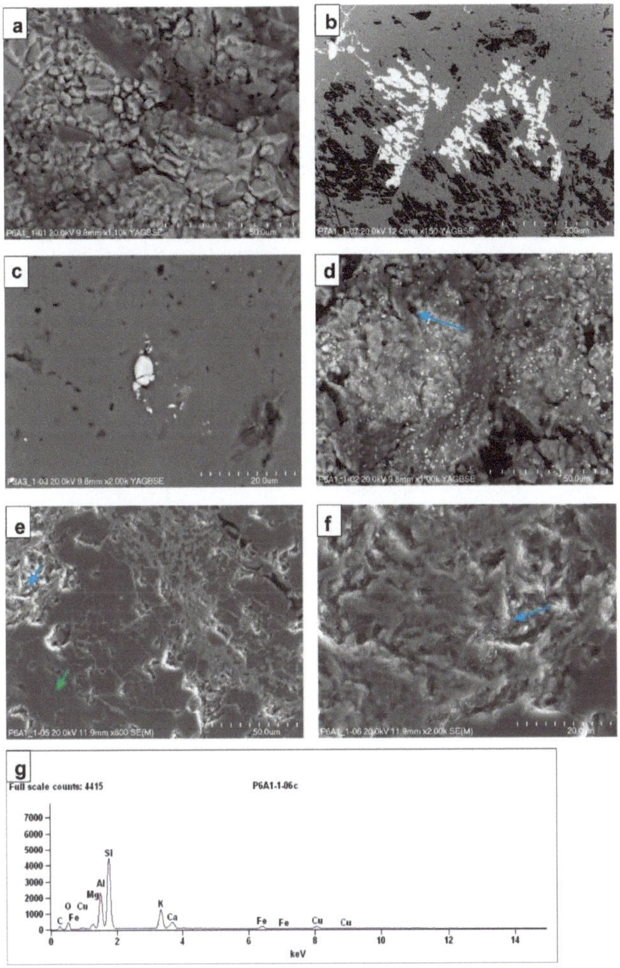

Figure 9. SEM images of secondary minerals in the oxidised zones: (**a**) host Devonian limestones—calcite with admixture of magnesium, sample P6A1; (**b**) azurite nest (white) in limestone (gray), (black is an artefact), sample P7A1; (**c**) scattered hematite crystals with admixture of V_2O_5 0.6 wt.% (limit detection for EDS $\pm$ 0.05) in creme-pinkish calcite, sample P3A3; (**d**) host Devonian limestones—Fe-Mg illite nests (blue arrow) and scattered Fe oxides (light) in calcite, sample P6A1; (**e**) Mg-Fe illite nests (blue arrow) in calcite (green arrow), sample P6A1; (**f**) calcite with Fe-Mg illite nests (blue arrow); blue arrow also marks the EDS analyses, shown in (**g**), sample P6A1, (**g**) microprobe analyses of Fe-Mg illite, taken from site marked with blue arrow in (**f**).

5.2. Characteristics of Minerals from the Teresa Adit

5.2.1. Primary Minerals—Ore Mineralisation

In the samples studied from the Teresa adit only relics of primary chalcopyrite are visible in sample P6A1-1 (Figure 8c,d). This mineral has a yellow colour in reflected light. In the other samples (P1A1, P3A3, P7A1), the pseudomorphoses after primary sulphides are filled by Fe oxides (hematite), calcite, Fe-Mg illite (Figure 8g,h (sample P1A1) and 10a (sample P3A3)), as well as azurite (Figure 10b sample P7A1).

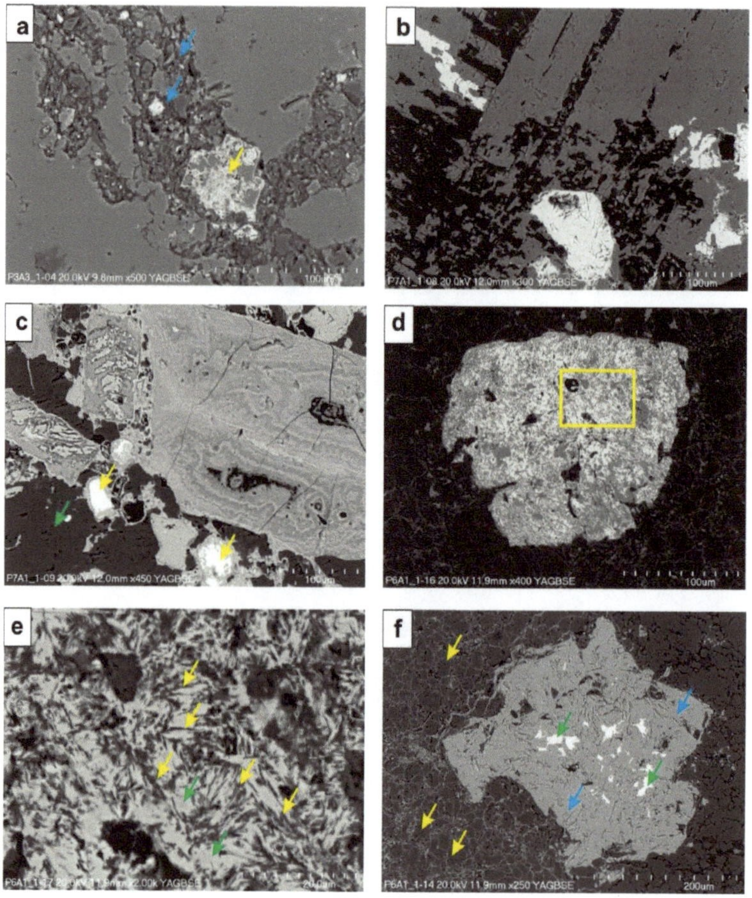

Figure 10. SEM images of primary and secondary sulphide minerals in the oxidised zones: (**a**) pseudomorphosis after primary sulphide (pyrite) (marked by yellow arrow) filled with a mixture of calcite, hematite, and Fe-Mg illite is present in cracks (blue arrows), sample P3A3; (**b**) pseudomorphosis after primary sulphide (pyrite) filled with azurite, sample P7A1; (**c**) pseudomorphosis (right side) after primary sulphide (pyrite) and sulfosalt (Zn-rich tennantite-Cu) filled with a mixture of Fe oxide and conichalcite. Covellite (yellow arrows), calcite (green arrow), sample P7A1; (**d**) replacement of Cu sulphide (chalcocite) by azurite, enlarged in (**e**), sample P6A1; (**e**) enlarged view from (**d**) of replacement of Cu sulphide (chalcocite, green arrows) by azurite (yellow arrows), sample P6A1; (**f**) secondary sulphide relics (covellite, green arrows) in azurite (blue arrows) pseudomorphosis and host calcite (yellow arrows), sample P6A1.

5.2.2. Secondary Minerals in the Oxidised Zones

During the research performed in the present study the minerals Cu sulphide minerals chalcocite (Cu_2S) and covellite (CuS) were detected. They were identified in the studied samples (P6A1-1 and P7A1) using micro-area chemical analysis (Figure 10c–f). The size of these minerals is small (a few microns to 10–25 μm). They appear as relics in the secondary Cu carbonates (mainly azurite). We have to remark here that the primary or secondary origin of chalcocite in this Miedzianka deposit is still under debate (see Section 2.3).

Cuprite—Cu_2O. This mineral forms octahedral crystals or less often cubic crystals; more common are massive, compact varieties (Figure 8c,d). Its colour is dark red to cochineal red, sometimes black; compact varieties are brown-red. In transmitted light, thin and small crystals of cuprite are translucent and have cochineal red colour (Figure 8c). It has exceptionally high relief (n = 2.849), is isotropic, and has no birefringence but sometimes exhibits anomalous birefringence [47]. Its colour in reflected light is bluish-white with internal blood-red reflections (Figure 8d). Cuprite is commonly formed as an oxidation product of copper sulphides in the upper zone of veins. In the studied samples it is often associated with azurite (Figure 8c). In the Teresa adit this mineral occurs in sample P6A1-1. Earthy and massive clusters of cuprite with Fe hydroxides formed the so-called "orange-red (brick colour) copper ore" [50], while the mixture of earthy clusters of cuprite, tenorite, and Fe oxide and hydroxide with silica formed the so-called "black (tarry colour) copper ore" [50] (Figure 8b).

Azurite—$Cu_3(CO_3)_2(OH)_2$. This mineral is typically found as tabular to prismatic crystals of a deep blue ("azure blue") colour [18]. In the Teresa adit it occurs as fine-grained massive aggregates or earthy clusters (Figure 8c) as well as fibrous aggregates (Figures 8e and 10d). In transmitted light, azurite is transparent or translucent and has a blue colour (Figure 8c,e). The mineral has high surface relief (n_α = 1.730, n_β = 1.758, n_γ = 1.830) and high birefringence (δ = 0.108). Its pleochroism is visible in blue tones (in plane-polarised light) [49]. Azurite often fills pseudomorphoses after primary and secondary sulphides (Figure 10d–f) and secondary Cu oxides (Figure 8c). This mineral also occurs in veinlets (Figure 8e). XRD data of azurite (the sample P6A2-1 blue) are presented in Tables 1 and 2, and Figure 7. Optimized/calculated unit cell parameters of azurite from the Teresa adit (sample P6A2-1 blue) follow, and are identical to the parameters of the standard from the AMCSD database:

- a = 5.0150 ± 0.0063
- b = 5.8482 ± 0.0092
- c = 10.349 ± 0.012
- β = 92.48° ± 0.10

These data indicate a lack of significant atomic substitutions in the mineral structure.

Malachite—$Cu_2(CO_3)(OH)_2$. Similar to azurite, malachite is formed as a result of the oxidation of copper ore. In the samples studied from the Teresa adit, malachite is rare compared to azurite. This mineral occurs as massive, compact, or earthy clusters in veinlets in Devonian limestones (Figure 8f). Crystalline, radiate clusters of malachite are uncommon in the studied samples. Small crystals are rarely observed [6]. In transmitted light it is transparent and has a green colour (Figure 8f). Pleochroism is visible: X—nearly colourless, Y—yellowish green, Z—deep green. Surface relief is very high (n_α = 1.655, n_β = 1.875, n_γ = 1.909); birefringence is also very high (δ = 0.254) [49].

XRD data of malachite (samples P6A1-1 green and P6A1-1 black) are presented in Tables 1 and 2, and Figure 7. Calculated unit cell parameters of malachite from the Teresa adit (samples P7A2-1 green and P7A2-1 black) are as follows:

Sample P7A2-1 green:	Sample P7A2-1 black:
a = 9.492 ± 0.027	a = 9.493 ± 0.038
b = 11.946 ± 0.025	b = 11.943 ± 0.033
c = 3.2441 ± 0.0082	c = 3.243 ± 0.012
β = 98.56° ± 0.20	β = 98.55° ± 0.30

Both are very close to the standards from AMCSD. Small admixtures of bigger atoms of Ag or Au in the structure can lead to slight deformations of the unit cell and thus a decrease in a and b parameters with an increase in the c parameter in comparison to the values from AMSCD.

Hematite—Fe_2O_3. Fe oxide is an opaque mineral that has brown to reddish-brown colours (Figure 8g,h). This mineral occurs as fibrous masses, sometimes radiating or in reniform masses with a smooth fracture "kidney ore" (Figure 8g). It is frequently in earthy masses in veinlets or in stylolite seams in Devonian limestones. Hematite has very high surface relief (n_ω = 3.150–3.220, n_ε = 2.870–2.940) and high birefringence (δ = 0.280) [49]. In transmitted light, pleochroism is weak from brownish-red to yellowish-red (Figure 8g).

In reflected light this mineral is white to greyish-white with a bluish tint (Figure 8h). The mineral filled the pseudomorphoses after primary sulphides and occurs in veinlets (Figures 8g,h and 10a). It also occurs in the pseudomophoses after primary sulphides accompanied by primary sulphide pyrite and sulfosalt Zn-rich tennantite-Cu (historical name "miedziankite") Figure 10c. XRD investigations showed the presence of hematite in sample P4A2-1 (Table 1, Figure 7). Very small crystals of hematite are scattered in limestones (sample P6A1,Figure 9d) and in creme-pinkish calcite veinlets (sample P3A3, Figure 9c). Sometimes, hematite has the admixture of V_2O_5 0.6 wt.% (limit of detection for EDS $\pm$ 0.05) as shown in the microprobe chemical analysis of sample P3A3.

Illite flakes occur in nests in Devonian limestones (Figure 9e,f). They are very small in size, up to 3 μm. Illite together with calcite and micro-crystalline quartz/silica is found in fractures in limestones (Figure 10a). XRD investigations showed the presence of illite in sample P4A2-1 (Table 1, Figure 7). Based on the microprobe chemical analysis, it was found that the composition of this mineral includes FeO 3.3 wt.% (limit of detection for EDS $\pm$ 0.03) and MgO 3.0 wt.% (limit of detection for EDS $\pm$ 0.04). The occurrence of this mineral in nest and karst fractures in Devonian limestones is connected to the weathering processes and the development of karst on the exposed limestone surface during the Permian/Early Triassic.

6. Discussion

The Teresa adit is one of the oldest adits in Miedzianka Mountain. It was hollowed in 1805 by the Austrians, then expanded many times and was finally connected to the natural cave system. A preliminary description of the Teresa adit exploration was presented by Borowiecki et al. [9], which focused on technical issues related to safety and adaptation for tourism purposes. Furthermore, in the work of Ciurej et al. [10] there is a description of the adit, itspavements, and the natural corridors of the cave.

In the Miedzianka Mountain copper deposit, 40 minerals, both primary and secondary, have been described [8]. Paragenesis of primary ore minerals was described in the works of researchers for over 100 years [6,7,22,24–27,33,37] and include the chalcopyrite, tennantite variety ("miedziankite"), and galena.

In the samples collected in the Teresa adit, only one mineral of this paragenesis was found, unfortunately in the form of relics of the primary chalcopyrite in a pseudomorphosis filled with secondary cuprite (Figure 8d). Relics of primary chalcopyrite surrounded by cuprite in the Miedzinka deposits was for the first time described in this paper. In this study, the copper sulphides covellite (Figure 10f) and chalcocite (Figure 10e,d) were found.

The origin of the chalcocite is still not clear, whether it is a primary or secondary mineral, as was mentioned in Section 2.3. According to Rubinowski [6,51], the formation of these minerals is related to the stage of weathering of the primary deposit, the so-called weathering stage. Balcerzak et al. [7] described that chalcocite occurs in the veins in the primary zone, which could suggest its primary origin, although they do not exclude the possibility of the formation of this mineral as a result of chalcocite migration from the altered ores (oxidised and karstic secondary ores) and by distention solutions. Balcerzak et al. [7] determined that chalcocite was formed at a temperature not higher than 103 °C. Chalcocite is formed in reducing conditions as the first sulphide mineral in the secondary sulphide cementation zone [41]. Covellite is also related to the secondary sulphide paragenesis, but it is formed later during cementation processes.

During this stage, apart from the formation of secondary copper sulphides, copper oxides, mainly cuprite, were formed, which was described in our work. In sample P6 (Figure 8c,d) a pseudomorphosis filled with cuprite with relics of primary chalcopyrite was found. Cuprite is formed at a lower partial pressure of O_2 than tennorite [41].

The release of iron during primary chalcopyrite weathering is indicated by the presence of veins of goethite (Figure 8f) and hematite (Figures 7 and 8g,h) in the limestones. Also, the two generations of rosa-like calcite (older (red) and younger (creme-pinkish) veins) found in limestones are coloured with diffused hematite pigment (Figures 8a and 9c).

Copper carbonate mineralisation (azurite and malachite) is visible in many thin veins (Figure 6c,d), and can also be found in the creme-pinkish veins of calcite (Figure 6d). The formation of these minerals is related to the last stage of cupriferous mineralisation—the adsorption stage [6,51]. Then, the clay residue, formed after the dissolution of the limestones in the karsting process, became a medium carrying copper ions, which was facilitated by the network of cracks in the limestones. It was then that nests were created in unchanged/partially karstic Devonian limestones filled with copper carbonates (azurite and malachite—Figure 6c,d). Azurite forms in the oxidation zone initially. This mineral precipitates under slightly acid conditions (pH ca. 6.95 [52]), and then locally turns into malachite. We can observe such a process in Figure 8e. Malachite forms under neutral to slightly basic pH conditions [52] during the neutralization of acidic supergene solutions by Devonian limestones. In samples taken from the Teresa adit it can be observed that azurite not only forms nests in the limestones (Figure 6d), but also occurs in pseudomorphoses, where relics of copper sulphides can still be observed: chalcocite (Figure 10d,e) and covellite (Figure 10f) as well as the copper oxide cuprite (Figure 8c). Malachite, similarly to azurite, occurs in the form of fibrous aggregates (Figure 8e,f).

Devonian limestones in the Miedzianka Mountain deposit are strongly tectonically fractured and cut by numerous veins of different colours (Figure 6a). Fragments of red veins strongly stained with hematite can be observed (Figure 6a,b). Calcite veins with hematite in limestones and their mechanism of formation were described by Bąk et al. [53] and Bąk et al. [54]. Rubinowski [6] described two generations of veins. The oldest generation of calcite is of a rosa-like calcite type. Creme-pinkish calcite veins can be observed in the studied Devonian limestones from the Teresa adit (Figures 6a and 8a), and belong to the younger generation of calcite [6]. According to Migaszewski et al. [55], it is the so-called younger post-Variscan mineralisation. This calcite is the most common in the Miedzianka Mountain deposit and was formed in conditions typical of the terrestrial environment, characterised by strongly developed karst processes, as determined by isotope studies (C, O, and Sr) carried out by Migaszewski et al. [55]. Micro-area chemical analysis of the calcite from the creme-pinkish vein revealed the presence of scattered very small grains of hematite with V_2O_5 admixture. The presence of hematite containing vanadium, which occurs in creme-pinkish calcite veins in Devonian limestones, was for the first time described in this paper. The colour of calcite is related to various amounts of disseminated hematite. The oxidation of iron minerals in limestones occurs during the processes of weathering and karsting, and results in the appearance of a red colour in limestones and veins caused by the presence of small grains of hematite [56].

Rubinowski [6] described the occurrence of clay residuum in Devonian limestones in the Miedzianka Mountain deposit, and suggested that it was a medium that transferred copper ions into fractured Devonian limestones and was responsible for the formation of copper carbonates. This author also believed that this process was related to the adsorption stage in the evolution of the copper deposit on Miedzianka Mountain [6]. During our study of Devonian limestones we managed to find the clay residuum in limestone microcracks, and we have defined its mineral composition. Our research showed that the residuum, formed after the dissolution of limestones during the karsting process, is composed of a mixture of Fe-Mg illite and micro-crystalline quartz/silica (Figures 7, 9f and 10a). In addition, isolated azurite nest was found in limestones (Figure 10b), which might be evidence of the transfer by clay minerals (illite) of copper ions in the limestones. Taking into account Rubinowski's hypothesis [6] concerning the adsorption stage of copper-bearing mineralisation as well as our results, we proposed a model of the formation of copper carbonates in the adsorption stage of copper-bearing mineralisation in the Miedzianka Mountain deposit, presented in Figure 11.

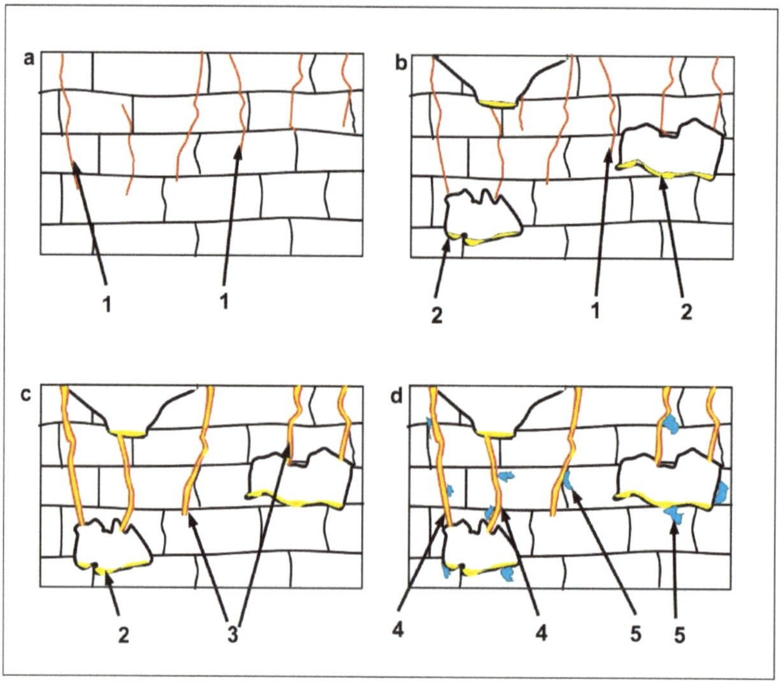

Figure 11. Scheme of formation of azurite nests, Fe-Mg illite, and fine crystalline silica within microcracks in Devonian limestones in the adsorption stage (see [6]). Not to scale. (**a**,**b**)—stages of dissolution of limestones during karstification; (**c**,**d**)—stages of adsorptive mineralisation. Legend: 1—microcracks in limestones; 2—karst voids with residuum of clay minerals and microcrystalline silica; 3—microcracks with residuum of clay minerals (illite) + solution with ions of Cu^{2+} and Fe^{2+}; 4—Fe-Mg illite and microcrystalline silica in microcracks (see Figure 10a); 5—nests of azurite (see Figure 9b).

7. Summary

Our investigation of the Teresa adit shows that mineralisation occurs as veins of irregular shapes with various minerals. The minerals detected in the studied samples are copper minerals (primary and secondary) and rock-forming minerals.

The primary minerals occurring only as relics of chalcopyrite were developed during hydrothermal mineralisation (the first stage of mineralisation). Secondary minerals are developed during: (a) weathering and karstification processes (the second stage of mineralisation), namely, secondary copper sulphides, copper oxides, and iron oxides including chalcocite, covellite, cuprite, and hematite; and (b) adsorptive mineralisation (the third stage of mineralisation), namely, azurite and malachite.

Our investigations show for the first time the presence of secondary cuprite with relicts of primary chalcopyrite in the studied deposit. Moreover, the disseminated hematite with admixture of vanadium, which occurs in younger creme-pinkish calcite veins, was not described previously in this deposit. Although clay minerals (illite) are common minerals in sedimentary rocks, it has not been described in limestones in the Miedzianka deposit previously. We found Fe-Mg illite together with fine-grained silica in cracks in limestones (Figures 9f and 10a). We proposed a model of the formation of copper carbonates in the adsorption stage of copper-bearing mineralisation in the Miedzianka Mountain deposit (Figure 11). In addition, for the first time unit cell parameters of malachite and azurite from Miedzianka Mountain were determined. The data indicated very low substitutions of atoms other than Cu in their structures (Table 1), probably silver and gold.

The presence of the minerals described in the Miedzianka deposit suggest that it is a deeply weathered deposit of copper sulphide with Pb-Zn hydrothermal veins. These minerals' parageneses are known from the Mississippi Valley deposit types.

The collected samples together with the detailed results described in this paper will be donated to the Museum Chamber of Ore Mining in Miedzianka located in Miedzianka village. The aim of our work was to enlarge the museum's collection of mineralogical specimens from the Teresa adit in the Miedzianka Mountain deposit. We are convinced that the set of samples collected in the Teresa adit and studied in this paper will certainly enrich the scientific and educational activity of this public institution.

The described minerals from samples in situ collected at seven points in the Teresa adit may constitute the basis for the graphic design of information boards in the planned underground geotourist route in the Teresa adit on Miedzianka Mountain [9].

Our research leads to the conclusion that, despite the fact that the Miedzianka Mountain deposit is already exhausted and not exploited, and has been studied in various aspects for almost 200 years, it still has great scientific potential. Findings of paragenesis of various minerals in heaps and in adits are still being described. Therefore, the study of old deposits usually extends knowledge with new scientific aspects, as presented in many works on various ore and historical deposits in Europe, e.g., Lavrion Mines in Greece [57], the Sulcis, Sardinia, Italy [58], Eastern Liguria, Italy [59], Wieliczka Salt Mine [60], Jachymoc, Czech Republic [61], Freiberg, Germany [62], and Zemplín, Slovakia [63].

Author Contributions: Conceptualization, A.C. and A.W.; Methodology, A.C., A.W., M.S. (Monika Struska), M.S. (Marek Szczerba) and J.O.; Investigation and Data Curation, A.C., A.W., M.S. (Monika Struska) and M.S. (Marek Szczerba); Writing—Original Draft Preparation, A.C. and A.W.; Writing—Review and Editing, A.C., A.W., M.S. (Monika Struska), M.S. (Marek Sczerba) and J.O. All authors have read and agreed to the published version of the manuscript.

Funding: This research was financed from statutory research of the Pedagogical University of Krakow: Project BN. 711-171/PBU/2021 and WPBU/2021/05/00168.

Data Availability Statement: The samples are archived at the Institute of Geography, Pedagogical University of Krakow, Podchorażych 2, 30-084 Kraków, Poland; Collection reference: Góra Miedzianka, 2019/ P1-P7.

Acknowledgments: The authors thank the Regional Director for Environmental Protection in Kielce, Ing. Aldona Sobolak, for issuing permission for study of the Teresa adit. The authors are grateful to anonymous Reviewers, for their important remarks on this paper.

Conflicts of Interest: The authors declare no conflict of interest.

References

1. Kowalczewski, Z. Historia górnictwa świętokrzyskiego—Geologia złóż, dzieje eksploatacji, wykorzystanie kopalin. In Proceedings of the 42 Zjazd Polskiego Towarzystwa Geologicznego, Kielce, Poland, 27–30 June 1993; pp. 7–8, 24–32.
2. Staszic, S. O Ziemiorództwie Karpatów i Innych Gór i Równin Polski. Warszawa, w Drukarni Rządowei, 1815, 107s. + Atlas. Available online: https://jbc.bj.uj.edu.pl/dlibra/doccontent?id=267 (accessed on 15 June 2021).
3. Morozewicz, J. Staszycyt—Nowy minerał złoża kruszcowego na Miedziance (Staszicit, ein neues Mineral des Kupfererzvorkommens Miedzianka). In *Rozprawy Akademii Umiejętności-Rozprawy Wydziału Matematyczno-Przyrodniczego*; dział, A., Drukarnia, U.J., Eds.; Wielkopolska Biblioteka Cyfrowa: Poznań, Poland, 1919; Volume 18, pp. 123–134.
4. Morozewicz, J. O kobaltowym minerale Miedzianki, lubeckicie. In *Rozprawy Akademii Umiejętności-Rozprawy Wydziału Matematyczno-Przyrodniczego*; dział, A., Drukarnia, U.J., Eds.; Wielkopolska Biblioteka Cyfrowa: Poznań, Poland, 1919; Volume 18, pp. 205–209.
5. Czarnocki, J. Surowce mineralne w Górach Świętokrzyskich. Surowce skalne. *Pract. Państwowego Instytutu Geol.* **1958**, *21*, 128–160.
6. Rubinowski, Z. Z badań złóż kruszców w Polsce. *Biul. Inst. Geol.* **1971**, *247*, 27–46, 105–132.
7. Balcerzak, E.; Nejbert, K.; Olszyński, W. Nowe dane o paragenezach kruszcowych w żyłach siarczków pierwotnych złoża Miedzianka (Góry Świętokrzyskie). *Przegląd Geol.* **1992**, *40*, 659–663.
8. Kozłowski, A. Native gold from Miedzianka Mt., Świętokrzyskie Mts. In *Gold in Poland*; Kozłowski, A., Mikulski, S.Z., Eds.; Uniwersytet Warszawski, Wydział Geologii, Państwowy Instytut Geologiczny, Państwowy Instytut Badawczy: Warszawa, Poland, 2011; Volume 2, pp. 339–349.

9. Borowiecki, N.; Barzycki, M.; Chudzik, W. Możliwości zagospodarowania naturalnych i pogórniczych podziemi rezerwatu "Góra Miedzianka" do celów geoturystycznych. *Bezp. Pract. Ochr. Sr. Gór.* **2017**, *275*, 8–17.
10. Ciurej, A.; Struska, M.; Wolska, A.; Chudzik, W. The Miedzianka Mountain Ore Deposit (Swiętokrzyskie Mountains, Poland) as a Site of Historical Mining and Geological Heritage: A Case Study of the Teresa Adit. *Minerals* **2021**, *11*, 1177. [CrossRef]
11. Ciurej, A.; Struska, M.; Wolska, A.; Chudzik, W. Atrakcje geoturystyczne rejonu pogórniczego na przykładzie dawnej sztolni Teresa, Góra Miedzianka koło Chęcin. In Proceedings of the XXIX Konferencja Aktualia i Perspektywy Gospodarki Surowcami Mineralnymi, Nowy Sącz, Poland, 6–8 November 2019; pp. 15–17.
12. Łabęcki, H. *Górnictwo w Polsce: Opis Kopalnictwa i Hutnictwa Polskiego Pod Względem Technicznym, Historyczno-Statystycznym i Prawnym*; Drukarnia Jana Kaczanowskiego: Warszawa, Poland, 1841; Volume 1, pp. 1–538.
13. Morozewicz, J. O miedziankicie (On miedziankite). *Spraw. Pol. Inst. Geol.* **1924**, *2*, 1–3.
14. Manecki, A. *Encyklopedia Minerałów. Minerały Ziemi i Materii Kosmicznej*; AGH, Uczelniane Wydawnictwa Naukowo-Dydaktyczne: Kraków, Poland, 2004; p. 498.
15. Biagioni, C.; George, L.L.; Cook, N.J.; Makovicky, E.; Moëlo, Y.; Pasero, M.; Sejkora, J.; Stanley, C.J.; Welch, M.D.; Bosi, F. The tetrahedrite group: Nomenclature and classification. *Am. Mineral.* **2020**, *105*, 109–122. [CrossRef]
16. Łaszkiewicz, A. Mimetyt z Miedzianki (Mimetite from Miedzianka). *Arch. Mineral.* **1928**, *4*, 133–135.
17. Czarnocki, J. O budowie zachodniej części fałdu chęcińskiego. *Posiedz. Nauk. Państw. Inst. Geol.* **1928**, *21*, 52–55.
18. Swęd, M.; Urbanek, P.; Krechowicz, I.; Dworczak, P.; Wiecka, P.; Mleczak, M.; Tobys, P. Mineralogia hałd wietrzeniowych złoża Miedzianka (Góry Świętokrzyskie). *Przegląd Geol.* **2015**, *63*, 363–372.
19. Wróblewski, T. Rys historyczny górnictwa kruszcowego w Górach Świętokrzyskich. *Przegląd Geol.* **1962**, *10*, 414–417.
20. Czarnocki, J. *Ogólna Mapa Geologiczna Polski 1:100,000. Arkusz 4, Kielce*; Państwowy Instytut Geologiczny: Warszawa, Poland, 1938.
21. Wojciechowski, A. Hałdy dawnego górnictwa świętokrzyskiego jako źródło metali kolorowych i szlachetnych. *Przegląd Geol.* **2002**, *50*, 240–244.
22. Szunke, M.; Kowalczewski, Z. Zjazdy Polskiego Towarzystwa Geologicznego w regionie świętokrzyskim. *Przegląd Geol.* **2006**, *54*, 509–512.
23. Giełżecka-Mądry, D.; Fijałkowska-Mader, A.; Szczepanik, Z. Działalność Oddziału Świętokrzyskiego PIG-PIB w zakresie geologii środowiskowej, ochrony przyrody nieożywionej oraz edukacji geologicznej (Activity of the Holy Cross Mountains Branch of the Polish Geological Institute-NRI in the field of environmental geology, protection of inanimate nature and geological education). *Przegląd Geol.* **2021**, *69*, 250–258.
24. Wieser, T.; Żabiński, W. Copper arsenate and sulphate minerals from Miedzianka near Kielce (Poland). *Mineral. Pol.* **1986**, *17*, 17–39.
25. Bzowska, G.; Karwowski, Ł.; Juzaszek, A. New phosphate and arsenate minerals from Miedzianka near Chęciny (Holy Cross Mts.). *Pol. Mineral. Soc.–Spec. Pap.* **2003**, *22*, 201–204.
26. Lenik, P.; Salwa, S.; Bazarnik, J. Ni,Co-pyrites (bravoites) from the Holy Cross Mountains, Poland. *Mineral.–Spec. Pap.* **2019**, *49*, 58.
27. Lenik, P.; Salwa, S.; Bazarnik, J. Thiospinels (siegenite, fletcherite) from the Holy Cross Mountains, Poland—Preliminary results. *Mineral.–Spec. Pap.* **2019**, *49*, 59.
28. Rubinowski, Z. Wyniki badań geologicznych w okolicy Miedzianki świętokrzyskiej. *Biul. Inst. Geol.* **1958**, *126*, 143–153.
29. Rubinowski, Z. Nowe obserwacje okruszcowania na Miedziance świętokrzyskiej. *Przegląd Geol.* **1955**, *6*, 299–301.
30. Urban, J. Bloczne skały węglanowe w Paleozoiku zachodniej części Gór Świętokrzyskich. *Kwart. Geol.* **1986**, *31*, 238–239.
31. Ryka, W. Adinol contact in Bardo (Święty Krzyż Mountains). *Kwart. Geol.* **1957**, *1*, 163–167.
32. Kardymowicz, I. Studia petrologiczne lamprofirów Gór Świętokrzyskich. *Kwart. Geol.* **1962**, *6*, 271–318.
33. Rubinowski, Z. Lamprofiry okolic Daleszyc i związane z nimi przejawy mineralizacji. *Kwart. Geol.* **1962**, *6*, 245–269.
34. Samsonowicz, J. Lamprofiry okolic Iwanisk Łysogórskich i okolic Siewierza. *Pos. Nauk. Państw. Inst. Geol.* **1928**, *19–20*, 38–39.
35. Migaszewski, M.Z. Datowanie diabazu i lamprofirów świętokrzyskich metodą K-Ar i Ar-Ar. *Przegląd Geol.* **2002**, *50*, 227–229.
36. Jaśkowski, B.; Kowalski, B.; Sołtysik, R. Góra Miedzianka—Budowa geologiczna i historia eksploatacji rud miedzi. In Proceedings of the Konferencja Naukowa nt. Geomorfologia Gór i Wyżyn w Polsce—Kontrowersje i Nowe Spojrzenia, Kielce, Poland, 16–19 May 2000; pp. 90–94.
37. Morozewicz, J. O miedziankicie. *Spraw. Pol. Inst. Geol.* **1923**, *2*, 1–3.
38. Bolewski, A. *Mineralogia Szczegółowa*; Wydawnictwa Geologiczne: Warszawa, Poland, 1965; p. 796.
39. Czarnocki, J. Marmury świętokrzyskie. W: Materiały do znajomości skał w Polsce. *Biul. Inst. Geol.* **1952**, *80*, 27–48.
40. Wojciechowski, J. Minerały Miedzianki pod Chęcinami. *Pract. Muz. Ziemi* **1958**, *1*, 133–156.
41. Bolewski, A.; Manecki, A. *Mineralogia Szczegółowa*; Polska Agencja Ekologiczna: Warszawa, Poland, 1993; p. 663.
42. Konon, A. Regionalizacja tektoniczna Polski–Góry Świętokrzyskie i regiony przyległe. *Przegląd Geol.* **2008**, *56*, 921–926.
43. Recielski, K. Jaskinia w Sztolni Teresa na Miedziance. In *Jaskinie Regionu Świętokrzyskiego*; Urban, J., Ed.; PTPNoZ: Warszawa, Poland, 1996.
44. Belokoneva, E.L.; Gubina, Y.K.; Forsyth, J.B. The charge density distribution and antiferromagnetic properties of azurite $Cu_3[CO_3]_2(OH)_2$. *Phys. Chem. Miner.* **2001**, *28*, 498–507. [CrossRef]
45. Zigan, F.; Joswig, W.; Schuster, H.D.; Mason, S.A. Verfeinerung der struktur von Malachit, $Cu_2(OH)_2CO_3$, durch neutronenbeugung. *Z. Für Krist.* **1977**, *145*, 412–426. [CrossRef]

46. Kazmierczak, J. Morphogenesis and systematics of the Devonian Stromatoporoidea from the Holy Cross Mountains, Poland. *Palaeont. Pol.* **1971**, *26*, 1–150.
47. Narkiewicz, M.; Racki, G.; Wrzołek, T. Litostratygrafia dewońskiej serii stromatoporoidowo-koralowcowej w Górach Świętokrzyskich. *Kwart. Geolog.* **1990**, *34*, 433–456.
48. Racki, G. Evolution of the bank of reef complex in the Devonian of the Holy Cross Mountains. *Acta Palaeont. Pol.* **1993**, *37*, 87–182.
49. Borkowska, M.; Smulikowski, K. *Minerały Skałotwórcze*; Wydawnictwo Geologiczne: Warszawa, Poland, 1973; p. 477.
50. Kotański, Z. *Z Plecakiem i Młotkiem w Góry Świętokrzyskie. Przewodnik Geologiczny Dla Turystów*; Wydawnictwo Geologiczne: Warszawa, Poland, 1968; p. 227.
51. Rubinowski, Z. *Zarys Metalogenezy Paleozoiku Świętokrzyskiego. Przew. 35 Zjazdu. Pol. Tow. Geol. Kielce, 16–19 Września, 1962*; PIG: Warszawa, Poland, 1962.
52. Vink, B.W. Stability relations of malachite and azurite. *Mineral. Mag.* **1986**, *50*, 41–47. [CrossRef]
53. Bąk, K.; Kowalczyk, J.; Wolska, A.; Bąk, M.; Natkaniec-Nowak, L. Iron and silica enrichments in the Middle Albian neptunian dykes from the High-Tatric Unit, Central Western Carpathians: An indication of hydrothermal activity for an extensional tectonic regime. *Geol. Mag.* **2018**, *155*, 1–19. [CrossRef]
54. Bąk, M.; Górny, Z.; Bąk, K.; Wolska, A.; Stożek, B. Successive stages of calcitization and silicification of Cenomanian spicule-bearing turbidites based on microfacies analysis, Polish Outer Carpathians. *Ann. Soc. Geol. Pol.* **2015**, *85*, 187–203. [CrossRef]
55. Migaszewski, M.Z.; Hałas, S.; Durakiewicz, T. Wiek i geneza mineralizacji kalcytowej w Górach Świętokrzyskich w świetle badań litologiczno-petrograficznych i izotopowych. *Przegląd Geol.* **1996**, *44*, 275–281.
56. Bąk, K.; Bąk, M.; Gatlik, J.; Błachowski, A. Oscillating redox conditions in the Middle–Late Jurassic Alpine Tethys: Insights from selected geochemical indices and 57Fe Mössbauer spectroscopy. *Palaeogeogr. Palaeoclimatol. Palaeoecol.* **2020**, *537*, 109440. [CrossRef]
57. Voudouris, P.; Melfos, V.; Mavrogonatos, C.; Photiades, A.; Moraiti, E.; Rieck, B.; Zaimis, S. The Lavrion Mines: A Unique Site of Geological and Mineralogical Heritage. *Minerals* **2021**, *11*, 76. [CrossRef]
58. Gioncada, A.; Pitzalis, E.; Cioni, R.; Fulignati, P.; Lezzerini, M.; Mundula, F.; Funedda, A. The volcanic and mining geoheritage of San Pietro Island (Sulcis, Sardinia, Italy): The potential for geosite valorization. *Geoheritage* **2019**, *11*, 1567–1581. [CrossRef]
59. Marescotti, P.; Brancucci, G.; Sasso, G.; Solimano, M.; Marin, V.; Muzio, C.; Salmona, P. Geoheritage values and environmental issues of derelict mines: Examples from the sulphides mines of Gromolo and Petronio valleys (Eastern Liguria, Italy). *Minerals* **2018**, *8*, 229. [CrossRef]
60. Wiewiórka, J. Level of the oldest rock-salt in the stratified salt deposit of Wieliczka. *Stud. I Mater. Do Dziejów Żup Solnych W Polsce* **1974**, *3*, 17–38.
61. Aulický, R.; Iványi, K.; Kafka, J. *Rudné a Uranové Hornictví České Republiky*; Anagram: Ostrava, Czech Republic, 2003; p. 647.
62. Stockmann, M.; Hirsch, D.; Lippmann-Pipke, J.; Kupsch, H. Geochemical study of different-aged mining dump materials in the Freiberg mining district, Germany. *Environ. Earth Sci.* **2012**, *68*, 1153–1168. [CrossRef]
63. Sidor, C.; Kršák, B.; Štrba, Ľ.; Gajdoš, J.; Šebešová, A.; Kolačkovska, J. Examples of secondary online data for raising awareness about geo and mining heritage. *Acta Montan. Slovaca* **2020**, *25*, 116–126.

minerals

MDPI

Article

New Strategies in Archaeometric Provenance Analyses of Volcanic Rock Grinding Stones: Examples from Iulia Libica (Spain) and Sidi Zahruni (Tunisia)

Lluís Casas [1,*], Roberta Di Febo [1], Anna Anglisano [2], África Pitarch Martí [3], Ignasi Queralt [4], Cèsar Carreras [5] and Boutheina Fouzai [6]

1 Departament de Geologia, Universitat Autònoma de Barcelona (UAB), Edifici C, 08193 Cerdanyola del Vallès, Spain; roberta.difebo@uab.cat
2 Departament de Ciències Ambientals, Universitat de Girona (UdG), c/Maria Aurèlia Capmany i Farnés, 69, 17003 Girona, Spain; anna.anglisano@udg.edu
3 Departament d'Arts i Conservació-Restauració, Facultat de Belles Arts, Universitat de Barcelona (UB), c/Pau Gargallo, 4, 08028 Barcelona, Spain; africa.pitarch@ub.edu
4 Department of Geosciences, IDAEA-CSIC, Jordi Girona 18-26, 08034 Barcelona, Spain; ignasi.queralt@idaea.csic.es
5 Departament de Ciències de l'Antiguitat i de l'Edat Mitjana, Universitat Autònoma de Barcelona (UAB), 08193 Cerdanyola del Vallès, Spain; cesar.carreras@uab.cat
6 Département Génie Chimique Industriel et Minier, École Nationale D'ingénieurs de Gafsa (ENIGA), Campus Universitaire Sidi Ahmed Zarroug, Gafsa 2112, Tunisia; fouzaiboutheina11@gmail.com
* Correspondence: lluis.casas@uab.cat

Abstract: Archaeometry can help archaeologists in many ways, and one of the most common archaeometric objectives is provenance analysis. Volcanic rocks are often found in archaeological sites as materials used to make grinding tools such as millstones and mortars or as building materials. Petrographic characterization is commonly applied to identify their main mineralogical components. However, the provenance study of volcanic stones is usually undertaken by comparing geochemical data from reference outcrops using common descriptive statistical tools such as biplots of chemical elements, and occasionally, unsupervised multivariate data analysis like principal component analysis (PCA) is also used. Recently, the use of supervised classification methods has shown a superior performance in assigning provenance to archaeological samples. However, these methods require the use of reference databases for all the possible provenance classes in order to train the classification models. The existence of comprehensive collections of published geochemical analyses of igneous rocks enables the use of the supervised approach for the provenance determination of volcanic stones. In this paper, the provenance of volcanic grinding tools from two archaeological sites (Iulia Libica, Spain, and Sidi Zahruni, Tunisia) is attempted using data from the GEOROC database through unsupervised and supervised approaches. The materials from Sidi Zahruni have been identified as basalts from Pantelleria (Italy), and the agreement between the different supervised classification models tested is particularly conclusive. In contrast, the provenance of the materials from Iulia Libica remained undetermined. The results illustrate the advantages and limitations of all the examined methods.

Keywords: archaeometry; volcanic stone; grinding tools; provenance studies; supervised methods; machine learning; clustering; XRF

1. Introduction

Among the vast field of archaeometry, provenance studies stand out [1]. These studies involve the analysis of artifacts, raw materials, and geological sources to determine the origin and movement of ancient objects [2]. Provenance studies help unravel the mysteries of human history, trade, and cultural exchange. This type of research has been developed on a number of materials (see [2] and references therein), including pottery (and their corresponding clays) [3,4], stones [5–7], metals (e.g., [8,9]), mortars/plasters (e.g., [10,11]), glasses [12], glazes (e.g., [13,14]) and pigments (e.g., [15,16]). The strategy to trace the source of the corresponding raw materials is specific to every type of material. Importance can be focused on specific mineral inclusions, relevant chemical compositions, isotopic ratios, etc. The most convenient and useful characterization techniques vary depending on the material under study and the tracing strategy.

In the particular field of stone provenance studies, there is also a large diversity of approaches that simply reflect the variety of rock types [17–22]. Marbles and volcanic rocks are probably among the materials whose provenance has been most frequently investigated, and this is for both archaeological and analytical reasons. Obsidians (volcanic glasses) aside, volcanic rocks had specific uses. They were often used as grinding tools like millstones and mortars [23–25] or as building materials such as vault stones [26,27]. The number of plausible volcanic sources can be delimited by using geological, geographical and historical data. In some cases, the analyzed materials can bear a particular chemical and mineral content, enabling the ability to trace them back to a specific origin among the possible sources [28–31]. These provenance studies are of capital importance to infer trading routes and to assess the status of a given archaeological site.

In the present paper, we petrographically and geochemically characterize volcanic stones retrieved from two Roman archaeological sites. We explore and compare conventional and new strategies to infer their corresponding volcanic sources. The novelty lies (i) in the exhaustive quantification of petrographic data as opposed to the common non- or poorly quantified petrographic descriptions and, (ii) more importantly, in the innovative use of supervised machine learning algorithms instead of the common unsupervised exploratory geochemical data analysis to tackle the provenance of the characterized materials. The studied materials were retrieved from Iulia Libica, a Roman municipium (2nd–3rd century CE) located in the Eastern Pyrenees (between Spain and France) and Sidi Zahruni, a non-excavated late Roman large pottery center (5th–7th century CE) located in northern Tunisia (Figure 1a). The two sites have been selected for their role as trading centers [32,33] and their proximity to a known geological source of volcanic stone (Olot Volcanic Field and Pantelleria island, respectively).

Figure 1. (**a**) Location of the archaeological sites of Iulia Libica (Spain) and Sidi Zahruni (Tunisia). (**b**) The three sampled millstones from Iulia Libica. (**c**) The seven small volcanic samples retrieved from Sidi Zahruni.

2. Materials and Methods

2.1. Archaeological Materials

2.1.1. Three Roman Millstones from Iulia Libica (Eastern Pyrenees)

The Roman city of Iulia Libica (currently Llívia) was placed in the most important route crossing the Eastern Pyrenees, at more than 1200 m high. Its most splendorous period was during the 1st and 2nd centuries CE. Archaeological research in the area started in the 1970s, and regular excavation campaigns have been carried out since 1997. Since 2013, one of the excavated areas has been confirmed as the central square (forum) of the Roman city [34]. In 2019, excavation campaign works focused on the external part of the eastern wall of the forum reaching its foundations. Within the infilling material of a ditch that runs parallel to the eastern wall, abundant construction materials (tegulae and imbrices) and pottery were retrieved, with the latter indicating a chronology around the change of era. Also, within this ditch, three rotary millstones were retrieved along with a terracotta artifact interpreted as a small sundial [35].

The three millstones (Figure 1b) correspond to three thin catilli, the biconical funnel that rotates on a fixed bell-shaped millstone (meta) in rotary hand-mills. The morphology of the millstones corresponds to type Lattes B2g, dating from the last decades of 1st century BCE to the late 1st CE [36]. Most of the similar millstones found at Lattes and other sites in Southern France were produced from local basalts.

Small chips from every retrieved millstone were carefully collected from fracture surfaces using a small hammer and a chisel. The three samples (labeled L1, L2 and L3) were split into two parts, one to be used for thin-section preparation and the other one to be pulverized for chemical analysis.

2.1.2. Volcanic Stone from Sidi Zahruni (Nabeul, NE Tunisia)

Sidi Zahruni is a non-excavated site located near the modern town of Beni Khiar, 6 km northeast of Nabeul (ancient Neapolis), one of the best-documented amphora production areas in Roman Africa Proconsularis. The site covers an area of 13 ha and hosts a huge pottery center where amphorae, in particular African Keay 25 type, were massively produced [37], as attested by the tilled fields containing an excessive number of pottery shards. In 2012, a systematic surficial survey was performed on the site to determine the different levels of shard concentration and their corresponding identifiable pottery typologies [38]. The chronology of the site, largely based on the known periods of production of the identified pottery typologies, can be placed within the 5th–7th centuries CE.

Apart from pottery shards, marble, volcanic scoria, mosaic and cocciopesto mortar fragments were also occasionally found. These were interpreted as evidence of at least two areas with luxurious houses. Regarding the vesiculated volcanic scoria blocks, they are clearly non-local stones but part of the archaeological site. The volcanic blocks on site were fragmented and irregular, with the larger ones having an approximate volume of 3 dm^3. A curved surface on some of them indicated a probable use as millstone, though other uses cannot be excluded. Pottery fragments, including small embedded volcanic chips, were also found, indicating their use as mortarium.

Seven small samples of highly vesiculated volcanic scoria (labeled Z1 to Z7) were obtained in the form of irregular fragments from the largest blocks found on the surface of the site, using a small hammer and a chisel. Similarly to the samples from Iulia Libica, the seven samples from Sidi Zahruni were split into two parts to be used for thin-section preparation and chemical analysis, respectively (Figure 1c).

2.2. Characterization Methods

Standard thin sections of the samples were prepared at the Laboratori de Preparació de Làmines Primes, Universitat Autònoma de Barcelona (UAB). Petrographic descriptions were performed using a Nikon Eclipse E400POL microscope; pictures in plane-polarized light (PPL) and cross-polarized light (XPS) modes were taken using a DS-Fi2 high-definition color camera head mounted on the microscope. Thin section observation was used to determine the fabric, to identify the mineralogy of phenocrysts and to describe their shapes and features. High-resolution (4800 ppp) full scans of the analyzed thin sections were obtained using an EPSON 4990 scanner, including both PPL and XPL modes. The scans were treated with image processing software ImageJ [39] to quantify porosity, matrix and phenocryst ratios, including discrimination between the different minerals that appear as phenocrysts. The modal relative proportions of the components were formally measured as area percentages. However, the relative surface measurements are representative of the volumetric relative proportions, considering that the analyzed samples are relatively isotropic and fine-grained. The different phenocryst compositions were analyzed separately using the "analyze particles" command within ImageJ to obtain size-frequency distributions. The crystals were fit to ellipses, and several shape parameters were computed; among others, circularity (C) defined as $4\pi \cdot \text{area}/(\text{perimeter})^2$.

Petrographic information was complemented with elemental point-composition measurements on the phenocrysts imaged using a Merlin Field Emission Scanning Electron Microscope (FE-SEM) (Zeiss, Oberkochen, Germany) equipped with an energy-dispersive X-ray spectrometer (EDS) at the Servei de Microscòpia (UAB). Average elemental composition over different areas of the microcrystalline matrices was also measured. Measuring conditions were accelerating voltage of 15 kV, 1 nA beam current and an overall counting time of 60–100 s.

Multielemental bulk analysis was accomplished by means of energy-dispersive X-ray fluorescence (EDXRF) for the largest remaining sample parts after thin-section preparation. For EDXRF sample preparation, 5 g of previously powdered sample was mixed with 0.4 g of a binding agent (Elvacite™) and then homogenized in an agate mortar [40]. The mixture was pressed at 10 tons to achieve a homogeneous pressed powder pellet of 4 cm in diameter. Different pellets were produced for the three samples from Iulia Libica and five of the Sidi Zahruni samples (Z1, Z2, Z4, Z5, Z7). The pellets were analyzed using a benchtop spectrometer S2 Ranger EDXRF system (Bruker AXS, GmbH, Karlsruhe, Germany) with a Pd X-ray tube (max. power 50 W) and XFLASHTM Silicon Drift Detector (SDD) with a resolution of <129 eV Mn-Kα1 at 100,000 cps. Samples were evaluated under vacuum at different excitation voltage conditions to properly excite elements of low, medium and high atomic number. Multielemental quantitative determination for Si, Al, Fe, Mg, Ca, Na, K, Ti, P, Mn, V, Cr, Ni, Cu, Zn, Rb, Sr, Y, Zr and Nb was made using a calibration designed for volcanic rocks and operated with the SPECTRA.EDX package (Bruker AXS, GmbH, Karlsruhe, Germany). These elements were always present above detection limits in all the analyzed samples. Nine certified reference materials (CRMs) were used, comprising several basalts (BR–CRPG, BE-N–IWG-GIT, BHVO2–USGS, DC73303–NCS), an andesite (AGV-1–USGS), a diorite (DR-N–ANRT), a dolerite (WSE–IWG-GIT), a microgabbro (PM-S–IWG-GIT) and a syenite (STM-1–USGS). Calibration lines indicate very good correlation ($r^2 > 0.997$) between measured and expected concentrations, with absolute errors ranging from 2 to 9% for all the elements. Achieved precision is higher than the intrinsic compositional variation of the analyzed rocks.

Reference geochemical data corresponding to volcanic materials were extracted from the GEOROC database [41]. A starting dataset was obtained by setting a query by geography (latitude from 25° N to 54° N and longitude from 12° W to 45° E) to cover the Mediterranean Basin, North Africa and a great part of Europe and then selecting only the records that declare the rock name as basalt, excluding modern eruptions. This produced a database of around 3600 records that was additionally restricted to chemical analyses that really correspond to the basalt area in a total alkali–silica (TAS) diagram [42] (with data recalculated on an anhydrous basis). Data regarding Fe were standardized, and all the analyses were recalculated to total iron as Fe_2O_3. This left 1855 records that were further reduced to smaller homogeneous datasets containing a given set of analyzed chemical elements. This produced datasets G90, G60 and G30, containing roughly 90%, 60% and 30% of the original records. The data were labeled into different classes according to the corresponding tectonic setting and location (Figure 2). The original database

(Table S1) and the three reduced datasets (Tables S2–S4) can be downloaded from the Supplementary Materials section.

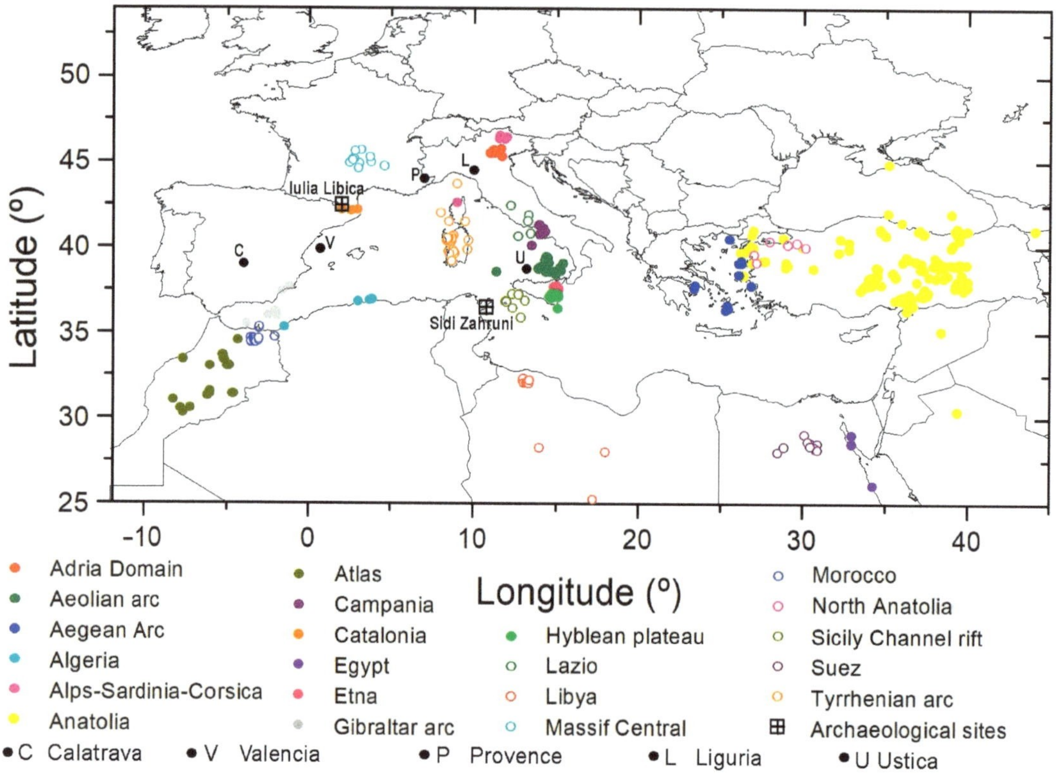

Figure 2. Map of the Mediterranean area, including the location of all the selected data from the geochemical database GEOROC, colored or labeled according to the corresponding data group. The archaeological sites where the studied basalts were retrieved have also been highlighted.

Several statistical methods were applied combining both the geological reference geochemical values as well as the data obtained from the analyzed archaeological materials. Firstly, principal component analysis (PCA) was used to check the structure of the data. This method, widely used in provenance studies, reduces the number of relevant variables by defining new variables (or components) as a linear combination of the original ones (i.e., the compositional variables). The two components that accumulate the largest percentages of variance (PC1 and PC2) were represented in a biplot to investigate if any clustering appeared and to check if the clusters correlate with any of the reference fields. As usual, data were standardized to zero mean and unit variance before computing the analysis. PCA is categorized as an unsupervised method, meaning that the procedure does not consider the class labels. In contrast, supervised machine learning methods use labeled reference datasets to train models to predict the class of unlabeled data (i.e., the archaeological samples). Training is usually performed using 80% of the reference data, and then the performance of the trained model is tested using the remaining 20%. The split between train and test subsets is made by using a random seed. This procedure is well-known in machine learning, but to our knowledge, it has never been applied to provenance analyses of grinding stones. All the compositional values were divided by the SiO_2 concentration as a way to standardize them. Different 8:2 splits were tested (using different random seeds) following the train–test approach. The supervised approach was performed using the "Supervised Provenance Analysis" R Markdown files from Anglisano et al. [43], including different classification models: generalized linear models (GLM), random forest (RF), artificial neural network (ANN), weighted k-nearest neighbors (kkNN), linear discriminant analyses (LDA) and a stack of all these models as described in [44]. Four additional classification models were tested using Python code (see Supplementary Materials section) under the Jupyter Notebook interactive computing platform: gradient boosting (GB), Gaussian process classifier (GPC), Gaussian naive Bayes (GNB) and linear support vector machine (LSVM).

3. Results

3.1. Petrography

The three millstone samples from Iulia Libica exhibit a porphyritic to glomerophyric and vesiculated texture with rather idiomorphic phenocrysts of two compositions embedded within a microlitic groundmass (Figure 3a). The most abundant and largest phenocrysts correspond to clinopyroxene (cpx), with augite composition as analyzed by SEM-EDS, see Figure 4 and Table 1. Cpx often appears as clusters of polygonal idiomorphic crystals, some of them exhibiting occasional hourglass sector zoning (Figure 3b).

Figure 3. Petrographic images of the samples from Iulia Libica illustrating their volcanic origin and basaltic composition. (**a**) PPL (**left**) and XPL (**right**) micrograph of the same area, showing abundant clinopyroxene (cpx) and olivine (ol) phenocrysts within a microlitic, vesiculated groundmass. (**b**) XPL micrograph of a cpx aggregate including hourglass sector zoning crystals. (**c**) XPL micrograph of a subhedral, corroded ol crystal, including an iddingsite alteration rim. (**d**) XPL micrograph including a basal section of cpx showing typical two cleavage directions at near 90° (**left**) and a corroded and altered ol crystal (**right**).

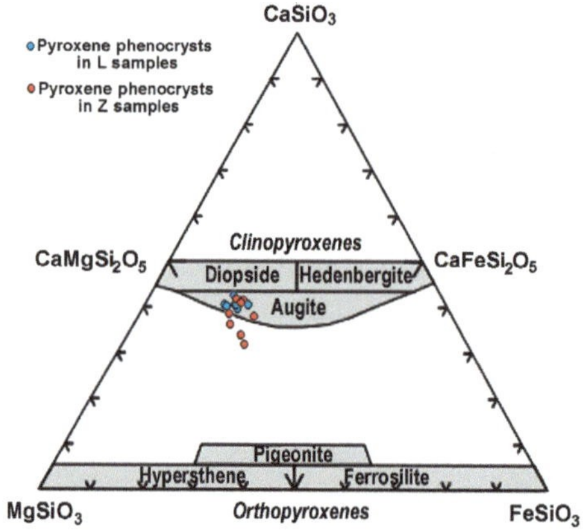

Figure 4. Pyroxene ternary diagram [45] showing the compositions of the pyroxene phenocrysts in samples from Iulia Libica (blue) and Sidi Zahruni (red).

Table 1. Average structural formulae of the analyzed (SEM-EDS) clinopyroxenes calculated on the basis of six oxygens. Eight augite crystals were measured for L samples and Z samples, as well as a single additional augite–aegirine crystal, which was found in sample Z6.

	Iulia Libica (L) Samples	Sidi Zahruni (Z) Samples	
	Augite	Augite	Augite–Aegirine
Si	1.85 ± 0.05	1.87 ± 0.05	2.03
Ti	0.04 ± 0.01	0.04 ± 0.01	0.01
Al	0.17 ± 0.03	0.16 ± 0.06	0.02
Fe	0.33 ± 0.03	0.38 ± 0.05	0.55
Mn	n.d.	n.d.	0.03
Mg	0.78 ± 0.04	0.82 ± 0.05	0.57
Ca	0.76 ± 0.01	0.74 ± 0.07	0.75
Na	0.03 ± 0.01	0.01 ± 0.01	0.12

Smaller phenocrysts of olivine (ol) are also very common in the three samples. These are anhedral (often rounded and occasionally corroded, showing that they partly dissolved into the liquid now solidified as the groundmass). The ol phenocrysts systematically exhibit an iddingsite alteration rim, made of iron oxides and clay minerals, resulting from post-magmatic hydrothermal alteration (typical for olivine in many volcanic rocks) (Figure 3c,d). The fresh olivine composition measured by SEM-EDS is $Fo_{75\pm2}$, while the obtained composition in the altered rims is still essentially olivine but with larger Fe contents (19–29 wt% instead of 18 wt%). Possibly, the measured spectra, even if focused on the rims, have contributions from iddingsite and nearby olivine. Small amounts (<1 wt%) of Ca and Al are also occasionally detected in the altered rims. The groundmass is a fine-grained matrix formed by mostly unoriented lath-shaped plagioclase microlites, clinopyroxene and opaque equidimensional minerals (possibly ilmenite and Ti-bearing magnetite).

All the collected samples from Sidi Zahruni exhibit a porphyritic and vesiculated texture with anhedral to idiomorphic phenocrysts and microphenocrysts within a microlitic groundmass (Figure 5). The mineral association is uniform and consists of the following: (i) Zoned and twinned plagioclase (pl) crystals with rectangular and quadratic sections (Figure 5a). SEM-EDS analyses indicate that they have compositions ranging from 46 to 78 An% (labradorite and bytownite). (ii) Clinopyroxene (cpx) with a slight brown color (augite as analyzed by SEM-EDS; see Table 1 and Figure 4). These augite crystals appear particularly idiomorphic, exhibiting mainly hexagonal sections or octagonal basal sections (Figure 5e), often showing the diagnostic two cleavage directions at nearly 90° angles.

Hourglass sector zoning also appears in some crystals (Figure 5f). (iii) Olivine (ol) with rare polygonal sections and more common rounded shapes (Figure 5g), sometimes significantly corroded. Its composition by SEM-EDS is $Fo_{72\pm4}$. Apart from these, in one of the samples (Z6), a large green-colored (PPL) rounded crystal was spotted (Figure 5h). The SEM-EDS analysis revealed Na- and Fe-rich clinopyroxene, a member of the aegirine–augite solid solution (see Table 1). The different phenocrysts are also embedded in a fine-grained matrix mainly formed by lath-shaped plagioclase crystals, clinopyroxene and opaque equidimensional minerals (possibly ilmenite or Ti-bearing magnetite).

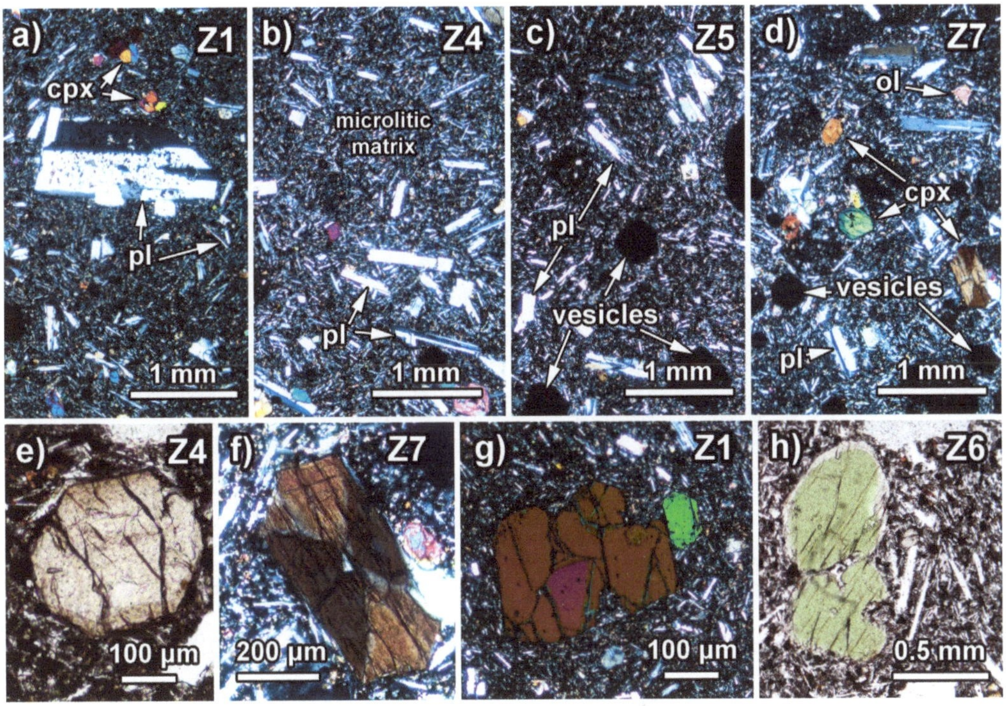

Figure 5. Petrographic images of the samples from Sidi Zahruni illustrating their volcanic origin and basaltic composition. (**a–d**) General aspect of representative samples showing plagioclase (pl) phenocrysts (**a**) and microphenocrysts (**a–d**), vesicles (mostly on **c**,**d**), olivine (ol) and clinopyroxene (cpx) microphenocrysts (mostly on **a**,**d**) within a microlithic groundmass viewed in XPL. (**e**) Basal section of a cpx microphenocryst, PPL. (**f**) cpx hourglass sector zoning microphenocryst, XPL. (**g**) Detail of an ol microphenocryst, XPL. (**h**) Section of a sodic, partly corroded, cpx macrocryst showing a brighter reaction rim with the groundmass, PPL.

The treated images from the scanned petrographic thin sections (e.g., Figure 6) allowed the discrimination between porosity, matrix and the different types of phenocrysts and their quantification. Porosity is around 24%–39% for Iulia Libica (L) samples and usually lower (6%–30%) for Sidi Zahruni (Z) samples, although, for sample Z5, it exceeds 50%. Pore-size histograms reveal that small pores (with areas below 1 mm^2) are the most abundant in all the analyzed samples (Figure 7). The three samples from Iulia Libica display a wide range of pore sizes similar to some of the samples from Sidi Zahruni. However, samples Z3 and Z4 have narrower pore size distributions. The circularity of the pores is within the range of 0.9 to 0.5 for all the samples, and the mean circularity values are around 0.7 or a bit lower for the samples with irregular or elliptic pore shapes (L2, Z2, Z6 and Z7).

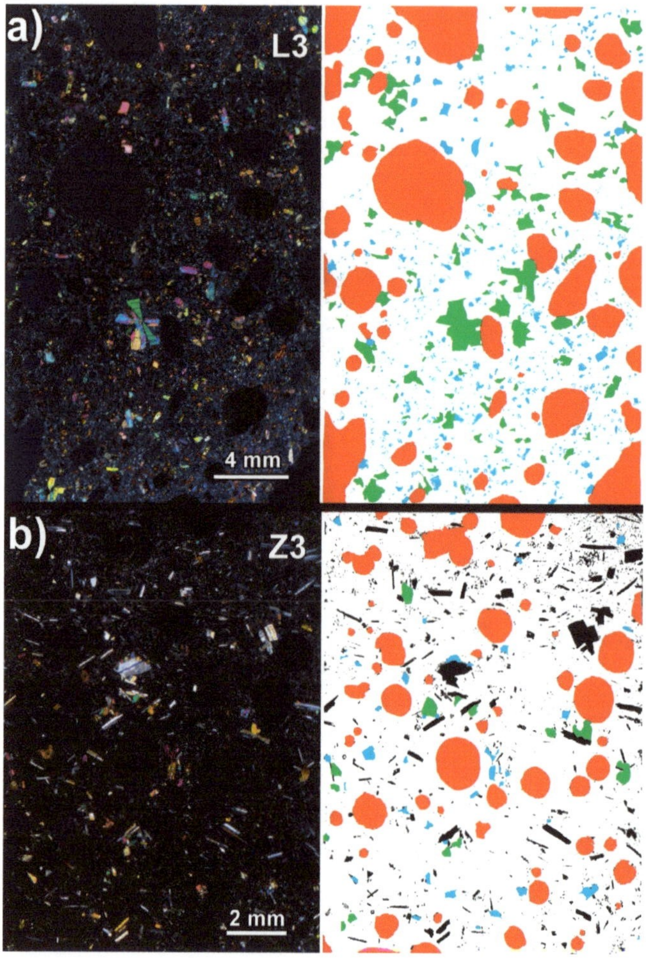

Figure 6. (**a**) L3, Iulia Libica, and (**b**) Z3, Sidi Zahruni, thin section images. To the left are the original XPL scans, and to the right are the corresponding colored images, including porosity (red), clinopyroxene (green), olivine (blue), plagioclase (black) and the microcrystalline matrix (white).

The mineral part of the samples is mainly formed by the fine-grained microlitic matrix in all the samples, and the phenocrysts are only a minor component (3%–12%), see Table 2. The matrix/phenocrysts content ratio is higher for the Sidi Zahruni samples (constituting 20 ± 8) compared to the Iulia Libica samples (8 ± 3).

The two types of phenocrysts that appear in Iulia Libica samples have a relative content ratio of about 2:1 between clinopyroxenes (cpx) and olivine (ol). In contrast, in the basalts from Sidi Zahruni, the abundance of these components is reversed and surpassed by plagioclase (pl) phenocrysts (Table 2). The pl:ol:cpx relative ratio in samples from Sidi Zahruni is around 4:1.5:1. The quantification of the crystals indicates that regardless of the sample and considered a mineral, the frequency of sizes increases towards small values forming a continuous range of sizes that connect the microphenocrysts (area < 0.25 mm^2) to the microlites of the matrix (Figure 8). The area of the phenocrysts only rarely reaches values above 1 mm^2.

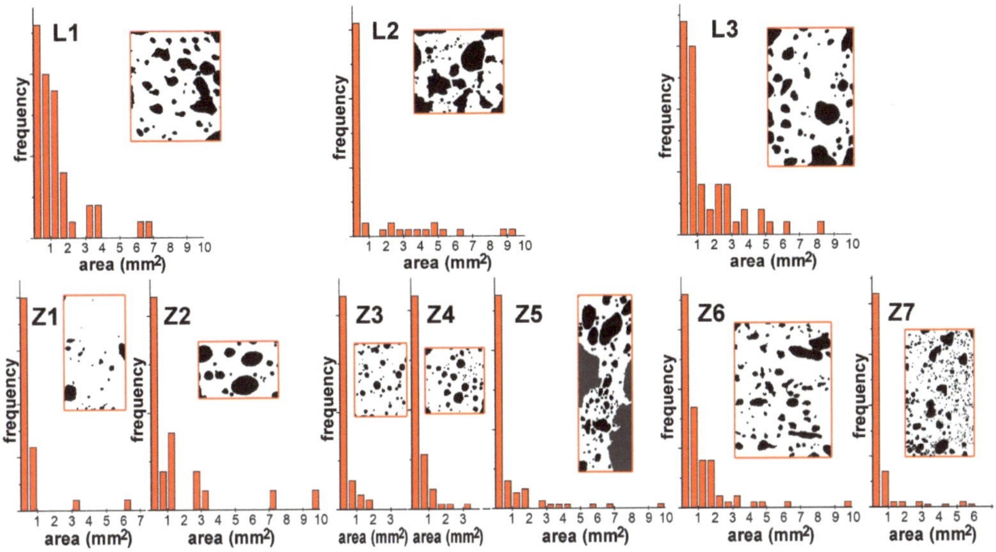

Figure 7. Porosity size histograms from the samples from Iulia Libia (**top**) and Sidi Zahruni (**bottom**). Next to each histogram, an inset shows the treated image with pores in black.

Table 2. Quantification of the different components identified in samples from Iulia Libica (L) and Sidi Zahruni (Z). Average values are also shown. Percentages were measured as area %, but they can also be read as vol%.

Sample	Phenocrysts			Matrix	Porosity
	cpx (%)	ol (%)	pl (%)	(%)	(%)
L1	7.5	4.6	0.0	64.3	23.6
L2	3.4	1.9	0.0	56.1	38.6
L3	5.8	2.4	0.0	66.3	25.5
L	6 ± 2	3 ± 2	-	62 ± 6	29 ± 8
Z1	1.3	2.1	3.5	86.9	6.2
Z2	0.5	1.4	4.1	64.1	29.9
Z3	1.1	1.1	3.6	81.0	13.2
Z4	0.5	0.7	1.4	78.9	18.5
Z5	0.5	0.5	2.2	46.0	50.8
Z6	0.7	0.8	3.0	76.5	19.0
Z7	0.4	0.7	1.5	74.0	23.4
Z	0.7 ± 0.4	1.0 ± 0.5	3 ± 1	73 ± 14	23 ± 14

The distribution of clinopyroxene and olivine crystals is less biased towards small crystals for the basalts from Iulia Libica, while the distribution corresponding to clinopyroxene exhibits a secondary peak centered around 0.7 mm^2 for these basalts. The distribution corresponding to olivine is relatively similar for all the samples, and there are also (regardless of the archaeological site provenance) hints of a secondary peak for sizes around 0.4 mm^2. The crystal size of plagioclase crystals (only present as phenocrysts in samples from Sidi Zahruni) is possibly the one that fits better to a monotonous exponential decay (to a second-order exponential decay function, to be precise).

Another quantified parameter (circularity) indicates higher mean values (<C>) for clinopyroxenes and olivine crystals for Sidi Zahruni samples. However, plagioclase phenocrysts from these basalts deviate markedly from circularity consistently with their common lath-shaped morphologies (see <C> in Figure 8).

The petrographic characterization indicates that all the investigated samples (L and Z) are volcanic rocks exhibiting basaltic compositions and are classifiable as such according to the international classification of igneous rocks [46].

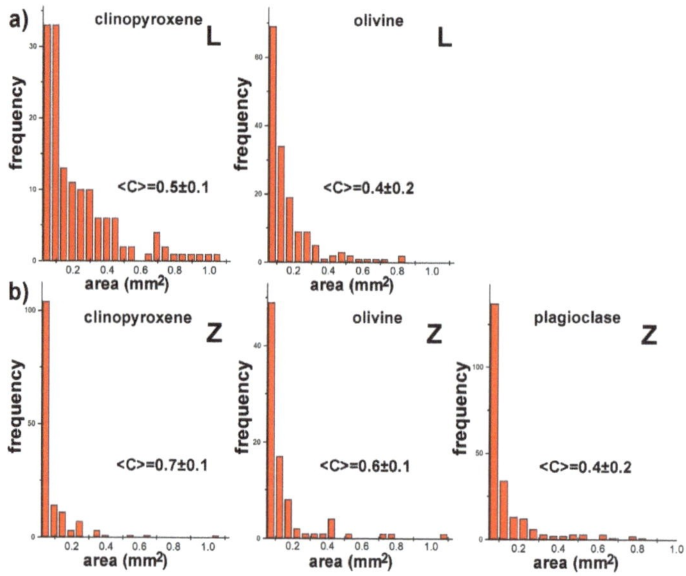

Figure 8. Crystal size (area) histograms for phenocrysts and microphenocrysts in basalts from (**a**) Iulia Libica (L) and (**b**) Sidi Zahruni (Z): the quantified minerals are clinopyroxene and olivine for L and Z, and plagioclase for Z samples. <C> indicates the average circularity values of the phenocrysts.

3.2. Geochemistry

The average bulk composition of the analyzed basalts is shown in Table 3. Each set of samples has a homogeneous composition in agreement with the observed petrographic texture and mineralogical composition. All the analyzed samples have a similar amount of silica (~46%–48%). In contrast, there are slight but significant differences between L and Z samples regarding the concentration of the other elements. For instance, L samples are richer in K, Fe and P and poorer in Al, Ca, Na and Ti. However, the sum of alkali oxides ($Na_2O + K_2O$) is similar for both types of samples, so all of the investigated samples can be classified as basalts (Figure 9) according to the TAS diagram [42]. The higher concentration of Al, Ca and Na in Z samples is the consequence of a magma richer in these elements and where plagioclase crystallized earlier, forming the pl phenocrysts only observed in the basalts from Sidi Zahruni. In contrast, the amount of Fe is slightly higher in the samples from Iulia Libica (~14% compared to ~12%). Finally, the trace element proportions are also dissimilar for both types of basalts. Those retrieved in Iulia Libica are richer in trace elements, except for the amount of V, which is higher in the basalts from Sidi Zahruni.

The tectonic setting of volcanic lavas can sometimes be geochemically inferred using certain compositional ratios (e.g., Zr vs. TiO_2 or Zr vs. Zr/Y) [47,48]. According to these criteria, the measured geochemical values indicate that the samples (both from Iulia Libica and Sidi Zahruni) correspond to intraplate basalts, which could help determine the provenance of the samples. However, the classification of volcanic districts as intraplate can be challenging for complex tectonic settings like those of some Mediterranean volcanic districts.

Table 3. Average composition (m) and corresponding standard deviations (σ) from N specimens from L and Z samples of the investigated archaeological basalts, including major (wt%) and trace elements (ppm).

	L1		L2		L3		Z1		Z2		Z4		Z5		Z7	
N	3		2		3		5		5		5		5		5	
Statistics	m	σ	m	σ	m	σ	m	σ	m	σ	m	σ	m	σ	m	σ
SiO_2 (%)	47.63	0.25	48.01	0.11	47.04	0.21	47.59	0.08	48.06	0.10	47.50	0.17	46.43	0.22	46.76	0.24
Al_2O_3 (%)	14.73	0.10	13.51	0.24	13.54	0.09	16.61	0.11	16.13	0.08	16.02	0.04	15.58	0.12	15.21	0.17
Fe_2O_3 (%)	14.73	0.10	13.51	0.24	13.54	0.09	11.82	0.03	11.89	0.03	12.89	0.04	12.30	0.12	12.32	0.06
MgO (%)	5.61	0.17	5.60	0.40	6.23	0.10	5.59	0.02	6.11	0.05	5.10	0.05	4.99	0.03	4.79	0.08
CaO (%)	9.48	0.08	9.74	0.21	10.24	0.21	11.89	0.02	11.57	0.06	10.90	0.03	11.79	0.06	10.81	0.03
Na_2O (%)	2.80	0.14	2.74	0.08	2.30	0.04	3.47	0.04	3.28	0.03	3.80	0.05	3.39	0.03	3.42	0.03
K_2O (%)	1.77	0.03	1.77	0.06	1.64	0.03	1.02	0.01	1.01	0.01	1.06	0.01	0.92	0.01	1.01	0.01
TiO_2 (%)	2.37	0.02	2.27	0.05	2.31	0.04	2.63	0.01	2.55	0.01	3.01	0.02	2.69	0.05	2.83	0.01
P_2O_5 (%)	0.76	0.01	0.85	0.01	0.72	0.01	0.51	0.01	0.51	0.00	0.60	0.00	0.55	0.01	0.61	0.01
Mn (ppm)	1376	14	1415	47	1686	89	1324	9	1352	9	1440	5	1332	20	1385	8
V (ppm)	268	11	266	18	277	18	333	17	316	19	379	22	364	23	358	22
Cr (ppm)	201	1	221	13	337	27	136	4	128	8	82	6	95	3	84	5
Ni (ppm)	181	2	204	9	280	11	83	0	88	0	57	0	61	0	56	0
Cu (ppm)	60	2	68	3	59	5	33	2	84	2	38	3	54	1	65	2
Zn (ppm)	123	2	126	7	133	9	85	1	96	1	97	1	91	2	96	1
Rb (ppm)	43	2	42	3	39	1	14	1	13	1	15	1	14	1	13	1
Sr (ppm)	898	7	938	60	849	41	496	1	596	1	471	1	500	2	476	2
Y (ppm)	29	0	31	2	31	1	22	0	22	1	27	0	23	1	27	1
Zr (ppm)	239	1	252	15	227	12	135	1	136	1	148	1	151	1	150	1
Nb (ppm)	79	3	80	8	71	3	30	3	31	1	32	4	25	3	30	3

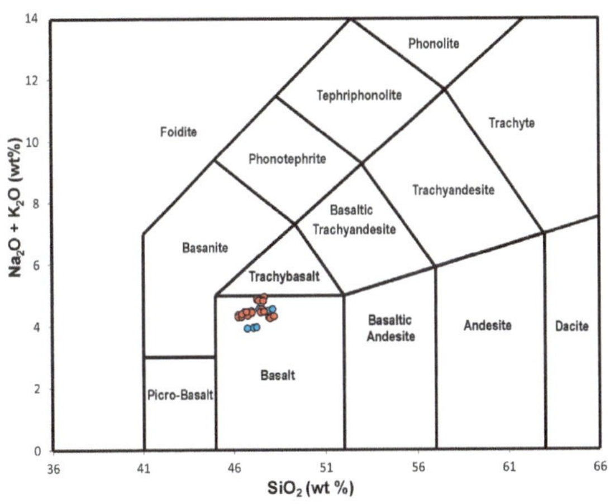

Figure 9. TAS diagram according to the IUGS classification scheme [42] showing the data plots for Iulia Libica (blue) and Sidi Zahruni (red) sample analyses.

3.3. Geochemical Comparison with Reference Materials

3.3.1. Elemental and PCA Biplots

Provenance determination using geochemical data is commonly undertaken using two-variable scatterplots that examine the relations between two chemical elements (or two simple chemical ratios). In the same plot, several reference variation fields are also depicted. The choice of the two elements (or chemical ratios) is arbitrary, and usually, the only criterion is to find a combination of variables that allows discrimination between the different possible provenances.

In Figure 10, several examples of biplots are shown. When only a few possible provenances are considered (Figure 10a), and the analyzed samples lie within a given reference field, the provenance seems well-determined. For instance, looking at Figure 10a, the geological provenance of the samples from Sidi Zahruni appears to be attributable to the Pantelleria island (strait of Sicily) and those from Iulia Libica to the Hyblean Plateau (southeastern Sicily). In contrast, for samples not lying within any reference field, it would be assumed that they do not belong to any of the considered provenances or, alternatively, that the reference fields perhaps are not well defined due to a lack of a fully representative set of reference samples. For instance, one of the samples from Sidi Zahruni strictly is not within any reference field.

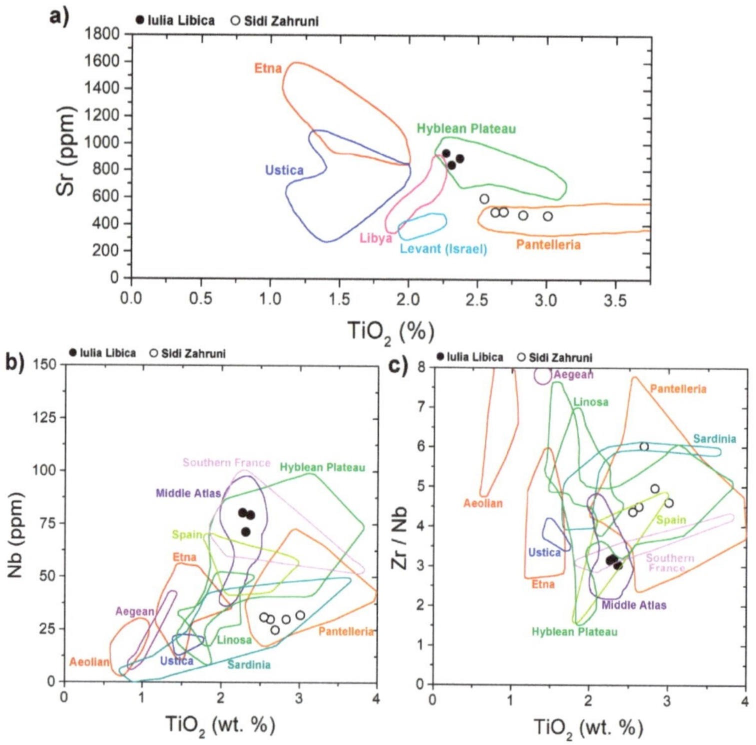

Figure 10. Average compositions of samples from Iulia Libica (black dots) and Sidi Zahruni (open dots) plotted in different discrimination diagrams: (**a**) Sr vs. TiO_2; (**b**) Nb vs. TiO_2; (**c**) Zr/Nb vs. TiO_2. Reference compositional fields from [28,49,50].

When the number of possible provenances is increased, the degree of overlap between reference fields increases inevitably, and it becomes very unlikely that the analyzed samples will be unambiguously assigned to a given provenance. Looking at Figure 10b, the basalts from Sidi Zahruni could have been quarried in Pantelleria but also in Sardinia, and those from Iulia Libica, apart from the suggested Hyblean Plateau, could also belong to the Southern France basalts or those from the Middle Atlas (Morocco). Also, using different biplots, divergent provenances are sometimes suggested. Using the biplot from Figure 10c, the samples from Sidi Zahruni appear more scattered within the reference fields of Pantelleria and partially within the Hyblean Plateau and Spanish basalts reference fields. On the other hand, the samples from Iulia Libica are now plotted within the reference fields of Middle Atlas, Southern France and Spain basalts but, strictly, not within the Hyblean Plateau reference field.

Using two-variable plots, a large number of compositional parameters are disregarded. Multivariate statistical analyses extend the examination of the data to all the analyzed elements. In particular, PCA condenses the relevant information in a few principal components that can also be represented in biplots. There are a few precedents of the use of this statistical tool in stone provenance studies (see, e.g., [30,51,52]). However, one major difficulty is to have a uniform set of data. The reference data from the GEOROC database is heterogeneous since samples were prepared using different methods and measured using different equipment [53]. The analyzed elements are not always the same set; in particular, the minor and trace elements can vary.

Two PCAs were computed using two different sets of data. For the PCA shown in Figure 11, a total of 1702 records (G90 dataset) from the preselected GEOROC database were combined with the 25 individual analyses from Sidi Zahruni samples and the 8 from Iulia Libica. To obtain this homogeneous dataset, the number of considered compositional variables was restricted to the most represented within the preselected database (SiO_2, TiO_2, Al_2O_3, CaO, MgO, Mn, K_2O, Na_2O and P_2O_5), all of them being main components of the basalt samples (both geological reference and archaeological materials). This allowed us to consider samples from 26 different sites/geological settings. The first two principal components have a joint explained variance ratio of ~50%. The obtained PCA reveals a positive correlation between SiO_2 and Al_2O_3 and a negative correlation between them with Mn, P_2O_5 and TiO_2. These variables are the main contributors to the first principal component (PC1). On the other hand, there is also a very good positive correlation between K_2O and CaO, and both are negatively correlated with Na_2O. Both Sidi Zahruni and Iulia Libica samples appear in a similar area of the biplot with slightly negative PC1 values. Concerning the PC2 values, they are slightly positive for Sidi Zahruni samples and almost zero or slightly negative for Iulia Libica samples. However, the degree of overlap between different provenances observed in the two main components biplot is very high; therefore, it is not possible to conclude on the origin of the analyzed samples. The samples from the most represented provenance (Anatolia, see yellow dots in Figure 11) spread, covering almost all the point cloud formed by the G90 dataset. Reference samples that are geographically close to Iulia Libica (L), such as those labeled Catalonia, Calatrava and Massif Central, are plotted near the L samples, but individual samples from many other reference sites are plotted in the same area. Instead of assigning provenances, it perhaps appears more clearly that some of the considered reference provenances can be excluded. Specifically, the Aeolian arc, the Campanian and the Lazio basalts seem to bear compositions that have no affinity with those sampled at Sidi Zahruni and Iulia Libica. The Aegean arc, Algeria, Alps–Sardinia–Corsica and Etna provenances could also be discarded, except for some occasional individual samples. Most of the discarded provenances agree with the fact that both Z and L basalts are intraplate. From the discarded volcanic districts, only Algeria is considered intraplate.

An additional PCA was computed (Figure 12) with a much more complete set of compositional variables (20, including, besides the previously considered, Fe_2O_3, Cr, V, Ni, Cu, Zn, Rb, Sr, Y, Zr and Nb). In this case, the preselected GEOROC database was reduced to 550 records from 18 different geological sites, constituting dataset G30. In this newly computed PCA, the degree of overlap between different classes is not as strong as in the previous PCA. Moreover, the samples from Sidi Zahruni (Z) and Iulia Libica (L) appear in a marginal area of the point cloud, and therefore, some provenances can be discarded with better confidence, especially for Iulia Libica basalts, which appear with higher PC1 values (see X symbols in Figure 12). Namely, the Aegean, Gibraltar and Tyrrhenian arcs form clusters without intersecting either the Z or L clusters (as expected since these volcanic arcs are not in an intraplate tectonic setting), and from the Aeolian arc samples, only one is in the area of Z and L samples; the Campanian, Etna, Liguria and North Anatolia samples are also far from the Z and L clusters, but these provenances contain very few samples in the processed database. Considering that Z and L clusters plot relatively close to each other, some provenances match the position of both clusters. This is the case for Catalonia, Egypt,

Atlas and Sicily Channel Rift clusters (all of them considered intraplate volcanic districts). Additionally, the samples from the Suez Rift province match particularly well with those from Sidi Zahruni. Moreover, other provenances cannot be completely excluded. Also, there are a number of issues that should be considered: the joint PC1 and PC2 explained variance ratio is only ~44%, some reference volcanic districts are underrepresented (e.g., <6 samples) within the G30 database (see Table S4), and they could be non-statistically representative of their location, and some potential provenance locations have actually disappeared from the PCA because GEOROC does not contain any analyzed sample including all the considered compositional variables.

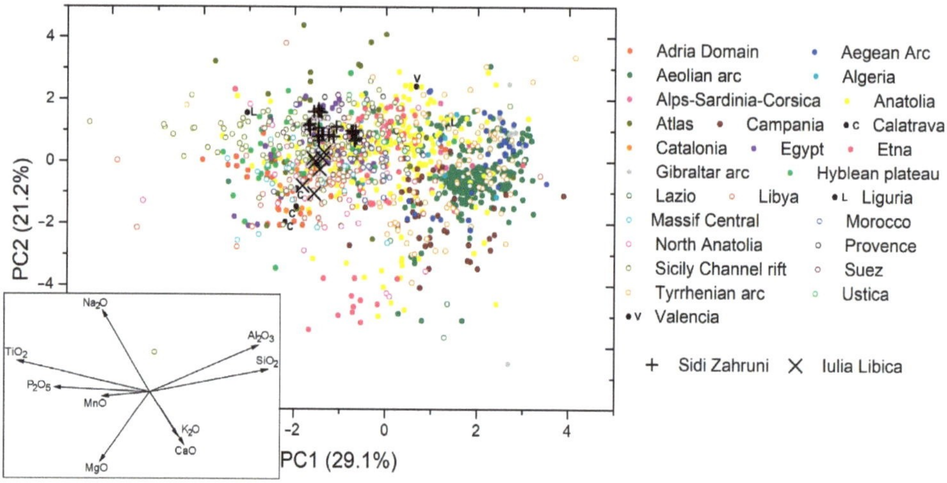

Figure 11. PCA biplot of factor scores for the first two principal components for the selected data from GEOROC (G90 dataset), Iulia Libica and Sidi Zahruni sample analyses have also been plotted. Inset: PCA biplot of the original variables.

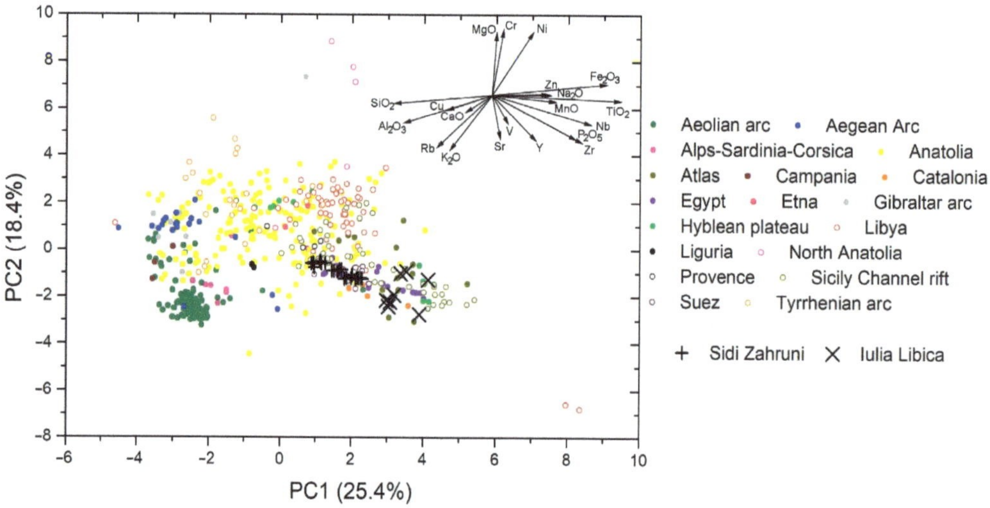

Figure 12. PCA biplot of factor scores for the first two principal components for the selected data from GEOROC (G30 dataset), Iulia Libica and Sidi Zahruni sample analyses have also been plotted. Inset: PCA biplot of the original variables.

3.3.2. Supervised Classification Methods

The supervised approach [43,44] was applied to the reference data obtained from the GEOROC database. Three different databases were tested (G90, G60 and G30, as described in Section 2.2). Datasets G30 and G60 only differ in two features (Cu and Zn values are not considered in G60; compare Tables S3 and S4). From the different classes (i.e., provenances) defined for the data in these three databases, only those bearing at least nine samples were considered. Classes with a smaller number of samples cannot be properly trained and predicted by the machine learning algorithms. As one of the classes relevant to the present study contained very few samples, additional basalt analyses [54,55] were added to the database to keep the corresponding provenance into consideration. Indeed, this class (X8), labeled Catalonia, corresponds to the Olot volcanic field, only about 50 km southeast of Iulia Libica. An additional reason to retain this class is that the corresponding provenance was, in fact, suggested from the results of some of the computed PCAs. Finally, 18, 16 and 12 different classes were considered from G90, G60 and G30 datasets, respectively.

Tables 4–6 present the results corresponding to individual runs of the supervised classification models using G90, G60 and G30 datasets, respectively. The tables display the classification accuracy of three different random test sets for each trained classification model and the overall classification predictions for the ensemble of Z samples and L samples. The results appear very consistently homogeneous and point to a given provenance for the Z samples. In contrast, for the L samples, the results are less monotonous, and a variety of provenances is suggested.

Table 4. Accuracies computed using the predictions of the trained models on three different test subsets (from dataset G90) and corresponding average probabilities of predicted class for Sidi Zahruni and Iulia Libica samples.

Model	Seed	Accuracy	Class [1] Predictions	
			Sidi Zahruni Samples	Iulia Libica Samples
GLM	1	0.6049	X21 (80%) X5 (20%)	X5 (62.5%) X1 (25%) X15 (12.5%)
	2	0.6292	X21 (68%) X5 (32%)	X5 (50%) X1 (37.5%) X15 (12.5%)
	3	0.5957	X21 (60%) X5 (40%)	X5 (50%) X1 (25%) X6 (12.5%) X15 (12.5%)
RF	1	0.7629	X21 (100%)	X5 (75%) X6 (25%)
	2	0.7508	X21 (100%)	X5 (62.5%) X6 (37.5%)
	3	0.7447	X21 (100%)	X6 (50%) X5 (37.5) X15 (12.5%)
ANN	1	0.6292	X21 (100%)	X15 (50%) X5 (37.5%) X1 (12.5%)
	2	0.5805	X21 (96%) X5 (4%)	X13 (75%) X21 (25%)
	3	0.5745	X21 (56%) X5 (44%)	X1 (62.5%) X13 (25%) X5 (12.5%)
kkNN	1	0.7660	X21 (100%)	X5 (75%) X8 (12.5%) X21 (12.5%)
	2	0.7264	X21 (100%)	X5 (37.5%) X8 (37.5%) X15 (12.5%) X21 (12.5%)
	3	0.7508	X21 (100%)	X5 (62.5%) X8 (37.5%)
LDA	1	0.5653	X21 (60%) X5 (40%)	X5 (87.5%) X15 (12.5%)
	2	0.5593	X21 (52%) X5 (48%)	X5 (62.5%) X1 (25%) X15 (12.5%)
	3	0.5471	X21 (52%) X5 (48%)	X5 (75%) X15 (12.5%) X21 (12.5%)
Stack	1	0.7538	X21 (100%)	X5 (100%)
	2	0.7295	X21 (100%)	X5 (87.5%) X15 (12.5%)
	3	0.7204	X21 (100%)	X5 (87.5%) X6 (12.5%)
GB	1	0.6770	X21 (92%) X5 (8%)	X5 (62.5%) X15 (37.5%)
	2	0.6794	X21 (88%) X10 (12%)	X13 (50%) X15 (37.5%) X5 (12.5%)
	3	0.6914	X21 (80%) X5 (16%) X10 (4%)	X13 (62.5%) X8 (25%) X5 (12.5%)

Table 4. *Cont.*

Model	Seed	Accuracy	Class [1] Predictions	
			Sidi Zahruni Samples	Iulia Libica Samples
GPC	1	0.6733	X21 (100%)	X5 (50%) X15 (50%)
	2	0.6962	X21 (92%) X5 (8%)	X5 (50%) X15 (37.5%) X21 (12.5%)
	3	0.6962	X21 (100%)	X5 (50%) X15 (50%)
GNB	1	0.8254	X21 (100%)	X5 (50%) X15 (50%)
	2	0.8014	X21 (100%)	X5 (50%) X15 (50%)
	3	0.8014	X21 (100%)	X5 (50%) X15 (50%)
LSVM	1	0.5670	X21 (80%) X5 (20%)	X1 (37.5%) X5 (37.5%) X8 (25%)
	2	0.5861	X21 (96%) X5 (4%)	X17 (87.5%) X1 (12.5%)
	3	0.5981	X21 (100%)	X8 (37.5%) X5 (25%) X17 (25%) X6 (12.5%)

[1] Class key: X1: Adria Domain; X5: Anatolia; X6: Atlas; X8: Catalonia; X10: Egypt; X13: Hyblean Plateau; X15: Lybia; X17: Massif Central; X21: Sicily Channel Rift.

Table 5. Accuracies computed using the predictions of the trained models on three different test subsets (from dataset G60) and corresponding average probabilities of predicted class for Sidi Zahruni and Iulia Libica samples.

Model	Seed	Accuracy	Class [1] Predictions	
			Sidi Zahruni Samples	Iulia Libica Samples
GLM	1	0.7282	X21 (88%) X5 (12%)	X6 (62.5%) X13 (37.5%)
	2	0.7538	X21 (80%) X5 (20%)	X6 (87.5%) X13 (12.5%)
	3	0.7795	X21 (88%) X5 (12%)	X6 (50%) X13 (50%)
RF	1	0.8359	X21 (100%)	X6 (87.5%) X13 (12.5%)
	2	0.8256	X21 (100%)	X6 (100%)
	3	0.8718	X21 (100%)	X6 (100%)
ANN	1	0.6974	X21 (100%)	X6 (62.5%) X13 (37.5%)
	2	0.7538	X21 (100%)	X6 (100%)
	3	0.7590	X21 (100%)	X6 (100%)
kkNN	1	0.8205	X21 (96%) X5 (4%)	X5 (50%) X8 (37.5%) X13 (12.5%)
	2	0.8154	X21 (100%)	X5 (50%) X8 (37.5%) X13 (12.5%)
	3	0.8718	X21 (96%) X5 (4%)	X5 (50%) X8 (37.5%) X21 (12.5%)
LDA	1	0.6564	X21 (100%)	X13 (100%)
	2	0.7231	X21 (100%)	X13 (100%)
	3	0.7231	X21 (100%)	X13 (100%)
Stack	1	0.8359	X21 (96%) X5 (4%)	X5 (50%) X6 (37.5%) X13 (12.5%)
	2	0.8205	X21 (100%)	X5 (50%) X6 (37.5%) X13 (12.5%)
	3	0.8718	X21 (96%) X5 (4%)	X5 (50%) X6 (37.5%) X21 (12.5%)
GB	1	0.8135	X21 (80%) X5 (20%)	X13 (62.5%) X6 (37.5%)
	2	0.8016	X21 (80%) X5 (20%)	X13 (100%)
	3	0.7619	X21 (80%) X5 (20%)	X13 (75%) X6 (25%)
GPC	1	0.7500	X21 (100%)	X21 (50%) X5 (37.5%) X15 (12.5%)
	2	0.7381	X21 (100%)	X21 (62.5%) X5 (25%) X10 (12.5%)
	3	0.7302	X21 (100%)	X5 (37.5%) X10 (37.5%) X21 (25%)
GNB	1	0.8413	X21 (100%)	X5 (37.5%) X10 (37.5%) X21 (25%)
	2	0.8690	X21 (100%)	X5 (37.5%) X10 (37.5%) X21 (25%)
	3	0.8651	X21 (100%)	X5 (37.5%) X10 (37.5%) X21 (25%)
LSVM	1	0.6825	X21 (100%)	X6 (87.5%) X1 (12.5%)
	2	0.6786	X21 (100%)	X6 (50%) X8 (37.5%) X13 (12.5%)
	3	0.6666	X21 (100%)	X6 (50%) X8 (37.5%) X13 (12.5%)

[1] Class key: X1: Adria Domain; X5: Anatolia; X6: Atlas; X8: Catalonia; X10: Egypt; X13: Hyblean Plateau; X15: Lybia; X21: Sicily Channel Rift.

Table 6. Accuracies computed using the predictions of the trained models on three different test subsets (from dataset G30) and corresponding average probabilities of predicted class for Sidi Zahruni and Iulia Libica samples.

Model	Seed	Accuracy	Class [1] Predictions	
			Sidi Zahruni Samples	Iulia Libica Samples
GLM	1	0.8416	X21 (100%)	X13 (75%) X8 (25%)
	2	0.8614	X21 (100%)	X15 (50%) X8 (37.5%) X13 (12.5%)
	3	0.8614	X21 (96%) X5 (4%)	X15 (87.5%) X8 (12.5%)
RF	1	0.8713	X21 (100%)	X6 (62.5%) X5 (12.5%) X13 (12.5%) X21 (12.5%)
	2	0.8416	X21 (100%)	X6 (87.5%) X21 (12.5%)
	3	0.8713	X21 (100%)	X6 (100%)
ANN	1	0.8119	X21 (100%)	X6 (75%) X5 (12.5%) X15 (12.5%)
	2	0.7525	X21 (100%)	X6 (100%)
	3	0.8317	X21 (100%)	X8 (50%) X6 (37.5%) X5 (12.5%)
kkNN	1	0.8911	X21 (92%) X5 (8%)	X5 (37.5%) X6 (25%) X8 (12.5%) X13 (12.5%) X21 (12.5%)
	2	0.8812	X21 (92%) X5 (8%)	X6 (50%) X5 (12.5%) X8 (12.5%) X13 (12.5%) X21 (12.5%)
	3	0.9010	X21 (92%) X5 (8%)	X5 (62.5%) X6 (25%) X13 (12.5%)
LDA	1	0.7921	X21 (100%)	X13 (100%)
	2	0.7723	X21 (96%) X22 (4%)	X13 (100%)
	3	0.8317	X21 (100%)	X13 (100%)
Stack	1	0.8713	X21 (100%)	X5 (37.5%) X6 (37.5%) X13 (12.5%) X21 (12.5%)
	2	0.8416	X21 (92%) X5 (8%)	X6 (50%) X5 (12.5%) X8 (12.5%) X13 (12.5%) X21 (12.5%)
	3	0.8812	X21 (92%) X5 (8%)	X5 (62.5%) X6 (25%) X13 (12.5%)
GB	1	0.8271	X21 (80%) X5 (20%)	X6 (50%) X5 (25%) X10 (12.5%) X15 (12.5%)
	2	0.8272	X21 (80%) X5 (20%)	X6 (50%) X5 (25%) X10 (12.5%) X15 (12.5%)
	3	0.7894	X21 (80%) X5 (12%) X22 (8%)	X15 (62.5%) X5 (37.5%)
GPC	1	0.7594	X21 (100%)	X5 (37.5%) X15 (37.5%) X21 (25%)
	2	0.7368	X21 (100%)	X15 (62.5%) X5 (37.5%)
	3	0.7820	X21 (100%)	X15 (62.5%) X5 (37.5%)
GNB	1	0.8571	X21 (100%)	X15 (62.5%) X5 (37.5%)
	2	0.8947	X21 (100%)	X15 (62.5%) X5 (37.5%)
	3	0.8571	X21 (100%)	X15 (62.5%) X5 (37.5%)
LSVM	1	0.7669	X21 (100%)	X6 (50%) X8 (37.5%) X13 (12.5%)
	2	0.7444	X21 (100%)	X6 (62.5%) X8 (37.5%)
	3	0.6617	X21 (100%)	X6 (37.5%) X8 (37.5%) X13 (25%)

[1] Class key: X1: Adria Domain; X5: Anatolia; X6: Atlas; X8: Catalonia; X10: Egypt; X15: Lybia; X21: Sicily Channel Rift; X22: Suez.

The different models tested, regardless of the database used, indicate a very high probability that the samples from Sidi Zahruni (Z) come from the Sicily Channel Rift (i.e., Pantelleria island), which is not far from the archaeological site (~115 km). Using the G90 database and combining all the Z samples and all the trained models, the overall probability for class X21 (Sicily Channel Rift) reaches 88.4%, and only two more classes are suggested, Anatolia (X5) with 11.1% and Egypt (X10) with only 0.5%. Using the G60 database, only two provenances are suggested: X21 captures an overall probability of 96%, and X5 captures the remaining 4%. Finally, using G30, class X21 holds almost all the probability (96.4%) again, and the rest splits between Anatolia (X5) with 3.3% and Suez (X22) withonly 0.3%.

Contrastingly, the different models do not show a similar agreement to assign provenance to the samples from Iulia Libica (L), and there are even significant differences using different seeds to define the train/test splits (see, for instance, L-sample predictions using the ANN model in Table 4). In any case, using the G90 database and combining all the results, the probability spreads into eight different provenances, including Anatolia (X5), 50.4%; Libya (X15), 17.5%; Adria Domain (X1), 7.9%; Hyblean Plateau (X13), 7.1%; Catalonia (X8), 5.8%; Atlas (X6), 5.0%, Massif Central (X17), 3.8%; and Sicily Channel Rift (X21), 2.5%.

One could think that at least one of the classes (X5) captures a bit more than 50%. However, after checking the results obtained using G60 (Table 5), we see that the predominance of class X5 is lost. The overall results using the G60 database indicate even more fractionated probabilities: Atlas (37.1%); Hyblean Plateau (25.4%); Anatolia (17.1%); Sicily Channel Rift (7.9%); Catalonia (6.3%); Egypt (5.4%); Adria Domain; and Libya (both 0.4%). Finally, using the database containing the highest number of compositional variables (G30), the overall probabilities spread again among different provenances: Atlas (30.8%); Anatolia (19.2%); Hyblean Plateau (17.1%); Catalonia (9.2%); Sicily Channel Rift (3.3%); and Egypt (0.8%).

4. Discussion

Regardless of the properties or techniques used, the attribution of provenance to archaeological volcanic samples requires accurate quantification of a set of parameters and a dataset containing the same quantification for a representative collection of reference geological samples.

In this paper, a quantification effort has been undertaken for both petrographic and compositional properties. Some of the quantifiable petrographic properties could arguably not be representative of a given provenance, e.g., the porosity or the shape of the pores could vary enormously even within the rocks of a given volcanic eruption. However, the petrographic characterization has possibly the potential to combine a set of discerning elements, including mineralogy of the phenocrysts, relative ratios, size, morphology, presence of alterations, texture/mineralogy of the matrix, etc. The problem with petrography is that, presently, there is a lack of large reference datasets that should comprise a standard set of quantified parameters. Therefore, the petrographic approach is not ready to fulfill the requirements to be applied as a routine technique for provenance determination. However, it has the potential to be developed as such. In any case, for well-delimited cases such as a binary classification problem, the petrographic approach can be applied because the self-production of the required reference dataset is feasible. Besides this, the petrographic characterization is also very helpful to determine whether the set of sampled materials from a given archaeological site is homogeneous (likely from a single provenance) or not. It is worth mentioning that a set of volcanic samples that are mineralogically and petrographically different could exhibit chemical homogeneity.

In the present investigation, the samples from Sidi Zahruni (Z) appear petrographically homogenous, indicating a single geological supply of volcanic materials, and the same can be said for the three millstones from Iulia Libica (L). The minerals that have been identified are those common in basalts, such as plagioclase (absent as phenocryst in L samples), clinopyroxenes and olivine. The absence of plagioclase phenocrysts (as the absence of leucite in both L and Z samples) could help to constrain the provenance. Other particular petrographic features that could be helpful are the alteration rims of olivine (L samples) or the occasional occurrence of green aegirine–augite (Z samples). Alteration rims in olivine are actually a common feature in many olivine basalts worldwide and have many potential provenances, including Sardinia, Agde (to the south of Massif Central), Olot (Catalonia), Ustica (Italy) and Middle Atlas (Morocco) basalts [56]. The presence of aegirine–augite crystals has been described in volcanic rocks from Pantelleria and is typical of more felsic rocks from this island [57], with which the basalts are intimately associated [58].

The chemical approach is currently much more ready than the petrographic approach to determine the provenance of archaeological volcanic materials. The quantification of the chemical composition is routinely applied by many researchers who publish their data, and some initiatives have been undertaken to systematically collect these data to create huge geochemical reference databases like GEOROC. However, some difficulties arise from the use of different equipment and sample preparation methods by the contributors to the reference database. Nevertheless, part of the solution is to build larger and larger datasets, which contributes to reducing random errors (improving precision), and the average of different systematic errors from different laboratories could also increase accuracy. Also, an extra complication of large datasets is that the geochemical data can be expressed in different ways (as oxides or elements, wt%, ppm, including volatiles or not, assuming a given valence state for elements like Fe or distinguishing different states, etc.). All these points could imply additional transformation steps before using the data, but they should not represent a real issue for well-managed databases.

Despite carefully selecting and standardizing relevant sets of data, the limitations of the common geochemical classification tools have been illustrated. On the one hand, biplots representing only two chemical elements (or two chemical elemental ratios) can be misleading because they disregard data that could be relevant. On the other hand, multivariate data analysis methods like PCA, despite considering all the features (i.e., the analyzed chemical elements), reveal that as the number of considered predefined classes increases, it is more likely that they form overlapping clusters. For the studied archaeological samples retrieved from Iulia Libica (L), the suggested provenances using certain biplots are the Hyblean plateau, Middle Atlas, Southern France and Spain. For those retrieved in Sidi Zahruni (Z), the suggested provenances are Pantelleria (recurrently), Sardinia and also the Hyblean Plateau. PCA can also be used to discard some of the multiple considered provenances. Provenances that can be discarded both for the samples from Iulia Libica and Sidi Zahruni include the basalts from the Aeolian, Aegean, Gibraltar and Tyrrhenian arcs and those from Campania, Lazio and possibly Algeria, Alps–Sardinia–Corsica and Etna. The fact that the suggested provenances correspond to those of intraplate basalts (intraplate nature was also geochemically deduced for L and Z samples) and that the discarded reference sites are essentially those linked to volcanic arcs (i.e., linked to a subduction zone) attest the robustness of both GEOROC database and the PCA method. However, PCA is not really useful in deciphering which, among the suggested intraplate provenances, is the most probable.

Finally, the supervised machine learning models are also multivariate methods that take into consideration a high number of compositional variables, but they have the additional advantage of being designed to discriminate the different predefined reference classes (i.e., the provenances). The capacity to discriminate between different classes can be measured using different statistical metrics computed using the test sets. Along this line, as was previously stated in a methodologically similar study [43], using supervised models is very important to check for the convergence between the results using different classification models and training several times using different splits. Reliable provenance determination requires a significant agreement between the results using these different strategies. In the present research, besides different models and splits, three different reference databases have been used (G90, G60 and G30) with accuracies ranging from 0.55–0.80 (using G90) to 0.66–0.90 (using G30).

In the case of the Sidi Zahruni (Z) samples, the results point very monotonously to the Sicily Channel Rift (X21 class) as the most probable provenance. This means that the Z basalts were extracted from Pantelleria or Linosa volcano islands, which are made of ocean island basalts known to have been exploited in antiquity for the manufacture of lava grinding tools like millstones [25,59]. This origin (especially Pantelleria) had appeared recurrently among the provenances suggested by the common geochemical classification tools (biplots and PCA). Moreover, the petrographic properties of Z samples also agree with those exhibited by the Pantelleria basalts, e.g., weakly porphyritic texture, mineralogy and relative abundance of the phenocrysts with pl > ol > cpx [59–61]. From the two basalt types described in Pantelleria [59], those with a higher affinity with the samples retrieved from Sidi Zahruni are the younger basalts with relatively low TiO_2 and P_2O_5 contents [60].

In contrast, for the samples from Iulia Libica, there is no agreement between the prediction models, and the predicted provenance usually appears statistically divided into different classes that hold fractions of provenance probability. Up to nine different classes (i.e., provenances) are suggested by the models, and, in general, the probability average percentages combining all the predictions are very scattered (the highest value is 50.4% for class X5 (Anatolia) using database G90). According to [43], this heterogeneity indicates that the provenance attribution is not robust. Indeed, in some cases, even the results from a given classification algorithm vary significantly when using different trained models (i.e., different train–test sets). Some classes (among others, incidentally X5, Anatolia) are overrepresented within the reference databases and may spread out, covering most of the point cloud in the corresponding PCA (e.g., yellow dots in Figures 11 and 12). Therefore, it is not surprising that such classes could capture substantial fractions of the provenance probability. Using geographical proximity as a criterion, classes X17 (Massif Central) and especially X8 (Catalonia) are the nearest locations to Iulia Libica. The models assign to these classes rather low probabilities, and both classes have a very low amount of data within the reference databases. Even if training is enabled, a low number of data may not be fully statistically representative of the class. In fact, class X17 has only been considered by the models using database G90 because the number of X17-class data in the other datasets was too low to train the models to discriminate this class. The same would have occurred for class X8 if we had not supplemented the original database (produced using GEOROC) with additional data for this class. Ideally, all classes with a low number of reference samples should be supplemented. In particular, it would have been interesting to increase the reference samples for class X17, as the Massif Central (in particular cap d'Agde) is the presumed origin for the basaltic millstones produced in Lattes [36], typologically similar to those found in Iulia Libica. Such a trading route would be consistent with the presence of pottery from the southern Gaulish area in Iulia Libica [32]. With the available data and comparing with the results obtained for Sidi Zahruni, the conclusion is that the provenance of Iulia Libica samples has not been successfully determined, and this implies that it corresponds to a class imperfectly represented or absent from the reference databases used.

To summarize, supervised machine learning methods have a number of advantages over other statistical methods. Conceptually, they are designed to learn the best way to discriminate the different classes. However, they are not foolproof, and they are subject to limitations linked to incomplete reference databases (unbalanced, underrepresented or missing classes). Despite the good indicators of overall performance (see accuracies in Tables 4–6) obtained after testing the trained models, a model will obviously never be able to predict a class not described in the training database and the performance of the model to predict underrepresented classes can be far from appropriate. One of the statistical metrics used to check how well a model performs for a particular class is sensitivity, which is defined as the ratio between well-predicted samples and the total samples within the class. For instance, from the confusion matrices obtained during the test step, it could be seen that classes X8 (Catalonia) and X17 (Massif Central) exhibit low average sensitivities per model (usually below 0.5 and sometimes even 0). Specifically, for class X8, the highest average sensitivity values are 0.78 and 0.69 using LSVM and kkNN models, respectively, and incidentally, these two models are the ones that statistically attribute higher provenance probabilities to class X8 compared to other models (see Tables 4–6). As for class X17 (only considered in runs using database G90), the highest average sensitivity is only 0.23, obtained using the LSVM model, and this is actually the only model that assigns a certain probability to this provenance.

5. Conclusions

Statistical approaches to establish the provenance of volcanic rock tools require quantification of a set of properties and a large reference database. The quantified properties could be geochemical, mineralogical, or petrographic. However, currently, only large reference geochemical datasets are available. Implementing the statistical approach to petrographic properties would first require an agreement on the parameters to be quantified and a collective effort to produce the corresponding datasets.

The limitations of the common geochemical and statistical classification tools have been illustrated. On the one hand, biplots using a few variables can be misleading because relevant data could be omitted. As a result of this, different biplots can suggest different provenances. On the other hand, multivariate data analysis methods, like PCA, often exhibit a single point cloud and not separated clusters, and then it is difficult to assign a given provenance to the unlabeled archaeological data.

The supervised machine learning approach applied to geochemical data has proven to be a powerful tool for discerning different reference clusters, and this enables provenance prediction for unlabeled samples. Among the 18 different considered provenances, the models assign systematically to all the samples from Sidi Zahruni a very high provenance probability to the class representing the basalts from the Sicily Channel Rift. The provenance of Sidi Zahruni samples has been successfully established, not only because of the strong agreement between the different supervised models but also because of the high petrographic consistency between the studied samples and the known characteristics of the Pantelleria basalts. Additionally, this provenance was among those suggested using common geochemical classification tools, and it lies only ~115 km from the archaeological site of Sidi Zahruni.

In contrast, the models assign different provenances to the samples from Iulia Libica, and this suggests that the corresponding provenance was missing in the considered reference datasets or that it was incompletely described within them. This highlights the limitations of the supervised approach and the need for a statistical approach making use of different sister samples, different trained models and different classification algorithms to be able to identify a homogeneous and, therefore, robust provenance prediction.

Supplementary Materials: The following supporting information can be downloaded at https://www.mdpi.com/article/10.3390/min14070639/s1. Table S1: Database contains 1855 records obtained by selecting analyzed basalt rocks within the area contained within latitude [25° N–54° N] and longitude [12° W to 45° E] from GEOROC database [41]. Table S2: Dataset G90 contains all records from Table S1, which includes the following set of analyzed elements: Si, Ti, Al, Ca, Mg, Mn, K, Na and P. Table S3: Dataset G60 contains all records from Table S2, which also includes the following set of analyzed elements: V, Ni, Rb, Sr, Y, Zr and Nb. Table S4: Dataset G30 contains all records from Table S3, which includes additional analyses on the Cu and Zn contents. The Python code for this study is attached: [https://www.mdpi.com/article/10.3390/min14070639/s1].

Author Contributions: Conceptualization, L.C.; fieldwork and sample retrieval, L.C., B.F. and C.C.; experimental work—petrographic analyses, L.C. and R.D.F.; experimental work—electron microcopy, R.D.F.; experimental work—geochemical analyses, Á.P.M., A.A., R.D.F. and I.Q.; code writing, A.A. and L.C.; formal analyses of data, L.C., I.Q. and Á.P.M.; writing—first draft preparation, L.C.; writing—review and editing, all authors. All authors have read and agreed to the published version of the manuscript.

Funding: IDAEA-CSIC is a Centre of Excellence Severo Ochoa (Spanish Ministry of Science and Innovation, Project CEX2018-000794-S).

Data Availability Statement: Most of the data presented in this study have been made publicly available. Any other data are available on request from the corresponding author.

Acknowledgments: We are grateful to the Tunisian Institut National du Patrimoine for allowing access and sampling at the Sidi Zahruni site. We thank Néjia Laridhi-Ouazaa, Siwar Baklouti, Aureli Álvarez, Marta Prevosti and Ramon Járrega for their support during fieldwork in Tunisia. Toni Lozano and Albert Ruiz are acknowledged for giving training on Python and machine learning. A.P.M. is a Serra Húnter Lecturer. L.C. and A.P.M. are members of the Grup de Recerca Conservació-Restauració de Patrimoni (AGAUR, 2021 SGR 00089). Finally, we would like to thank the editor as well as three anonymous reviewers for their insightful remarks and comments that definitely have helped to improve the paper.

Conflicts of Interest: The authors declare no conflicts of interest.

References

1. Liritzis, I.; Laskaris, N.; Vafiadou, A.; Karapanagiotis, I.; Volonakis, P.; Papageorgopoulou, C.; Bratitsi, M. Archaeometry: An Overview. *Sci. Cult.* **2020**, *6*, 49–98.
2. Kolb, C.C. Provenance Studies in Archaeology. In *Encyclopedia of Global Archaeology*; Smith, C., Ed.; Springer: New York, NY, USA, 2014; pp. 6172–6181, ISBN 978-1-4419-0465-2.

3. Waksman, Y. Provenance Studies: Productions and Compositional Groups. In *The Oxford Handbook of Archaeological Cerami Analysis*; Alice, H., Ed.; Oxford University Press: Oxford, UK, 2017; pp. 148–161.
4. Pollard, A.M.; Heron, C.; Gillard, R.D. The Geochemistry of Clays and the Provenance of Ceramics. In *Archaeological Chem istry*; Pollard, A.M., Heron, C., Gillard, R.D., Eds.; The Royal Society of Chemistry: Cambridge, UK, 2008; pp. 98–14 ISBN 978-0-85404-262-3.
5. Lancaster, L.; Sottili, G.; Marra, F.; Ventura, G. Provenancing of Lightweight Volcanic Stones Used in Ancient Roman Concret Vaulting: Evidence from Rome. *Archaeometry* **2011**, *53*, 707–727. [CrossRef]
6. Maniatis, Y.; Herz, N.; Basiakos, Y. *The Study of Marble and Other Stones Used in Antiquity: Asmosia III Athens: Transactions of the 3r International Symposium of the Association for the Study of Marble and Other Stones Used in Antiquity*; Archetype Publications Ltd. London, UK, 1995.
7. Flügel, E.; Flügel, C. Applied Microfacies Analysis: Provenance Studies of Roman Mosaic Stones. *Facies* **1997**, *37*, 1–48. [CrossRef
8. García de Madinabeitia, S.; Gil Ibarguchi, J.I.; Santos Zalduegui, J.F. IBERLID: A Lead Isotope Database and Tool for Meta Provenance and Ore Deposits Research. *Ore Geol. Rev.* **2021**, *137*, 104279. [CrossRef]
9. Levine, M.A. Determining the Provenance of Native Copper Artifacts from Northeastern North America: Evidence fror Instrumental Neutron Activation Analysis. *J. Archaeol. Sci.* **2007**, *34*, 572–587. [CrossRef]
10. Falkenberg, J.; Kaplan, U.; Mutterlose, J. Understanding the Provenance and Production Process of Historic Mortars—A Nove Approach Employing Calcareous Nannofossils. *Archaeol. Anthr. Sci.* **2023**, *15*, 138. [CrossRef]
11. Miriello, D.; Barca, D.; Pecci, A.; De Luca, R.; Crisci, G.M.; López Luján, L.; Barba, L. Plasters from Different Buildings of th Sacred Precinct of Tenochtitlan (Mexico City): Characterization and Provenance. *Archaeometry* **2015**, *57*, 100–127. [CrossRef]
12. Gratuze, B. Provenance Analysis of Glass Artefacts. In *Modern Methods for Analysing Archaeological and Historical Glass*; John Wile & Sons: Hoboken, NJ, USA, 2013; pp. 311–343. ISBN 9781118314234.
13. Di Febo, R.; Casas, L.; Capelli, C.; Cabella, R.; Vallcorba, O.; Di Febo, R.; Casas, L.; Capelli, C.; Cabella, R.; Vallcorba, O. Catalar Imitations of the Ligurian Taches Noires Ware in Barcelona (18th–19th Century): An Example of Technical Knowledge Transfe *Minerals* **2018**, *8*, 183. [CrossRef]
14. Fornacelli, C.; Briano, A.; Chiarantini, L.; Bianchi, G.; Benvenuti, M.; Giamello, M.; Kang, J.S.; Villa, I.M.; Talarico, F.M.; Hodges, R Archaeometric Provenance Constraints for Early Medieval Sparse Glazed Pottery from Donoratico (Livorno, Italy). *Archaeometry* **2021**, *63*, 549–576. [CrossRef]
15. Velliky, E.C.; MacDonald, B.L.; Porr, M.; Conard, N.J. First Large-Scale Provenance Study of Pigments Reveals New Comple Behavioural Patterns during the Upper Palaeolithic of South-Western Germany. *Archaeometry* **2021**, *63*, 173–193. [CrossRef]
16. Doménech-Carbó, A.; Doménech-Carbó, M.T.; Valle-Algarra, F.M.; Gimeno-Adelantado, J.V.; Osete-Cortina, L.; Bosch-Reig F. On-Line Database of Voltammetric Data of Immobilized Particles for Identifying Pigments and Minerals in Archaeometry Conservation and Restoration (ELCHER Database). *Anal. Chim. Acta* **2016**, *927*, 1–12. [CrossRef]
17. Ouazaa, N.L.; Casas, L.; Álvarez, A.; Fouzai, B.; Moreno-Vide, M.; Vidal, L.; Sihem, R.; Sonzogni, C.; Borschneck, D. Provenance of Marble Sculptures from the National Museum of Carthage (Tunisia). *J. Archaeol. Sci.* **2013**, *40*, 1602–1610. [CrossRef]
18. Casas, L.; Di Febo, R.; Parcerisa, D. Petrographic Markers for Archaeometric Identification of Montjuïc Sandstone, the Flagship Stone of Barcelona (NE Spain). *Minerals* **2020**, *10*, 154. [CrossRef]
19. Bigazzi, G.; Ercan, T.; Oddone, M.; Özdoğan, M.; Yeğingil, Z. Application of Fission Track Dating to Archaeometry: Provenance Studies of Prehistoric Obsidian Artifacts. *Nucl. Tracks Radiat. Meas.* **1993**, *22*, 757–762. [CrossRef]
20. Re, A.; Giudice, A.L.; Angelici, D.; Calusi, S.; Giuntini, L.; Massi, M.; Pratesi, G. Lapis Lazuli Provenance Study by Means of Micro-PIXE. *Nucl. Instrum. Methods Phys. Res. B* **2011**, *269*, 2373–2377. [CrossRef]
21. Malfilatre, C.; Hallot, E.; Boulvais, P.; Poujol, M.; Chauvin, A.; Gapais, D.; Dabard, M.-P.; Bourquin, S.; Pallix, D. Fingerprinting the Provenance of Building Stones: A Case Study on the Louvigné and Lanhélin Granitic Rocks (Armorican Massif, France). *Bull. Société Géologique Fr.* **2014**, *185*, 13–31. [CrossRef]
22. Malfilatre, C.; Boulvais, P.; Dabard, M.-P.; Bourquin, S.; Hallot, E.; Pallix, D.; Gapais, D. Petrographical and Geochemical Characterization of Comblanchien Limestone (Bourgogne, France): A Fingerprint of the Building Stone Provenance. *Comptes Rendus Geosci.* **2012**, *344*, 14–24. [CrossRef]
23. Williams-Thorpe, O.; Thorpe, R.S. Geochemistry and Trade of Eastern Mediterranean Millstones from the Neolithic to Roman Periods. *J. Archaeol. Sci.* **1993**, *20*, 263–320. [CrossRef]
24. Peacock, D.P.S. The Roman Millstone Trade: A Petrological Sketch. *World Archaeol.* **1980**, *12*, 43–53. [CrossRef]
25. Santi, P.; Foresta Martin, F.; Spatafora, F.; de Vita, S.; Renzulli, A. Volcanic Grinding Tools in Ustica Island (Tyrrhenian Sea, Italy): Local Production vs. Import of Morgantina-Type Millstones in the Hellenistic-Roman Period. *Minerals* **2020**, *10*, 389. [CrossRef]
26. Lancaster, L.C.; Sottili, G.; Marra, F.; Ventura, G. Provenancing of Lightweight Volcanic Stones Used in Ancient Roman Concrete Vaulting: Evidence from Turkey and Tunisia. *Archaeometry* **2010**, *52*, 949–961. [CrossRef]
27. Jackson, M.; Marra, F. Roman Stone Masonry: Volcanic Foundations of the Ancient City. *Am. J. Archaeol.* **2006**, *110*, 403–436. [CrossRef]
28. Antonelli, F.; Lazzarini, L. Mediterranean Trade of the Most Widespread Roman Volcanic Millstones from Italy and Petrochemical Markers of Their Raw Materials. *J. Archaeol. Sci.* **2010**, *37*, 2081–2092. [CrossRef]

29. Di Bella, M.; Italiano, F.; Martinelli, M.C.; Mazzoleni, P.; Quartieri, S.; Tigano, G.; Tripodo, A.; Sabatino, G. Archeometric Characterization of Prehistoric Grindstones from Milazzo Bronze Age Settlement (Sicily, Italy). *Archaeol. Anthr. Sci.* **2018**, *10*, 1571–1583. [CrossRef]
30. Falcone, F.; Dionisio, A.; Castorina, F.; Tufo, A.; Francis, R.E.; Stoppa, F. Archaeometry of a Roman Millstone from Santa Maria Arabona, Manoppello (Abruzzo, Central Italy). *Minerals* **2021**, *11*, 948. [CrossRef]
31. Santi, P.; Renzulli, A.; Gullo, R. Archaeometric Study of the Hopper-Rubber and Rotary Morgantina-Type Volcanic Millstones of the Greek and Roman Periods Found in the Aeolian Archipelago (Southern Italy). *Eur. J. Mineral.* **2013**, *25*, 39–52. [CrossRef]
32. Guàrdia, J.; Carreras, C.; de Soto, P. Commercial Circuits of Iulia Lybica (Llívia): Notions Gleaned from Finds of Pottery and Marble. *Rev. D'arqueologia Ponent* **2017**, 147–168. [CrossRef]
33. Járrega Domínguez, R.; Prevosti Monclús, M. *La Producción de Ánforas En El Territorium de Neapolis Durante La Antigüedad Tardía El Taller de Sidi Zahruni (Nabeul, Túnez)*; BAR Publishing: Oxford, UK, 2022.
34. Guàrdia Felip, J. The City and the Forum of Iulia Libica (Llívia, Cerdanya). *Cuad. Arqueol. Univ. Navar.* **2021**, *29*, 111–128.
35. Carreras, C.; Farré, E. Un Model de Rellotge Solar Trobat a Iulia Libica (Llívia). *Rev. D'arqueologia Ponent* **2021**, *31*, 115–128.
36. Py, M. Meules d'époque Protohistorique et Romaine Provenant de Lattes. In *Lattara, Recherches sur l'Économie Vivrière des Lattarenses*; Py, M., Ed.; Association Pour La Recherche Archéologique En Languedoc Oriental: Lattes, France, 1992; Volume 5, pp. 183–232.
37. Baklouti, S.; Maritan, L.; Casas, L.; Laridhi Ouazaa, N.; Jàrrega, R.; Prevosti, M.; Mazzoli, C.; Fouzaï, B.; Larabi Kassaa, S.; Fantar, M. Establishing a New Reference Group of Keay 25.2 Amphorae from Sidi Zahruni (Nabeul, Tunisia). *Appl. Clay Sci.* **2016**, *132–133*, 140–154. [CrossRef]
38. Prevosti, M.; Járrega, R.; Casas, L.; Fouzai, B.; Baklouti, S.; Laridhi-Ouazaa, N. Sidi Zahruni, Una Alfarería En El Ager de Neapolis (Túnez): Aglomeración o Vicus circa Villam? *Dialogues D'histoire Ancienne* **2022**, *481*, 265–313. [CrossRef]
39. Schneider, C.A.; Rasband, W.S.; Eliceiri, K.W. NIH Image to ImageJ: 25 Years of Image Analysis. *Nat. Methods* **2012**, *9*, 671–675. [CrossRef]
40. Marguí, E.; Queralt, I. Sample Preparation for X-ray Fluorescence Analysis. In *Encyclopedia of Analytical Chemistry*; John Wiley & Sons, Ltd.: Hoboken, NJ, USA, 2024; pp. 1–29. ISBN 9780470027318.
41. Lehnert, K.; Su, Y.; Langmuir, C.H.; Sarbas, B.; Nohl, U. A Global Geochemical Database Structure for Rocks. *Geochem. Geophys. Geosystems* **2000**, *1*, 1012. [CrossRef]
42. Le Bas, M.J.; Le Maitre, R.W.; Streickeisen, A.; Zanettin, B. A Chemical Classification of Volcanic Rocks Based on the Total Alkali-Silica Diagram. *J. Petrol.* **1986**, *27*, 745–750. [CrossRef]
43. Anglisano, A.; Casas, L.; Queralt, I.; Di Febo, R. Supervised Machine Learning Algorithms to Predict Provenance of Archaeological Pottery Fragments. *Sustainability* **2022**, *14*, 11214. [CrossRef]
44. Anglisano, A.; Casas, L.; Anglisano, M.; Queralt, I. Application of Supervised Machine-Learning Methods for Attesting Provenance in Catalan Traditional Pottery Industry. *Minerals* **2020**, *10*, 8. [CrossRef]
45. Morimoto, N.; Fabries, J.; Ferguson, A.K.; Ginzburg, I.V.; Ross, M.; Seifert, F.A.; Zussman, J.; Aoki, K.; Gottardi, G. Nomenclature of Pyroxenes. *Miner. Mag.* **1988**, *52*, 535–550. [CrossRef]
46. Le Maitre, R.W.; Streckeisen, A.; Zanettin, B.; Le Bas, M.J.; Bonin, B.; Bateman, P. *Igneous Rocks: A Classification and Glossary of Terms: Recommendations of the International Union of Geological Sciences Subcommission on the Systematics of Igneous Rocks*, 2nd ed.; Cambridge University Press: Cambridge, UK, 2002; ISBN 9780521619486.
47. Pearce, J.A. *Trace Element Characteristics of Lavas from Destructive Plate Boundaries*; Thorpe, R.S., Ed.; John Wiley and Sons: Chichester, UK, 1982.
48. Pearce, J.A.; Norry, M.J. Petrogenetic Implications of Ti, Zr, Y, and Nb Variations in Volcanic Rocks. *Contrib. Mineral. Petrol.* **1979**, *69*, 33–47. [CrossRef]
49. Weller, D.; Orlandini, O.; Benton, J.; Schirmer, C.; LoBue, L.; Culotta, S. Provenancing the Stone Tools of Volubilis, Morocco: A Socio-Economic Interpretation of Stonework Lithologies. *J. Archaeol. Sci. Rep.* **2021**, *38*, 103105. [CrossRef]
50. Antonelli, F.; Lazzarini, L.; Luni, M. Preliminary Study on the Import of Lavic Millstones in Tripolitania and Cyrenaica (Libya). *J. Cult. Herit.* **2005**, *6*, 137–145. [CrossRef]
51. Orange, M.; Le Bourdonnec, F.-X.; Bellot-Gurlet, L.; Lugliè, C.; Dubernet, S.; Bressy-Leandri, C.; Scheffers, A.; Joannes-Boyau, R. On Sourcing Obsidian Assemblages from the Mediterranean Area: Analytical Strategies for Their Exhaustive Geochemical Characterisation. *J. Archaeol. Sci. Rep.* **2017**, *12*, 834–844. [CrossRef]
52. Germinario, L.; Hanchar, J.M.; Sassi, R.; Maritan, L.; Cossio, R.; Borghi, A.; Mazzoli, C. New Petrographic and Geochemical Tracers for Recognizing the Provenance Quarry of Trachyte of the Euganean Hills, Northeastern Italy. *Geoarchaeology* **2018**, *33*, 430–452. [CrossRef]
53. Gluhak, T.M.; Schwall, C. Provenance Analyses of the Volcanic Rock Grinding Stones from the Greek Colony of Selinunte, Sicily (Italy)—Constraints and Possibilities. *Archaeometry* **2015**, *57*, 246–268. [CrossRef]
54. Gisbert Pinto, G. Estudi Petrològic i Geoquímic Del Vulcanisme Recent de La Garrotxa. Master's Thesis, Universitat de Barcelona, Barcelona, Spain, 2009.
55. Enrique, P.; Toribio, V. Subalkali Basalt Flows from the Ser Basin (Olot Volcanic Field). *Geogaceta* **2009**, *47*, 129–132.
56. Williams-Thorpe, O. Provenancing and Archaeology of Roman Millstones from the Mediterranean Area. *J. Archaeol. Sci.* **1988**, *15*, 253–305. [CrossRef]

57. White, J.C.; Parker, D.F.; Ren, M. The Origin of Trachyte and Pantellerite from Pantelleria, Italy: Insights from Major Element Trace Element, and Thermodynamic Modelling. *J. Volcanol. Geotherm. Res.* **2009**, *179*, 33–55. [CrossRef]
58. Ferla, P.; Meli, C. Evidence of Magma Mixing in the 'Daly Gap' of Alkaline Suites: A Case Study from the Enclaves of Pantelleria (Italy). *J. Petrol.* **2006**, *47*, 1467–1507. [CrossRef]
59. Renzulli, A.; Santi, P.; Gambin, T.; Serrano, P.B. Pantelleria Island as a Centre of Production for the Archaic Phoenician Trade in Basaltic Millstones: New Evidence Recovered and Sampled from a Shipwreck off Gozo (Malta) and a Terrestrial Site at Cádiz (Spain). *J. Archaeol. Sci. Rep.* **2019**, *24*, 338–349. [CrossRef]
60. Civetta, L.; D'Antonio, M.; Orsi, G.; Tilton, G.R. The Geochemistry of Volcanic Rocks from Pantelleria Island, Sicily Channel Petrogenesis and Characteristics of the Mantle Source Region. *J. Petrol.* **1998**, *39*, 1453–1491. [CrossRef]
61. Gail, A.M.; Don, R.B. Experimental Constraints on Depths of Fractionation of Mildly Alkalic Basalts and Associated Felsic Rocks Pantelleria, Strait of Sicily. *Contrib. Mineral. Petrol.* **1986**, *93*, 251–264. [CrossRef]

minerals

MDPI

Article

An Approach to Accurately Identifying Binders in Historic Mortars by the Combination of Microscopic and Microanalytical Techniques

Luís Almeida [1,2,*], António Santos Silva [3], Rosário Veiga [4] and José Mirão [1,5]

1 HERCULES Laboratory, University of Évora, Largo Marquês de Marialva, 8, 7000-809 Évora, Portugal
2 UNIARQ, Archaeology Center of the University of Lisbon, Faculty of Letters, Alameda da Universidade, 1600-214 Lisbon, Portugal
3 Materials Department, LNEC, National Laboratory for Civil Engineering, Avenida do Brasil, 101, 1700-066 Lisbon, Portugal; ssilva@lnec.pt
4 Buildings Department, LNEC, National Laboratory for Civil Engineering, Avenida do Brasil, 101, 1700-066 Lisbon, Portugal; rveiga@lnec.pt
5 Geosciences Department, Colégio Luís António Verney, University of Évora, Rua Romão Ramalho, 59, 7000-671 Évora, Portugal; jmirao@uevora.pt
* Correspondence: lfalmeida@uevora.pt

Abstract: Mortars are among the most important materials in building construction. They are generally obtained by mixing aggregates with an inorganic binder. The identification of mortar constituents, particularly the binder type in historic buildings, is one of the essential aspects of building conservation, considering that the new conservation materials must be chemically, mechanically, and physically compatible with the old masonries. Among other techniques used to characterise binders, those related to optical and electronic microscopy are particularly important. Microscopy and combined techniques may be the key to this identification since the classic mineralogical and chemical-based identification approaches are not conclusive enough in investigating the types of hydraulic binders in mortars. This work presents an analysis procedure to identify mortar binders by combining EDS microanalysis and petrography. Mortar samples of known composition were used as a reference for analyzing mortars from historic buildings. The proposed methodology made it possible to identify the type of binder or a mixture of binders based on the identification of the binder features by petrography together with analysis of the chemical composition of the paste by X-ray microanalysis under a scanning electron microscope.

Keywords: petrography; SEM-EDS; mortars; binders; characterization

1. Introduction

Mortars are an essential part of the construction of built structures and have become more sophisticated over time, evolving in close connection with manufacturing technologies and construction techniques.

Mortars have various functions, such as rendering walls, repointing and joints, covering and bedding masonry elements, or bonding ceramic tiles.

Over time, the functions assigned to mortars have mostly stayed the same. They are generally composite materials technologically characterized by a mixture of aggregates with one or more types of binders, water, and additions. The most typical traditional binders could be based on clay, lime, and gypsum [1,2]. The innovative changes were related to the binders' production technology, the mix's formulation, and the incorporation of different materials that granted characteristics or performances that increased durability and strength. Regarding innovation in formulation, pozzolanic materials in Roman times stand out, namely replacing volcanic ashes with crushed or powdered ceramic fragments when the former was unavailable [3]. Mortars with ceramic fragments (*cocciopesto*) were preferred for water-bearing structures and moisture protection, often used in baths, canals, and aqueducts [3–8]. This innovation in mortar production aligns with modern sustainability and recycling themes. Studies [9,10] indicate that the ceramic fragments in cocciopesto differ in optical activity (birefringence), petrographic characteristics, and texture. This diversity implies that they were made from recycled materials from various ceramic productions, potentially representing one of the earliest instances of reuse and recycling in history [9,10].

There have been considerable changes throughout history in the manufacture of binders, especially since the 18th century, when technologies were discovered that made it possible to produce various types of hydraulic binders and expand the application of hydraulic mortars in construction.

Lime mortars can be divided into air lime mortars and hydraulic lime mortars, depending on whether air or hydraulic lime is used. Compounds can also be incorporated to make mortars hydraulic (e.g., natural or artificial pozzolans, crushed bricks, silica, and amorphous alumina) even without a hydraulic binder.

Hydraulic mortars can also be based on natural or artificial types of cement, such as Portland cement. Hydraulic lime is obtained from a burnt natural rock (siliceous/argillaceous limestone or calcareous marl) below the sintering temperature (800–1200 °C). It must contain enough free CaO to be slaked with water and be capable of setting under water. The calcined product must contain a minimum amount of free CaO to reduce the entire mass to a powder when slaked [1].

Natural cement, also known as Roman cement, was patented by James Parker in England in 1796 [11]; its hydraulicity is due to the raw material used. Parker used calcareous nodules (septarian nodules) found in the London clay beds on the Isle of Sheppey, England, and calcined them in his domestic fire. He then ground the resulting clinker and mixed it with water. This process led to the creation of a quick-drying, nut-brown-colored cement [11–13].

Despite implied links to the Roman binders, Parker's 'Roman cement' was a proper hydraulic cement very different from the hydraulic binders used by the Romans in which pozzolanic materials, not cementitious in themselves, had combined with lime in the presence of water to form insoluble compounds possessing cementing properties [12].

The standardization of technological processes has led to the manufacture of various types of binders with characteristics that are now known in more recent historical periods. Air lime can currently be classified according to standard EN 459-1:2015 [14], which divides lime into calcium (CL) or dolomitic (DL), considering its chemical composition. The same standard also establishes the classification of natural hydraulic lime (NHL), which differs chemically and mineralogically, leading to different compressive strength ranges.

Air lime is divided into fat and lean. Fat air lime is derived from almost pure limestone with at least 99% carbonate content. Lean (generally greyish) is derived from limestones with clay and other impurity contents between 1 and 5% [15].

Roman cement is a type of cement made from marl or septaria that contains 25% or more clay. This cement is considered "natural" because all the necessary components, such as lime, silica, and alumina, are found in a single source material, unlike Portland cement, which is made from different sources [13].

Although they are natural hydraulic binders, according to Kozłowski (2010) [12], natural cements are distinct from hydraulic limes in that they have low levels of free lime, which means they need to be ground instead of slaked. They are also different from Portland cements due to their different chemical composition, resulting in significantly lower calcination temperatures. The primary hydraulic phase in natural cements is C_2S (dicalcium silicate or belite), whereas in ordinary Portland cement, it is C_3S (tricalcium silicate or alite).

Although both Roman cement and natural hydraulic lime are calcined at low temperatures, they differ in that the cement hardens quickly, usually in less than 15 min. Both materials contain substantial amounts of belite and have a prolonged strength development profile. Roman cements can be differentiated from Portland cements by the presence of residual quartz and calcite and the absence, or residual content, of tricalcium silicate

(C_3S—alite), which is responsible for the substantial strength development in Portland cements [12].

Historical cementitious or highly hydraulic mortar binders exhibit a lesser degree of chemical definition than their contemporary counterparts. Consequently, comprehensive characterization of these binders within mortar specimens requires meticulous examination of the residual unhydrated particles within their structural framework. This endeavor is optimally facilitated through microscopic techniques, potentially augmented by chemical point analysis [16].

As some authors recognize [16–18], instrumental bulk analyses such as X-ray diffraction or chemical analysis are less capable of tracing the binder constituents than imaging or microscopical analytical tools.

This study proves innovative as it combines elementary chemical analyses with polarized light microscopy. It aims to group different mortar types according to their binders, distinguishing mortars made with hydraulic binders, namely 20th-century Portland cement, natural cement (from the late 19th century) and NHL. Although NHL mortars were widely used in construction until the 19th century, the mass production of Portland cement, combined with its performance, likely meant that hydraulic cementitious binders were the choice of binders in the 20th century buildings studied. In order to cover the various types of binders, NHL mortar samples produced in the laboratory were also investigated.

It should also be mentioned that the ageing of hydraulic binders, such as Portland cement involves several chemical processes that can affect their long-term durability and performance. One critical process is carbonation, which involves the reaction of carbon dioxide (CO_2) from the air with the hydrated phases of the binder. The carbonation leads to calcium carbonate ($CaCO_3$) formation and subsequent decalcification, where calcium ions are leached out of the material [19]. Thus, the Si/Ca and Al/Ca ratios can increase with time.

Since the main factors in obtaining the various types of mortar binders depend on the raw material and the temperature of calcination and/or sintering, the latter depending on the manufacturing technology available at the time. The applied methodology and the results obtained can be used to characterize mortars from other historical periods.

The aspects identified above are even more critical regarding material compatibility. This greatly impacts conservation and restoration interventions, which must respect the original materials' physical, mechanical, and chemical characteristics [20,21] as much as possible to prevent future decay phenomena.

2. Materials and Methods

2.1. Samples and Analytical Background

Fifty-one samples were investigated from buildings constructed throughout the 20th century in Lisbon (Portugal), whose characteristics and sampling setup can be consulted elsewhere [22], as part of a thoroughly characterization study. Additionally, two natural cement samples were analyzed from cast decorative elements in Barcelona buildings (Spain), built between the end of the 19th century and the beginning of the 20th century [23].

Table 1 presents the type of binder and the binder-to-aggregate (b:a) ratio of the analyzed render and plaster samples and their construction or application period. The b:a ratio was relevant to investigate if this ratio would impact the chemical analysis results.

For lime mortars, the b:a ratio was calculated using the insoluble residue (IR) values obtained by wet chemical analysis and the CO_2 content obtained by thermogravimetry (TGA-DTA) [22]. The insoluble residue (IR) corresponds to siliceous aggregate content after acid dissolution of the binder by a nitric acid solution. The samples are previously ground until they completely pass through a 106 µm sieve. The samples are quartered until 2g is obtained and then etched with HNO_3 until the release of CO_2 stops. The IR is then determined by gravimetry. In contrast, the binder content was calculated from the CO_2 content (obtained by TGA-DTA in the weight loss range between 500–900 °C). The amount of CO_2 correspond to the amount of calcium carbonate in the binder, which can then be converted to $Ca(OH)_2$ [24], considering that all of the CO_2 came from the decomposition of carbonated lime ($CaCO_3$). For cementitious mortars, regardless of whether they contain other binders, the Portland cement content was obtained according to Arliguie's (2007) method [25]. This method does not consider the use of mineral additions in the binder, so the calculations of Portland cement as a binder were performed by default.

Table 1. Mortars from Lisbon (Portugal) and Barcelona (Spain) buildings.

Sample ID	Construction Period	Case Study/Building	Type of Building	Binder Type	Binder to Aggregate Ratio by Weight (b:a)
CVT1B	1902–1903	CVT (1903)/ *Ventura Terra* building (Lisbon)	Residencial	AL	1:5.4
CVT1C				AL	1:7.8
CVT3B				AL	1:4.3
AR49-2B	1920–1923	AR49 (1923)/ *Luiz Rau* building (Lisbon)	Residential	AL	(c)
AR49-6C				AL	1:5.8
AR49-7B				AL	1:3
AR49-8A				AL	1:2.9
AR49-8B				AL	1:11.2
AR49-11B				AL	1:6.7
AR49-15B				AL	1:7.1
AR49-15C				AL	1:7.9
IRF1B	1934–1938	IRF (1938)/*Nossa Senhora de Fátima* church (Lisbon)	Church	AL+OPC	1:0.1:7
IRF2A				AL	1:4.3
IRF2B				AL	1:9.9
IRF3A				AL	1:4.2
IRF3B				AL	1:8
IRF4A				AL+OPC	1:0.4:5.4
IRF7A				AL+OPC	1:0.4:3.8
IRF7B				AL+OPC	1:0.4:5.3
CBP1A	1938–1939	CBP (1939)/*Bernardo da Maia* house (Lisbon)	Residential	AL	1:8.4
CBP4B (d)				OPC	1:20.3
CBP6B				AL	1:8.7
CBP7B				AL	1:11.2
DN9A	1936–1940	DN (1940)/*Diário de Notícias* building (Lisbon)	Office and service	OPC	1:6.1
DN10A				OPC	1:7
DN11A				OPC	1:7.4
DN11B				OPC	1:4.2
DN12A				OPC	1:12.9
DN12B				AL+OPC	1:2.1:15.1
DN12C				OPC	1:4.2
DN12D				AL	1:4.3
DN19B				PCC	1:25.2 (a)
DN19C				OPC	1:8.9
DN19D				AL+PCC	1:1:6 (a)
AAC1A	1942–1944	AAC (1944)/*Cristino da Silva* building (Lisbon)	Residential	AL+OPC	1:0.2:6.1
AAC1B				AL+OPC	1:0.2:7
AAC2A (c)				AL+WPC	1:0.3:1.0
AAC2B				OPC	1:24.5
AAC3A (c)				OPC	1:3.0
AAC4A (c)				OPC	1:1.9
LIP1A	1955–1957	LIP (1954)/Laboratories of Pasteur Institute of Lisbon (Lisbon)	Laboratory and services	OPC	1:7.6
LIP9A				OPC	1:6.6
EUA53-2A (c)	1966–1969	EUA53 (1970)/America building (Lisbon)	Residential and commercial	WPC (e)	1:3.7
EUA53-2B				OPC	1:6.7
EUA53-3A (c)				WPC (e)	1:19.1
EUA53-3B				OPC	1:4.9
EUA53-4A (c)				WPC	1:11.1
EUA53-4B				OPC	1:11.5
FCG4A	1963–1969	FCG (1975)/Calouste Gulbenkian Foundation Headquarters and Museum (Lisbon)	Office and cultural facilities	OPC	1:10.2
JRP2A	1984–1987	JRP (1987)/Jacob Rodrigues Pereira Institute (Lisbon)	Educational/academic	PCC	1:4.9 (a)
UNL3A	2000–2002	UNL (2002)/New University of Lisbon Rectory (Lisbon)	Educational/academic	OPC	1:10.2
0-NC	1887	Hivernacle (Barcelona)	Greenhouse	NC	(b)
1-NC	1900	Villarroel (Barcelona)	Residential	NC	(b)

Legend: AL—air lime; OPC—ordinary Portland cement; WPC—white Portland cement; PCC—Portland composite cement; NC—natural cement; (a)—overestimated aggregate content, due to the presence of mineral additions (GGBS—samples DN19D and JRP2; FA—sample DN19D); (b)—not assessed; (c)—stone-imitating mortars; (d)—mortar with ceramic aggregates; (e)—stone imitating mortar with limestone filler.

The b:a ratio was obtained by point counting in thin sections for mortars with carbonate aggregates, according to RILEM Technical Committee TC167-COM recommendations [26].

The samples analyzed have the following binders: air lime (AL), natural hydraulic lime (NHL), natural cement (NC), ordinary Portland cement (OPC), Portland composite cement with mineral additions, like fly ash (FA) or ground granulated blast furnace slags (GGBS), and blended air lime and Portland cement (with OPC or white Portland cement—WPC). To investigate the different types of binders, a methodology involving a series of complementary analytical procedures was previously used, namely mineralogical investigation by XRD to characterize the overall and the binder-rich fraction and microstructural and optical microscopy analysis (SEM-EDS and petrography) in accordance with specialized literature to identify the key features of each type of binder [16,24–29].

In addition, mortar specimens formulated in the laboratory were also analyzed. These specimens were used as a control and a reference for elemental analysis. They were mainly formulated in the laboratory for previous research work and used in the present context [30–32]. Their formulations can be found in Table 2.

Table 2. Composition of laboratory formulated mortar specimens.

Specimens ID	Mortar Formulation	Binder Type	b:a (a)
CA-Sb-CP-360d	CL90-S air lime mortar with siliceous sand, 360 days laboratory curing	AL	1:11
CA-AL-CP-360d	CL90-S air lime mortar with washed siliceous sand, 360 days laboratory curing	AL	1:11
CH-AL-CP-360d	NHL 3.5 (1) mortar with washed siliceous sand, 360 days laboratory curing	NHL	1:5.6
CH-Sb-CP-360d	NHL 3.5 (1) mortar with siliceous sand, 360 days laboratory curing	NHL	1:5.6
CEM I (42.5)	Portland cement mortar (CEM I 42.5) with siliceous sand, 180 days laboratory curing	OPC	1:3
CEM I (52.5)	Portland cement type mortar (CEM I 52.5) with siliceous sand, 180 days laboratory curing	OPC	1:3
Li310-Ref-fib	Blended mortar with air lime and Portland cement (CEM I 42.5) with siliceous sand and organic fibers, 360 days laboratory curing	AL+OPC	1:2.6:26.3
Li310-Ref	Blended mortar with air lime and Portland cement (CEM I 42.5) with siliceous sand, 360 days laboratory curing	AL+OPC	1:2.6:26.3
CA-AL-CP-720d	CL90-S air lime mortar with washed siliceous sand, 720 days laboratory curing	AL	1:11
CH-AL-CP-720d	NHL 3.5 (1) mortar with washed siliceous sand, 720 days laboratory curing	NHL	1:5.6
124A	Blended mortars with air lime and white Portland cement (CEM I 42.5) with siliceous sand, 4 years laboratory curing	AL+WPC	1:1.3:19.7
134A		AL+WPC	1:0.9:17.5

Legend: AL—air lime; OPC—ordinary Portland cement; WPC—white Portland cement; NHL—natural hydraulic lime; AL+OPC (or WPC)—blended air lime and ordinary Portland cement or white Portland cement; (a)—binder to aggregate ratio; (1) The commercial SECIL NHL 3.5 is rated according to CEN EN 459-1:2015 [14].

2.2. SEM-EDS and Principal Component Analysis

Scanning electron microscopy with energy dispersive X-ray spectrometry (SEM-EDS) was performed in a TESCAN MIRA 3 field emission microscope (Tescan Group, a. s., Brno—Kohoutovice, Czech Republic) combined with a BRUKER XFlash 6 | 30 EDS system (Bruker Corporation, Billerica, MA, USA). Polished flat specimens or thin sections from each mortar sample were prepared as described in the following section and were analyzed in backscattered electron mode, with a chamber pressure of 20 Pa and 20 kV accelerating voltage, with an absorption current circa 300 pA, using a magnification over 500× in paste areas.

Over 50 EDS spot areas (<10 μm^2) were captured in each specimen, being the oxygen calculated by stoichiometry [33] and rejecting the values for Al/Si and Ca/Si atomic ratios higher than 10% of the coefficient of variation.

Principal component analysis (PCA) was performed using the results of the EDS analysis processed with OriginPro 9.0 software. PCA was performed to test and explore a mortar type's clustering structure to help understand how clusters are distributed and whether they are well-separated or overlapping.

2.3. *Optical Microscopy—Petrography*

The thin section technique, a significant advancement in the field of petrography, was developed by 19th-century geologists. It initially aided in studying rocks and minerals, and its application later expanded to include construction materials like concrete and mortar by the 1920s [34]. Today, this versatile technique is widely employed across various materials, including brittle substances like concrete and mortar. The samples were stabilized through embedding in epoxy resin, enabling cutting and grinding to a final thickness of 20–30 μm.

Petrographic observations were then conducted on thin sections using an Olympus BX60 (Tokyo, Japan) petrographic microscope featuring magnification lenses of 5×, 10×, 20×, and 40×. Residues within the mortar matrix necessitate microscopic tools to pinpoint the binder type [27], especially when distinguishing between the various hydraulic binders, which is why it became essential to apply this technique. It should be noted that petrography was only used to distinguish between mortars with hydraulic binders whenever the SEM-EDS methodology was inconclusive. Both methods (SEM-EDS and petrography) are complementary but should be used together.

3. Results and Discussion

3.1. *SEM—EDS and Principal Component Analysis*

EDS analysis and observations in backscattered mode were performed to mark out and group each type of binder used in terms of chemical composition. The analyses helped detect whether there were significant chemical differences between blended air lime Portland cement mortars and the other mortar types, considering the variability in the mix-design properties that could influence the composition, namely the binder-to-aggregate ratio.

The chemical compositions obtained by EDS were used to calculate the Al/Ca and Si/Ca atomic ratios (Table 3).

The hydraulicity of binders is highly variable [35] and was first determined by Louis Vicat (1818) [36] using the Hydraulicity Index. Vicat synthesized the available knowledge and directly linked hydraulicity to the SiO_2 and Al_2O_3 content. Subsequently, Eckel (2005) [37] introduced a novel index known as the Cementation Index. This index incorporates the influence of Fe_2O_3 and MgO on hydraulicity, presuming that all available SiO_2 combines with CaO to produce C_3S (Ca_3SiO_5) and that all Al_2O_3 combines to produce C_3A ($Ca_3Al_2O_6$). MgO is equated with CaO and Fe_2O_3 with Al_2O_3.

Nevertheless, this representation oversimplifies matters, as the mineralogy of hydraulic binders is more intricate than assumed [35]. The combustion temperature and duration indirectly influence the hydraulic characteristics, impacting the product's mineralogy [38]. Despite acknowledging this observation, the selection of Al/Ca and Si/Ca elemental atomic ratios is rationalized due to the importance of the CaO–SiO_2–Al_2O_3 system in evaluating the hydraulicity of mortars and binders, as evidenced by Mertens et al. 2008 [39], who scrutinized the composition of historical calcareous hydraulic binders within that chemical system.

Figure 1 shows an expected increase in hydraulic binders' Si/Ca ratio, while the Al/Ca ratio is variable within the specified clusters. Although there are few laboratory test specimens, finding a chemical composition concordance between them and the other analyzed samples is possible. The only discrepancy found is the composition of the mixed mortars, which has such a dispersion that some compositions are encompassed in the cluster of cement mortars or even at the limit between natural cement and NHL mortars.

Table 3. Al/Ca and Si/Ca atomic ratios obtained by EDS of the mortars from the buildings analyzed and of laboratory formulated test specimens.

Sample ID	Case Study		Binder Type	Al/Ca	Si/Ca
CVT1B	Lisbon buildings	CVT (1903)	AL	0.03	0.11
CVT1C			AL	0.06	0.19
CVT3B			AL	0.02	0.15
AR49-2B		AR49 (1923)	AL	0.05	0.13
AR49-6C			AL	0.05	0.12
AR49-7B			AL	0.02	0.10
AR49-8A			AL	0.03	0.11
AR49-8B			AL	0.05	0.19
AR49-11B			AL	0.07	0.16
AR49-15B			AL	0.05	0.14
AR49-15C			AL	0.05	0.17
IRF1B		IRF (1938)	AL+OPC	0.12	0.28
IRF2A			AL	0.04	0.09
IRF2B			AL	0.04	0.12
IRF3A			AL	0.06	0.12
IRF3B			AL	0.05	0.14
IRF4A			AL+OPC	0.11	0.35
IRF7A			AL+OPC	0.13	0.32
IRF7B			AL+OPC	0.11	0.30
CBP1A		CBP (1939)	AL	0.07	0.20
CBP4B			OPC	0.23	0.44
CBP6B			AL	0.03	0.13
CBP7B			AL	0.09	0.21
DN9A		DN (1940)	OPC	0.11	0.39
DN10A			OPC	0.09	0.34
DN11A			OPC	0.10	0.40
DN11B			OPC	0.10	0.33
DN12A			OPC	0.12	0.44
DN12B			AL+OPC	0.13	0.32
DN12C			OPC	0.12	0.45
DN12D			AL	0.05	0.15
DN19B			PCC	0.23	0.41
DN19C			OPC	0.07	0.31
DN19D			AL+PCC	0.13	0.37
AAC1A		AAC (1944)	AL+OPC	0.05	0.24
AAC1B			AL+OPC	0.09	0.27
AAC2A			AL+WPC	0.05	0.22
AAC2B			OPC	0.09	0.36
AAC3A			OPC	0.10	0.31
AAC4A			OPC	0.09	0.34
LIP1A		LIP (1954)	OPC	0.13	0.35
LIP9A			OPC	0.07	0.32
EUA53-2A		EUA53 (1970)	WPC	0.05	0.22
EUA53-2B			OPC	0.09	0.30
EUA53-3A			WPC	0.03	0.21
EUA53-3B			OPC	0.08	0.32
EUA53-4A			WPC	0.08	0.30
EUA53-4B			OPC	0.09	0.32
FCG4A		FCG (1975)	OPC	0.11	0.33
JRP2A		JRP (1987)	PCC	0.17	0.49
UNL3A		UNL (2002)	OPC	0.08	0.27

Table 3. *Cont.*

Sample ID	Case Study	Binder Type	Al/Ca	Si/Ca
CA-Sb-CP-360d	Laboratory formulated test specimens	AL	0.06	0.11
CA-AL-CP-360d		AL	0.05	0.07
CH-AL-CP-360d		NHL	0.15	0.26
CH-Sb-CP-360d		NHL	0.12	0.21
CEM I (42.5)		OPC	0.09	0.33
CEM I (52.5)		OPC	0.08	0.45
Li310-Ref-fib		AL+OPC	0.05	0.25
Li310-Ref		AL+OPC	0.05	0.27
CA-AL-CP-720d		AL	0.06	0.17
CH-AL-CP-720d		NHL	0.12	0.28
124A		AL+WPC	0.04	0.21
134A		AL+WPC	0.08	0.27
0-NC	Barcelona buildings	NC	0.12	0.23
1-NC		NC	0.16	0.35

Legend: AL—air lime; OPC—ordinary Portland cement; WPC—white Portland cement; PCC—Portland composite cement; AL+OPC (or WPC, or PCC)—blended air lime and Portland cement (for each type); NHL—natural hydraulic lime; NC—natural cement.

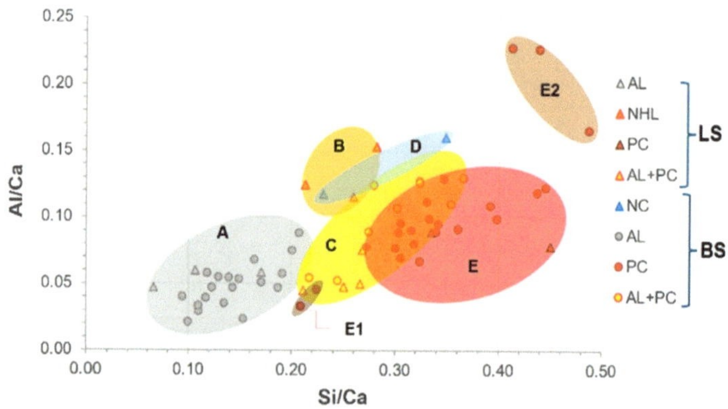

Figure 1. Si/Ca vs. Al/Ca plot of laboratory test specimens (LS) and building samples (BS) with clustering. Group A: air lime mortars (AL). Group B: NHL mortars (B). Group C: blended air lime Portland cement mortars (AL+PC). Group D: natural cement mortars (NC). Group E: Portland cement mortars (PC). Sub-group E1: Portland cement stone imitating mortars with limestone filler. Sub-group E2: Portland cement mortars with mineral additions or with pozzolanic additives. PC includes all types of Portland cement.

The presence of limestone filler in the paste of stone-imitating mortars reduced both ratios, as can be seen in cluster E1. Cluster E2 stands out from the others as it includes binders with mineral or pozzolanic additions rich in silica or alumina.

For comparative purposes and to attempt to improve the discrimination of this group of samples, principal component analysis (Figure 2) was carried out considering the analytical results presented in Table 3. The by-mixed mortars have high dispersion, and their cluster overlaps with the PC cluster. However, the air lime and Portland cement mortars were distinctively separated. A plot in the plane of the two principal components (PC1 and PC2) is shown in Figure 3. PC1 explains 43.73% of the variation and is controlled in the positive sense primarily by the contents in O, Si, Al, and Fe (somehow related to hydraulicity, which depends primarily on the clay content of the raw material) and in the opposite sense by the content in Ca (air lime binder's main element) and Cl. PC2 explains 16.38% of the variation and is controlled in the positive sense primarily by the contents in Ca and Na; in the opposite sense, no element has influence (as seen in the loadings plot of Figure 3). Although Cl may derive from the composition of the epoxy resin, which cannot be proved for all the analyzed samples, the influence of Na (PC2) raises questions regarding the contamination by salts and their possible influence on the analytical results of the binders.

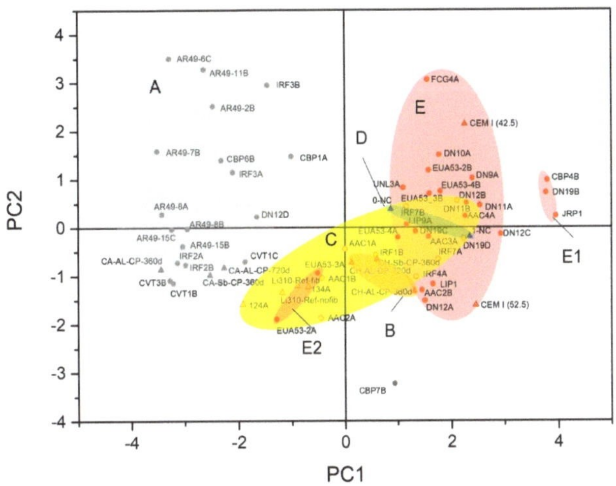

Figure 2. Score plot of the PCA of the analyzed samples, including the laboratory test specimens. Legend: Group A: air lime mortars (AL). Group B: NHL mortars (B). Group C: blended air lime with Portland cement mortars (AL+PC). Group D: natural cement mortars (NC). Group E: Portland cement mortars (PC). Sub-group E1: Portland cement stone imitating mortars with limestone filler. Sub-group E2: Portland cement mortars with mineral additions or with pozzolanic potential. PC includes all types of Portland cement.

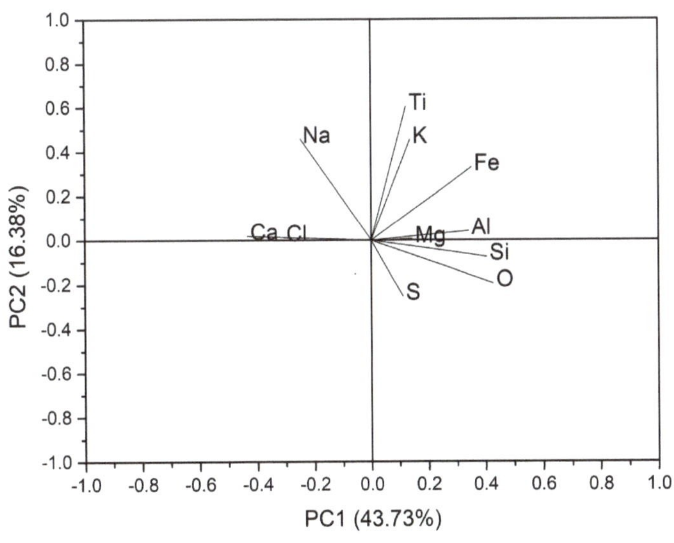

Figure 3. Loadings plot of the PCA of the experimental and the laboratory test specimens.

Almeida et al. (2023) [22] characterized the binder-rich fraction by X-ray diffractometry (XRD) of the Lisbon mortars, which was obtained by extracting the fines, passing through a 106 μm sieve directly from the bulk mortar. They also obtained the overall fraction corresponding to the samples as collected, obtained by crushing and grinding to pass through a 106 μm sieve and then analyzed by XRD. XRD detected Halite (NaCl) in samples CVT1C, CVT3B, AR49-7B, and AR49-11B (Figure 4), though only AR49-7B and AR49-11B are influenced by both contents. Nonetheless, sodium may be derived from the sodium feldspars (i.e., albite) from the aggregates, as the same authors stated elsewhere [22] (see Figure 4).

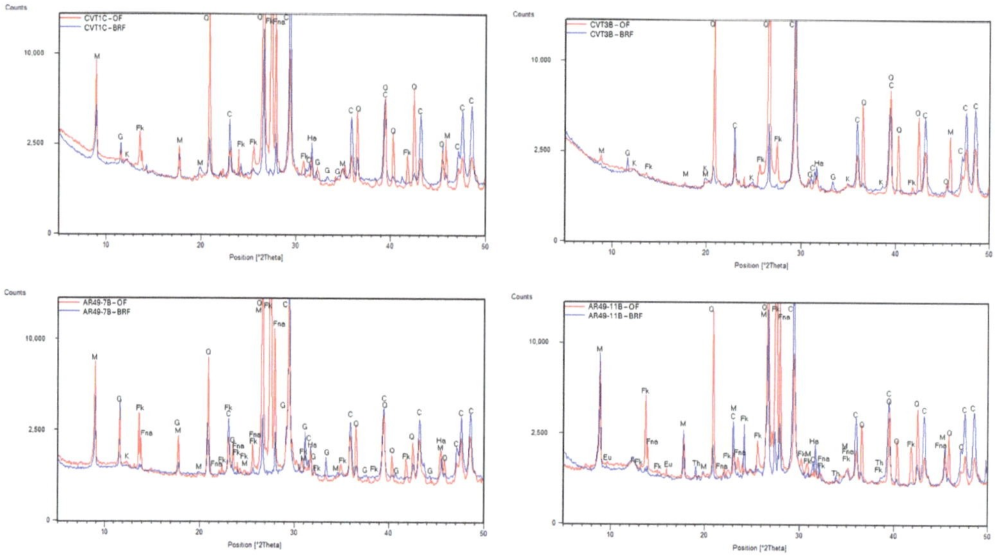

Figure 4. XRD patterns of the samples CVT1C, CVT3B, AR49-7B, and AR49-11B. Notation: OF—overall fraction; BRF—binder-rich fraction; M—muscovite; G—gypsum; Q—quartz; Fna—albite; Fk—microcline; C—calcite; Ha—halite; Eu—eugsterite; Th—thenardite; K—kaolinite.

Similarly to Figure 1, Figure 2 shows a cluster (E1) of samples with mineral additions and pozzolanic material and another, cluster E2, which refers to stone imitating mortars with limestone filler. As an outlier, sample CBP7B has a high influence of sulfur derived from gypsum in the binder [22]. Figure 5 shows a detail of the binder paste of this sample, confirming the presence of gypsum (needle-shaped crystals), in which can be seen the distribution of the elements Ca and S. This sample stands out from the others since it is a supporting mortar of a crown molding plaster element [22]. The incorporation of gypsum must, therefore, be related to the construction technique.

According to the EDS results (see Figures 1 and 2), mortars in which lime has been mixed with Portland cement are chemically similar to cement mortars. The plot in Figure 6 shows no tendency for the Si/Ca ratio to increase as the binder content (in this case, Portland cement) increases. This finding has significant implications, as it suggests that the Portland cement content in the mix does not considerably influence the chemistry of the paste, which can be explained by the fact that the pastes were poorly mixed or there may be an influence from the composition of the raw materials from different origins to contribute to this compositional variability.

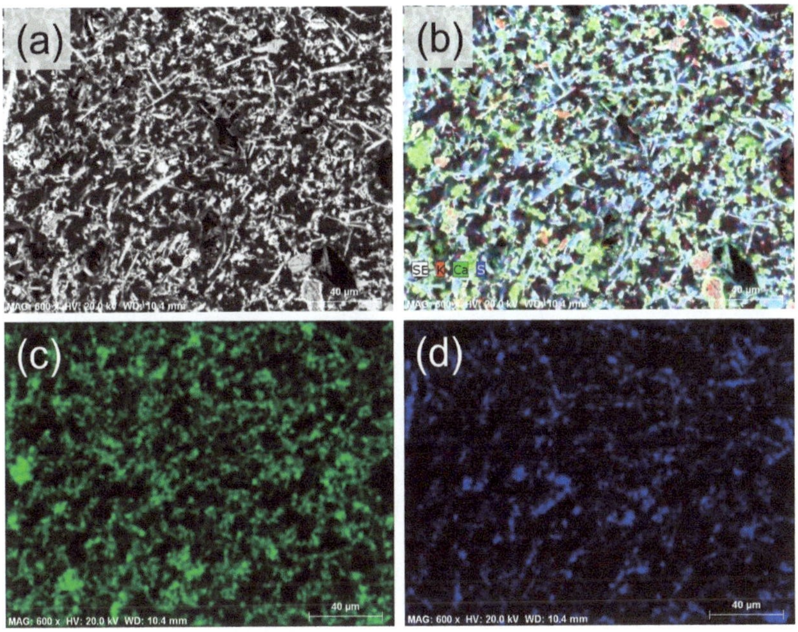

Figure 5. Backscattered scanning electron images of the binder of the sample CBP7B. (**a**) Needle-shape gypsum crystals forming a loose-like binder network; (**b**) X-ray map of elements K (red), Ca (green), and S (blue); X-ray maps of elements Ca (**c**) and S (**d**).

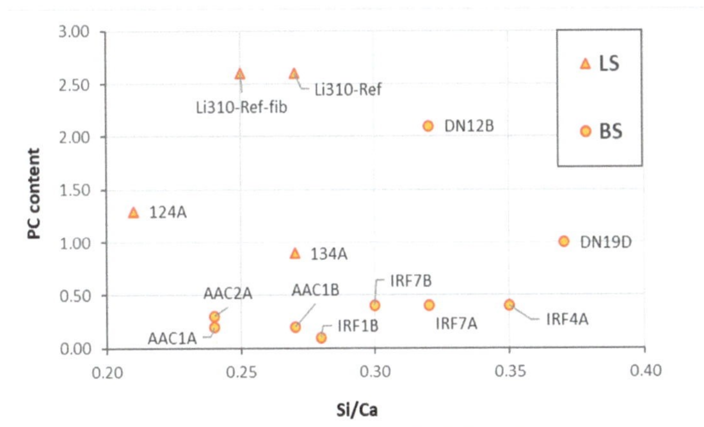

Figure 6. Si/Ca to Portland cement content (from b:a ratio) in the lime-mixed Portland cement mortars. Legend: LS—laboratory mortar specimens; BS—building mortar samples.

3.2. Optical Microscopy—Petrography

Considering the elemental analysis results, more specifically the Si/Ca ratio vs. Al/Ca ratio, there is a noticeable overlap between groups B (NHL mortars) and D (natural cement mortars) and between C (blended air lime Portland cement mortars), E (Portland cement mortars), and E1 (Portland cement stone imitating mortars with limestone filler). A complementary approach is needed to identify the binders unequivocally. A literature review shows that each type of binder has its characteristics with its compounds. Much of the identification using optical microscopy is carried out to find residues resulting from the manufacturing process, i.e., 'fossil' traces of the raw material, residues (or 'relicts') of the binder that have not produced chemical and physical reactions, or even neoformation products resulting from interaction with the binder matrix.

Air lime, which has been extensively studied, is often characterized by the presence of lumps. Lime lumps are fragments macroscopically showing a whitish color and sometimes an inconsistent appearance [40–43]. Among several authors, Cantisani et al. 2022 [42], observing the thin section of lumps under a polarized light microscope, stated that the presence of lime lumps is fundamental to recognize the lime binder nature, allowing information about the stone used to produce lime. According to Pavia and Caro (2007) [28], binder properties, including reactivity, shrinkage, cohesion, and fineness, can also be analyzed with petrographic microscopy, providing essential clues on lime calcination and slaking. The same authors point out that the accumulation of lime lumps often appears fractured under the microscope because pure calcium lime is a non-hydraulic binder with a high retraction coefficient. Other authors state that the presence of lime lumps can indicate incomplete calcination [44], poor mixing [45], or the practice of "hot lime mixing" [46] when they are abundant. Regarding the mortars in this study, particularly those from the Lisbon buildings, the lime binders are clearly identified by the presence of the lime lumps, as shown in Figure 7. In the case of blended binders, i.e., Portland cement with a lime mixture, the identification of lime lumps also makes it possible to conclude that lime was used in the mortar formulation (Figure 8b).

Figure 7. AR49-6C (**a**) and IRF2B (**b**) thin section micrographs in transmitted cross-polarized light, showing lime lumps (llu).

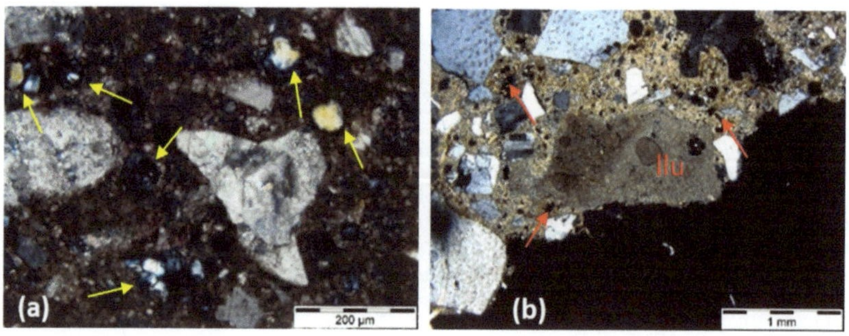

Figure 8. Thin-section micrograph of EUA53-4A in transmitted cross-polarized light (**a**), and thin section micrograph of IRF4A in transmitted cross-polarized light (**b**). Legend: unhydrated white Portland cement clinker grains (yellow arrows) showing no C_4AF phase in the residual cement grains; lime lump (llu) and Portland cement clinker grains (red arrows) coexisting in the same sample, indicating air lime with mixed ordinary Portland cement as binders.

When employing appropriate petrographic techniques, there is minimal risk of confusing a lime with Portland cement. Therefore, we can confine ourselves to the differences between the various types of hydraulic binders, as they present chemical overlaps in the elemental relationships studied and recognized by other features.

Whether in hot-mixed lime, fat lime putty, or eminently hydraulic lime, the essential constituent of the raw material remains calcium hydroxide. The manufacturing procedures involved are notably distinct from those of Portland cement. A limestone of varying purity undergoes a process known as "calcination", which is heated to a temperature sufficient to expel the carbon dioxide in the original limestone. This temperature threshold is significantly below the clinkering point, usually below 1000 °C. The primary output is calcium oxide or free lime; however, this compound is unstable, necessitating an additional step known as slaking [27].

The most eminent difficulty is the differentiation between NHL, natural cement, and air lime mixed with materials that promote hydraulicity, such as natural or artificial pozzolans. That difficulty is due to several factors, such as the optical similarity between natural cement and natural hydraulic lime. As far as natural cements are concerned, the poor quality of the milling technology means that the calcined marl fragments remain unhydrated during the setting process, making it easier to identify unhydrated belite crystals inside the lumps [23].

Table 4 compiles some key compounds according to several authors, and it is intended to be a guide for identifying the hydraulic binders from the compounds observed. Given the affinity between the NHL binder and the hydraulic mortars, both materials are combined in the exact identification batch. Despite using a transmitted light microscope, it should be kept in mind that using reflected (incident) light is beneficial for observing some of these components. Incident light microscopy proves helpful for mortars containing Portland cement (Figure 9), while polarized microscopy in transmitted light is advantageous for all types of hydraulic mortars [16].

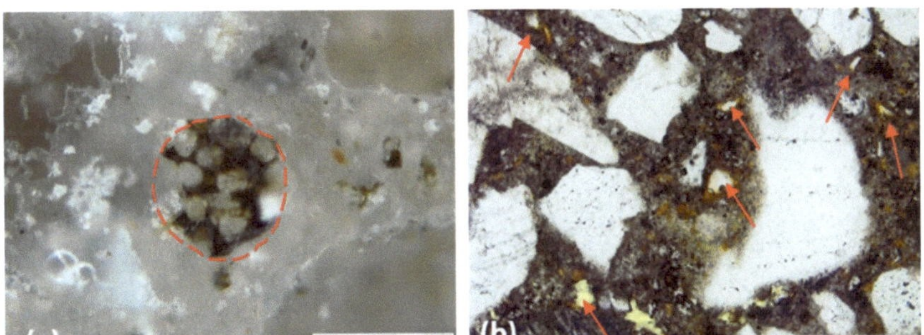

Figure 9. Polished section micrograph of Li310-Ref test specimen in incident light, dark field (**a**), and thin section micrograph of JRP2A sample in transmitted plane polarized light (**b**). Legend: unhydrated Portland cement clinker grain (red doted circle) containing C_3S (elongated), C_2S (rounded crystals) and a brownish phase corresponding to C_4AF; GBFS grains in a Portland cement matrix, showing typical orange to red hydration rims (red arrows).

Figures 8–10 show some key characteristics described in Table 4 and found in the samples studied, except the NHL mortar samples due to the inexistence of thin sections. The images show the presence of compounds that indicate different types of hydraulicity according to the binder used.

Table 4. Main characteristics for identifying binders by optical microscopy according to several authors.

<table>
<tr><th></th><th colspan="7">Authors [by Reference]</th></tr>
<tr><td rowspan="3">NHL and other hydraulic lime mortars/binders</td><td>[17]</td><td>[27]</td><td>[47]</td><td>[48]</td><td>[49,50]</td><td>[40,49,51]</td><td>[51]</td></tr>
<tr><td>Large C2S and C3S phases in a small amount of a brown matrix consisting of C3A and C4AF.</td><td>Small clusters of C2S may be detected in trace quantities in hydraulic limes.</td><td>Dominant C2S grains can present striations in different directions</td><td>Presence of "hot spots" as part of the hydraulic phases that result from local higher temperatures in the lime kiln.</td><td>If reaction rims did not develop, hydraulic phases dispersed in the binder can nevertheless still be observed both as veins and as pore filling.</td><td>Presence of small dark inclusions (non-hydrated relicts of C2S) that can be easily recognised.</td><td>C3S can be formed due to an overheating of the raw materials during the production of NHL</td></tr>
<tr><td></td><td>C3S can also be present in smaller quantities.</td><td>Hydration rims can be observed around individual hydraulic phases as a colourless rim.</td><td>The presence of gehlenite is a valuable indicator of the distinction between cement and natural hydraulic lime.</td><td></td><td>C2S: dark brown crystals medium-high relief, shaded contours, and sub idiomorphic habit.</td><td></td></tr>
<tr><td rowspan="4">Natural cement</td><td colspan="2">[16]</td><td>[23]</td><td colspan="2">[27]</td><td>[13]</td><td>[52]</td></tr>
<tr><td colspan="2">The NC residual nodules exhibit strong zoning which is best visible in transmitted light with parallel polars.</td><td>Presence of lumps of calcined marls in the natural cement (due to the calcination and to the deficient milling technology).</td><td colspan="2">Under polarised light, the matrix of natural cements often appears spotted with dark isotropic areas broken by bright but dense carbonated regions.</td><td>High proportion of non-reactive 'nodules' or relicts (under-burned, well burned, and over-burned).</td><td>Under-burned, well burned, and over-burned marl fragments.</td></tr>
<tr><td colspan="2">NC mortars can be identified equally well at low magnifications as their binders contain residual compounds that are larger than the clinker of historic Portland cement.</td><td>Presence of unhydrated C2S crystals.</td><td colspan="2">The dispersed quartz grains and other sand-sized silicate minerals also aid in distinguishing natural cements. These tend to retain most of their original texture when burned at calcining temperatures.</td><td>The largest 'nodules' are of millimetre size.</td><td>NHL does not show calcined marl fragments and have a more homogeneous structure.</td></tr>
<tr><td colspan="2"></td><td>Absence of unhydrated C3S crystals.</td><td colspan="2"></td><td>Contain greater percentage of relicts than either the hydraulic limes or the Portland cements.</td><td>Lime nodules can be observed in the NHL samples, while the natural cements lack these nodules.</td></tr>
<tr><td rowspan="2">OPC</td><td colspan="2">[23]</td><td colspan="2">[27]</td><td colspan="3">[53–55]</td></tr>
<tr><td colspan="2">Large size of the unhydrated C3S crystals in historic Portland cements. The calcination process in vertical kilns was longer than in a modern horizontal kiln and it allowed the C2S and C3S crystals to acquire a larger size than in modern Portland cement.</td><td colspan="2">Portland cement pastes are homogeneously isotropic or dark-colored where cementitious gels have formed, broken only by thin dispersed grains of calcium hydroxide that appear bright-colored.</td><td colspan="3">The unhydrated cement grains are identified by the optical properties of the cement minerals, primarily C3S and C2S.</td></tr>
<tr><td rowspan="2">WPC</td><td colspan="7">[27]</td></tr>
<tr><td colspan="7">The C2S phases are not surrounded by a brown-colored crystallographically indistinct ferrite phase (absence of brownish C4AF in the residual cement grains).</td></tr>
<tr><td rowspan="2">PCC</td><td colspan="7">[53,54]</td></tr>
<tr><td colspan="7">Unhydrated fly ash particles; GBFS—glassy slag particles with reddish hydration rims.</td></tr>
</table>

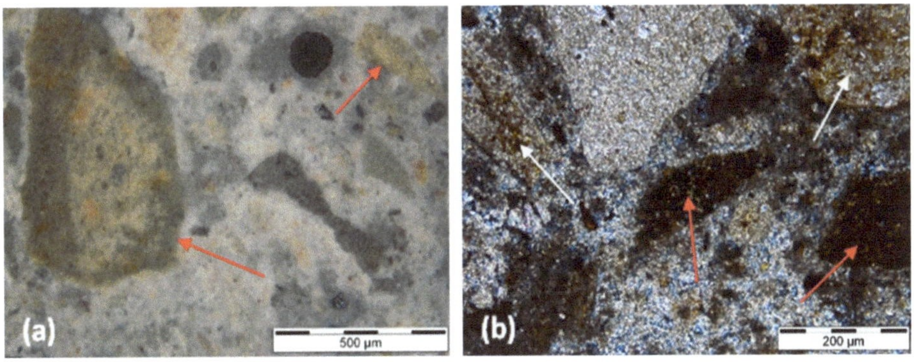

Figure 10. 0-NC thin section micrographs. Legend: (**a**)—fragments of calcined marl from the natural cement sample in incident light, dark field. These fragments are characterized by a hydrated outer ring (red arrows); (**b**)—detail of marls with non-hydrated belite inside lumps (red arrows); residual, under-burned relicts exhibiting low reactivity (white arrows) and the original rock texture observed in transmitted cross-polarized light.

4. Conclusions

The main conclusions of this work are:

- The interval ranges for the chemical compositions of the binders were obtained with precision through EDS analyses. The following interval of elemental ratios Al/Ca and Si/Ca were established for the types of binders identified in Table 5, ensuring the accuracy of the results.

Table 5. Ranges for chemical composition ratios (Al/Ca and Si/Ca) of binders.

	Al/Ca		Si/Ca	
Binder Type	**Min.**	**Max.**	**Min.**	**Max.**
Air lime	0.02	0.09	0.09	0.20
Natural hydraulic lime	0.12	0.15	0.21	0.28
Natural cement	0.12	0.16	0.23	0.35
Blended air lime and Portland cement (including ordinary, white, and Portland composite cement types)	0.04	0.13	0.21	0.37
Portland cement (including ordinary and white types)	0.03	0.23	0.21	0.45
Portland composite cements	0.17	0.23	0.41	0.49

- PCA analysis did not significantly refine the clusters because the samples have similar representations of the principal components. This suggests that PCA should not be applied to samples with elements that exhibit the same weight in their chemical composition.
- When dealing with overlapping grouped chemical results, optical microscopy is a critical step in identifying the key compounds of the binder. When consistently applied by a skilled petrographer, this practice ensures the accuracy and reliability of the results.
- Of the key compounds mentioned above, some stand out in helping to distinguish between historical hydraulic binders, such as: natural cement binders (calcined marl fragments, under-burned, well-burned, and over-burned marl fragments, presence of unhydrated C_2S crystals and the absence (or minimal presence) of alite); NHL binders (presence of small dark inclusions (non-hydrated relicts of C_2S), hydration rims around individual hydraulic phases, presence of gehlenite); and portland cement

binders (large size of the unhydrated C_3S crystals in historic Portland cement and ubiquity of C_3S in all types of Portland cement).

- In blended mortars, the pastes' homogeneity, or compositional variability, which may be attributable to the influence of the raw materials' composition, varying according to their provenance, points to being influential in EDS analyses.
- Although some of the samples analyzed were manufactured in the laboratory, namely the natural hydraulic lime, Portland cement, and air lime mixed with Portland cement test specimens, the chemical results can be extrapolated to these types of binders, historically applied in other periods, since the raw materials and calcination temperatures are similar to those obtained in the past, even though today there is greater quality control in the manufacturing processes.

Author Contributions: Conceptualization, L.A. and A.S.S.; methodology, L.A. and A.S.S.; software, L.A.; validation, A.S.S., R.V. and J.M.; investigation, L.A., A.S.S., R.V. and J.M.; writing—original draft preparation, L.A.; writing—review and editing, L.A., A.S.S., R.V. and J.M.; supervision, A.S.S., R.V. and J.M. All authors have read and agreed to the published version of the manuscript.

Funding: This research was funded by the Portuguese Foundation for Science and Technology—Fundação para a Ciência e a Tecnologia—FCT, grant SFRH/BD/112809/2015.

Data Availability Statement: Data is contained within the article.

Acknowledgments: The authors would like to acknowledge FCT through the PO-CI-01-0145-FEDER-031612 research project: 'CEMRESTORE: Mortars for early 20th century buildings' conservation: compatibility and sustainability' and the collaborative researcher Judith Ramirez-Casas for making available thin sections of natural cement mortars from the Barcelona buildings. The authors would like to thank researcher Sílvia Pereira for her support in software computing. The authors also acknowledge the buildings' owners for study authorizations and the National Laboratory for Civil Engineering for its support through the projects 'DUR-HERITAGE—Durability and characterisation of historical interest construction materials' and 'PRESERVe—Preservation of renders from built heritage with cultural value: identification of risks and contribution of traditional knowledge and new materials for conservation and protection'.

Conflicts of Interest: The authors declare no conflicts of interest.

References

1. Elsen, J.; Van Balen, K.; Mertens, G. Hydraulicity in historic mortars: A review. In Proceedings of the 2nd Historic Mortars Conference HMC2010 and RILEM TC 203-RHM Final Workshop, Prague, Czech Republic, 22–24 September 2010.
2. Furlan, V.; Bissenger, P. Les mortiers anciens. Histoire et essai d'analyse scientifique. *Rev. Suisse D'art D'archéol.* **1975**, *32*, 2–14.
3. Rispoli, C.; Montesano, G.; Verde, M.; Balassone, G.; Columbu, S.; De Bonis, A.; Di Benedetto, C.; D'Uva, F.; Esposito, R.; Graziano, S.F.; et al. The key to ancient Roman mortars hydraulicity: Ceramic fragments or volcanic materials? A lesson from the Phlegrean archaeological area (southern Italy). *Constr. Build. Mater.* **2024**, *411*, 134408. [CrossRef]
4. Eramo, G.; Spalluto, L.; Laviano, R. Paving stones of the Via Traiana in Egnazia (Brindisi, 2nd A.D.): Provenance of stones [Il basolato della Via Traiana nel sito archeologico di Egnazia (Fasano, Br): Provenienza dei materiali lapidei]. *Rend. Online Soc. Geol. Ital.* **2008**, *3*, 357–358.
5. Lezzerini, M.; Legnaioli, S.; Lorenzetti, G.; Palleschi, V.; Tamponi, M. Characterization of historical mortars from the bell tower of St. Nicholas church (Pisa, Italy). *Constr. Build. Mater.* **2014**, *69*, 203–212. [CrossRef]
6. Miriello, D.; Bloise, A.; Crisci, G.M.; De Luca, R.; De Nigris, B.; Martellone, A.; Osanna, M.; Pace, R.; Pecci, A.; Ruggieri, N. New compositional data on ancient mortars and plasters from Pompeii (Campania—Southern Italy): Archaeometric results and considerations about their time evolution. *Mater. Charact.* **2018**, *146*, 189–203.
7. Lezzerini, M.; Spampinato, M.; Sutter, A.; Montevecchi, N.; Aquino, A. Petrographic characteristics of the mortars from the Pisa's Cathedral apse. In Proceedings of the IMEKO TC4 International Conference on Metrology for Archaeology and Cultural Heritage, MetroArchaeo, Calabria, Italy, 19–21 October 2022; pp. 459–463.
8. Sitzia, F.; Beltrame, M.; Columbu, S.; Lisci, C.; Miguel, C.; Mirão, J. Ancient restoration and production technologies of Roman mortars from monuments placed in hydrogeological risk areas: A case study. *Archaeol. Anthropol. Sci.* **2020**, *12*, 147. [CrossRef]
9. La Russa, M.F.; Ruffolo, S.A.; Ricca, M.; Rovella, N.; Comite, V.; De Buergo, M.A.; Crisci, G.M.; Barca, D. Archaeometric approach for the study of mortars from the underwater archaeological site of Baia (Naples) Italy: Preliminary results. *Period. Miner.* **2015**, *84*, 553–557. [CrossRef]

10. Rispoli, C.; De Bonis, A.; Guarino, V.; Graziano, S.F.; di Benedetto, C.; Esposito, R.; Morra, V.; Cappelletti, P. The Ancient Pozzolanic Mortars of the Thermal Complex of Baia (Campi Flegrei, Italy). *J. Cult. Herit.* **2019**, *40*, 143–154. [CrossRef]
11. Parker, J. A Certain Cement or Terras to Be Used in Aquatic and Other Buildings and Stucco Work. British Patent n. 2120, 27 July 1796.
12. Kozłowski, R.; Hughes, D.; Weber, J. Roman cements: Key materials of the built heritage of the 19th Century. In *Materials, Technologies and Practice in Historic Heritage Structures*; Dan, M.B., Přikryl, R., Török, A., Eds.; Springer: Dordrecht, The Netherlands, 2010. [CrossRef]
13. Hughes, D.; Swann, S.; Gardner, A. Roman cement. *J. Arch. Conserv.* **2007**, *13*, 21–36. [CrossRef]
14. *CEN EN 459-1*; Building Lime—Part 1: Definitions, Specifications, and Conformity Criteria. European Committee for Standardization: Brussels, Belgium, 2015.
15. Coutinho, A.S. *Fabrico e Propriedades do Betão*; Laboratório Nacional de Engenharia Civil (LNEC): Lisboa, Portugal, 1988; Volume 1.
16. Weber, J.; Köberle, T.; Pintér, F. Methods of microscopy to identify and characterise hydraulic binders in historic mortars—A methodological approach. In *Historic Mortars*; Hughes, J., Válek, J., Groot, C., Eds.; Springer: Cham, Switzerland, 2019. [CrossRef]
17. Callebaut, K.; Elsen, J.; Van Balen, K.; Viaene, W. Nineteenth century hydraulic restoration mortars in the Saint Michael's Church (Leuven, Belgium): Natural hydraulic lime or cement? *Cem. Concr. Res.* **2001**, *31*, 397–403. [CrossRef]
18. Gadermayr, N.; Pintér, F.; Weber, J. Identification of 19th century roman cements by the phase composition of clinker residues in historic mortars. In Proceedings of the 12th International Congress on the Deterioration and Conservation of Stone, New York, NY, USA, 22–25 October 2012.
19. Taylor, H.F.W. *Cement Chemistry*; Academic Press: New York, NY, USA, 1990.
20. Veiga, M.R.; Aguiar, J.; Santos Silva, A.; Carvalho, F. *Conservação e Renovação de Revestimentos de Paredes de Edifícios Antigos*; LNEC: Lisboa, Portugal, 2004; p. 126.
21. Veiga, M.; Velosa, A.; Magalhães, A. Evaluation of mechanical compatibility of renders to apply on old walls based on a restrained shrinkage test. *Mater. Struct.* **2006**, *40*, 1115–1126.
22. Almeida, L.; Santos Silva, A.; Veiga, R.; Mirão, J. Composition of renders and plasters of award-winning buildings in Lisbon (Portugal): A contribution to the knowledge of binders used in the 20th Century. *Int. J. Arch. Herit.* **2023**, 1–31. [CrossRef]
23. Mayo, C.M.; Ramírez-Casas, J.; Sanz, D.; Ezquerra, A.N.; Rosel, J.R. Methodology of identification of natural and historic Portland cements. Application and study in mortars of Madrid and Barcelona. In Proceedings of the 5th Historic Mortars Conference, Pamplona, Spain, 19–21 June 2019.
24. Middendorf, B.; Hughes, J.J.; Callebaut, K.; Baronio, G.; Papayianni, I. Investigative methods for the characterisation of historic mortars—Part 2: Chemical characterisation. *Mater. Struct.* **2005**, *38*, 771–780. [CrossRef]
25. Arliguie, G.; Hornain, H. *Grandubé: Grandeurs Associées à la Durabilité des Bétons*, 1st ed.; Presses Ecole Nationale Ponts Chaussees: Paris, France, 2007; p. 438.
26. Lindqvist, J.E.; Sandström, M. Recommendations of RILEM TC 167-COM: Characterization of old mortars. Quantitative analysis of historical mortars using optical microscopy. *Mater. Struct.* **2000**, *167*, 612–617. [CrossRef]
27. Walsh, J.J. Petrography: Distinguishing Natural Cement from Other Binders in Historical Masonry Construction Using Forensic Microscopy Techniques. *J. ASTM Intern.* **2007**, *4*, 1.
28. Pavia, S.; Caro, S. Petrographic microscope investigation of mortar and ceramic technologies for the conservation of the built heritage. In Proceedings of the SPIE O3A: Optics for Arts, Architecture, and Archaeology, Munich, Germany, 16 July 2007; Volume 6618.
29. Mertens, G.; Madau, P.; Durinck, D.; Blanpain, B.; Elsen, J. Quantitative mineralogical analysis of hydraulic limes by X-ray diffraction. *Cem. Concr. Res.* **2007**, *37*, 1524–1530. [CrossRef]
30. Santos, A.R.; Veiga, M.R.; Santos Silva, A.; Brito, J.; Álvarez, J.I. Evolution of the microstructure of lime based mortars and influence on the mechanical behaviour: The role of the aggregates. *Constr. Build. Mater.* **2018**, *187*, 907–922. [CrossRef]
31. Santos, A.R.; Veiga, M.R.; Santos Silva, A.; Brito, J. Impact of aggregates on fresh mortars' properties. In Proceedings of the 5th Historic Mortars Conference, Pamplona, Spain, 19–21 June 2019.
32. Pederneiras, C.M. Improving Cracking Performance of Mortars with Selected Recycled Fibres for Non-Structural Uses. Ph.D. Thesis, Civil Engineering, Instituto Superior Técnico, University of Lisbon, Lisbon, Portugal, 2021.
33. Famya, C.; Brough, A.R.; Taylor, H.F.W. The C-S-H gel of Portland cement mortars: Part I. The interpretation of energy-dispersive X-ray microanalyses from scanning electron microscopy, with some observations on C-S-H, AFm and Aft phase compositions. *Cem. Concr. Res.* **2003**, *3*, 1389–1398.
34. Balksten, K.; Nitz, B.; Hughes, J.J.; Lindqvist, J.E. Petrography of historic mortar materials: Polarising light microscopy as a method for characterising lime-based mortars. In Proceedings of the 5th Historic Mortars Conference, Universidad de Navarra, Pamplona, Spain, 19–21 June 2019.
35. Elsen, J.; Mertens, G.; Snellings, R. Portland cement and other calcareous hydraulic binders: History, production and mineralogy. *Europ. Mineral. Uni. Notes Miner.* **2011**, *9*, 441–479. [CrossRef]
36. Vicat, L.J. *Recherches Expérimentales sur les Chaux de Construction, les Bétons et les Mortiers Ordinaires*; Goujon: Paris, France, 1818.
37. Eckel, E. *Cements, Limes and Plasters. Their Materials, Manufacture and Properties*; Routledge, Taylor & Francis Group: New York, NY, USA, 2015.

38. Mertens, G.; Elsen, J.; Laduron, D.; Brulet, R. Mineralogy of the calcium-silicate phases present in ancient mortars from Tournai. *Archeometrie* **2006**, *30*, 61–65.
39. Mertens, G.; Lindqvist, J.E.; Sommain, D.; Elsen, J. Calcareous Hydraulic Binders from a Historical Perspective. In Proceedings of the 1st Historical Mortars Conference, Characterization, Diagnosis, Conservation, Repair and Compatibility, Lisbon, Portugal, 24–28 September 2008; pp. 1–15.
40. Pecchioni, E.; Fratini, F.; Cantisani, E. *Atlas of the Ancient Mortars in Thin Section Under Optical Microscope*, 2nd ed.; Nardini: Florence, Italy, 2020; p. 78.
41. Bugini, R.; Toniolo, L. La presenza di grumi bianchi nelle malte antiche: Ipotesi sull'origine. *Arkos* **1990**, *12*, 4–8.
42. Cantisani, E.; Fratini, F.; Pecchioni, E. Optical and Electronic Microscope for Minero-Petrographic and Microchemical Studies of Lime Binders of Ancient Mortars. *Minerals* **2022**, *12*, 41. [CrossRef]
43. Hughes, J.; Cuthbert, S. The petrography and microstructure of medieval lime mortars from the west of Scotland: Implications for the formulation of repair and replacement mortars. *Mater. Struct.* **2000**, *33*, 594–600.
44. Pavia, S.; Fitzgerald, B.; Howard, R. Evaluation of properties of magnesian lime mortar. In *Structural Studies, Repair, and Maintenance of Heritage Architecture IX*; Brebbia, C.A., Torpiano, A., Eds.; Transactions on the Built Environment; WIT Press: Southampton, UK, 2005; Volume 83, pp. 375–384.
45. Leslie, A.B.; Hughes, J.J. Binder microstructure in lime mortars: Implications for the interpretation of analysis results. *Quart. J. Eng. Geol.* **2002**, *35*, 257–263.
46. Leslie, A.B.; Gibbons, P. Mortar analysis and repair specification in the conservation of Scottish historic buildings. In *Proceedings of the International RILEM Workshop on Historic Mortars: Characteristics and Tests*; Bartos, P., Groot, C., Hughes, J.J., Eds.; RILEM: Paisley, UK, 2000; pp. 273–280.
47. Furlan, V.; Pancella, R. Examen microscopique en lumière rèfléchie de ciments, bétons et mortiers. *Chantiers* **1982**, *13*, 25–30.
48. Gödicke-Dettmering, T.; Strübel, G. Mineralogische und technologische Eigenschaften von hydraulischen Kalken als Bindemittel für Restaurierungsmörtel in der Denkmalpflege. *Giessener Geol. Schr.* **1996**, *56*, 131–154.
49. Pecchioni, E.; Fratini, F.; Cantisani, C. *Le Malte Antiche e Moderne tra Tradizione e Innovazione*, 2nd ed.; Pàtron: Bologna, Italy, 2018; p. 231.
50. Moropoulu, A.; Cakmak, A.S.; Biscontin, G.; Bakolas, A.; Zendri, E. Advanced Byzantine cement based composites resisting earthquake stress: The crushed brick/lime mortars of Justinian's Hagia Sophia. *Constr. Build. Mater.* **2002**, *16*, 543–552.
51. Ingham, J. *Geomaterials under the Microscope: A Colour Guide*; John Wiley & Sons: Hoboken, NJ, USA, 2010; p. 192.
52. Mayo, C.; Sanz, D.; Pineda, J.I. Metodología simplificada de identificación mediante MOP de las cales hidráulicas y los cementos naturales. In Proceedings of the VI Jornadas FICAL—Forum Iberico de la Cal, Pamplona, Spain, 28–30 May 2018.
53. Jepsen, B.B.; Christensen, P. Petrographic examination of hardened concrete. *Bull. Intern. Assoc. Engin. Geol.* **1989**, *39*, 99–103.
54. Poole, A.; Sims, I. *Concrete Petrography. A Handbook of Investigative Techniques*, 2nd ed.; CRC Press: Boca Raton, FL, USA, 2020.
55. Pavía, S. A Petrographic study of the technology of hydraulic mortars at masonry bridges, harbours and mill ponds. In Proceedings of the Concrete Research and Bridge Infrastructure Symposium, Galway, Ireland, 4–5 December 2008; pp. 253–264.

minerals

MDPI

Article

Fast Elemental Analysis of Heavy Mineral Suites by Scanning Electron Microscopy (SEM-Unity BEX)

Jim Buckman [1,*], Amy Gough [1] and Max Webb [2]

1 Institute of GeoEnergy Engineering, School of Energy, Geoscience, Infrastructure and Society, Heriot-Watt University, Edinburgh EH14 4AS, UK; a.gough@hw.ac.uk
2 The Lyell Centre, Heriot-Watt University, Edinburgh EH14 4AS, UK; max.webb@hw.ac.uk
* Correspondence: j.buckman@hw.ac.uk

Abstract: Developments in scanning electron microscopy (SEM) have introduced instant live coloured SEM images based on elemental composition. Here, we use a technique utilising a Unity BEX detector system, with collection speeds up to 100 times faster than typical standard energy-dispersive X-ray (EDX) analysis systems, to obtain large area backscattered and elemental composition maps of heavy mineral (HM) suites from a sample from an Oligocene fluvio-deltaic system in the Central Myanmar Basin. The fast X-ray collection rate and a high-resolution backscattered (BSE) detector allow for rapid imaging of polished blocks, thin sections, and stubs. Individual HM species can be rapidly classified, allowing for the subsequent collection of compositional and morphological metrics. In addition, the identification of grains such as zircon and apatite allow for further analysis by cathodoluminescence (CL) to identify and record the presence of growth zonation, which is critical for further U-Pb geochronology and thermochronology, using fission track analysis of apatite, zircon, and titanite. The sample used in this study contains a diverse heavy mineral suite due to the complex tectonic history of Myanmar, juxtaposing multiple metamorphic basement terranes alongside volcanic arcs and obducted ophiolites. This, along with the textural and mineralogical immaturity of the sediments themselves (governed by short transport systems and the rapid weathering of the sources), means that a wide variety of heavy mineral species can be identified and tested using this new technique, which provides a time-efficient method in comparison to traditional optical techniques. As the Unity BEX detector is located at the polepiece, it is relatively insensitive to working distance; in addition, the geometry of paired X-ray detectors on either side of the polepiece (at 180°) means that the system is also capable of fully characterising individual particles, on uncut and unpolished grain mounts, without artefacts such as particle shadowing. The development of a more comprehensive heavy mineral EDX database (library) will improve the accuracy of this new technique, as will the correlation with other techniques such as Raman spectroscopy.

Keywords: automation; rapid mapping; heavy minerals; BEX

1. Introduction

Heavy minerals (HM) are those with a density greater than 2.8 to 2.9 g/m^3 [1] and commonly form around 1% of siliciclastic sediments. They are particularly useful in provenance studies [2–4] and can provide additional information on weathering, sediment transport, sediment mixing, and the degree of sediment recycling [3,5,6]. In such cases, heavy mineral species are typically paired (e.g., apatite/tourmaline (ATi), garnet/zircon (GZi), rutile/zircon (Ru/Zi) monazite/zircon (MZi) and chrome spinel/zircon (CZi)) and used as indexes for provenance, weathering, transportation, and degree of diagenetic alteration [7]. They also provide a method for dating sediment sources through the separation and study of mineral phases such as zircon and apatite [8–11] using SHRIMP or LA-ICP-MS U-Pb geochronology and fission track analysis [12–14]. Additionally, in some cases, high concentrations of HMs can be mined as valuable reserves of critical metals such as Nb, Ta, Sn, Zr, and Ti [15,16].

Due to the relative scarcity of heavy minerals in sediment, they normally require separation and concentration [16–19] and are traditionally sorted, hand-picked, and identified using optical microscopy [20–22]. The latter is time-consuming and requires a specialised knowledge of mineral optical properties. In recent years, it has been identified that optical analysis can produce varied results and occasionally introduce operator bias [23], which techniques for automated mineralogy can reduce. Combined SEM-EDX has been employed in the characterisation and composition of HM phases [15,17], which has included the use of automated SEM-EDX analysis using QEMSCAN [24], as well as more novel applications with machine learning applied to HM SEM-EDX data [25]. Other authors have used Raman spectroscopy [3,9,26,27] and ICP-AES [21].

Here, we illustrate a fast and efficient method that utilises SEM and BEX (combined backscattered electron and X-ray). This method can quickly collect high-resolution images (for size, shape, and distribution of heavy minerals) at the same time as elemental composition data. Once collected, data such as spatial distribution can be further extracted and used as the basis of additional analysis for Raman spectroscopy, cathodoluminescence, microprobe studies, and ICP analysis. The pros and cons of this technique are discussed, as are potential future improvements that would make automated HM analysis a valuable routine method.

2. Materials and Methods

The sample heavy mineral separate utilised herein was collected from the Oligocene fluvio-deltaic Padaung Formation from the Central Myanmar Basin (MCM-18-004). The Central Myanmar Basin (Figure 1) is an elongate basin containing ~15 km of Cenozoic sediment; it is bound by the Sino-Myanmar ranges (Jurassic to Eocene medium- to high-grade metamorphic rocks) in the east and the Indo-Myanmar ranges (Triassic to Eocene schists, turbidites, and ophiolitic rocks) in the west [9].

The methodology is illustrated in the form of a flow chart (Figure 2) and follows standard mineral separation processes [16–19,28]. Five hundred grams of medium-grained sandstone were partially disaggregated using a mortar and pestle before being milled using a Fritsch Pulverisette tungsten carbide mill (Fritsch, Idar-Oberstein, Germany) with a plate separation of 0.10 mm. The loose sediment was then sieved to a 63 μm to 250 μm fraction using a Fritsch Analysette 3 Spartan wet sieve (Fritsch, Idar-Oberstein, Germany) and cleaned using a sodium hexametaphosphate solution combined with sonic agitation. This 63 μm to 250 μm aliquot was then hydrodynamically separated into light, intermediate, and heavy fractions using a Holman–Wilfley 800 Laboratory Table (Holman–Wilfley, Redruth, UK) set to an angle of 15° with low amplitude vibrations. The retained heavy mineral fraction underwent an acetic acid bath to remove any carbonate content before using a hand magnet to remove any highly magnetic minerals. Eighty grams of the remaining sample was then separated using lithium heteropolytungstate (LST), a heavy liquid with a density of 2.89 $\pm$ 0.02 g/mL, to separate the heavy minerals. The concentrated HM separate was embedded into pucks using Araldite AY 103 resin and HY 956 hardener (Huntsman Advanced Materials, The Woodlands, TX, USA) and hand-polished for bulk SEM elemental analysis. In addition, individual HM grains were mounted onto a standard aluminium SEM stub to observe phases in their natural states. Both sample types were coated in gold, using an Emitech K550 gold sputter coater (Emitech Ltd., Ashford, UK), and analysed in high-vacuum mode with a Quanta 650 FEG scanning electron microscope (SEM) (FEI, Hillsboro, OR, USA) equipped with an Oxford Instruments X-Max^{N} 150 mm energy-dispersive X-ray (EDX) detector (Oxford Instruments, Abingdon-on-Thames, UK) in combination with an Oxford Instruments Unity backscattered electron and X-ray (BEX) system (Oxford Instruments, Abingdon-on-Thames, UK) (Figure 3). AZtec 6.1^{TM} BEX Mapping software was utilised to collect raw data and construct elemental maps; the parameters used during scanning are given in Table 1. AZtecFeature and AZtecMatch were also tested to extract numerical data for individual grains and identify mineral phases within the HM suite, respectively.

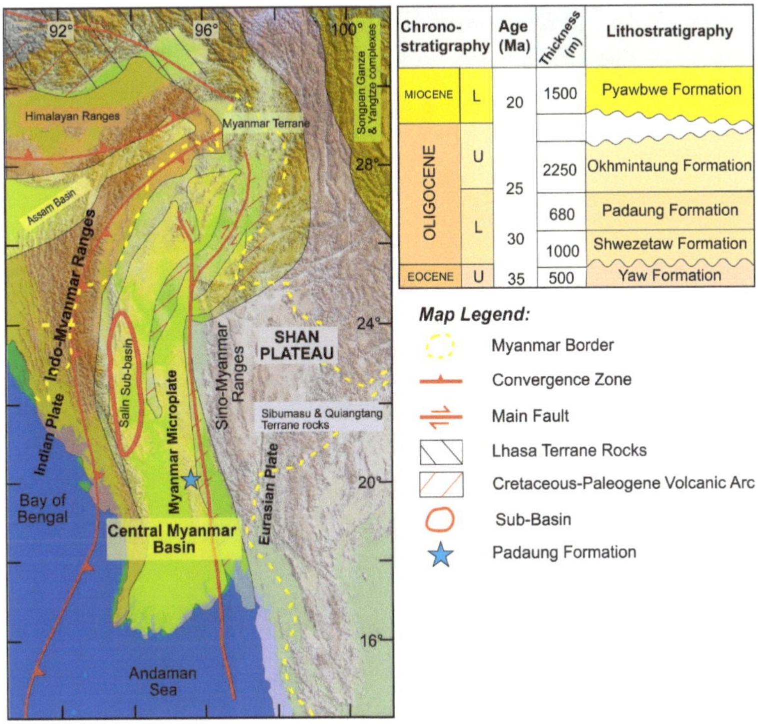

Figure 1. Digital elevation model (DEM) of Myanmar, showing the main geological terranes, alongside principal convergence zones and faults (**left**), alongside Oligocene lithostratigraphy of the Central Myanmar Basin, including the Padaung Formation (**right**). Modified from [29].

Table 1. Parameters used in mapping the surface of polished block and stub.

	Polished Block	Stub
Total number fields	164	84
Width per field (mm)	2.88	1.79
# pixels per field width	1024	1024
Montage pixel resolution	5560 × 5466	7413 × 7678
Scan time per field (s)	33	10
Magnification	145	233
Total scan time (h)	1.5	0.25
Dwell time (μs)	9.81	2.59
Operating voltage (kV)	20	20
Aperture strip position	1	1
Spot size	4.5	4.5
Working distance (mm)	9.6	10

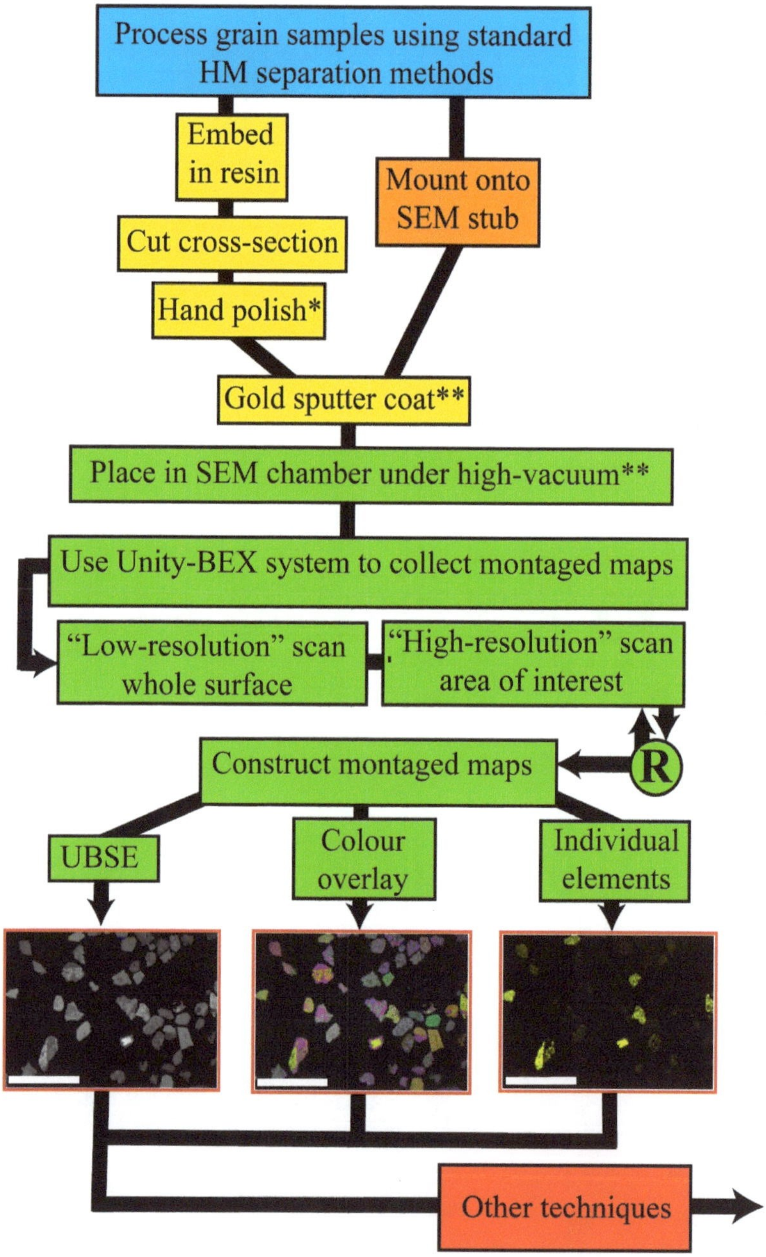

Figure 2. Schematic workflow of the methodology applied herein. R = repeat if required. Red outlined boxes at the bottom of the workflow respectively illustrate the BSE image, composite layered map, and elemental map for Fe, all for the same location. * Machine polishing will produce better images, but hand-polishing is adequate for most purposes. ** Samples can also be imaged and analysed under low-vacuum conditions (0.83 Torr), in which case no sputter coating is required. Solid red box indicates the potential for additional techniques (e.g., AZtecFeature, AZtecMatch, Raman microscopy, cathodoluminescence, in situ U-Pb geochronology, etc.).

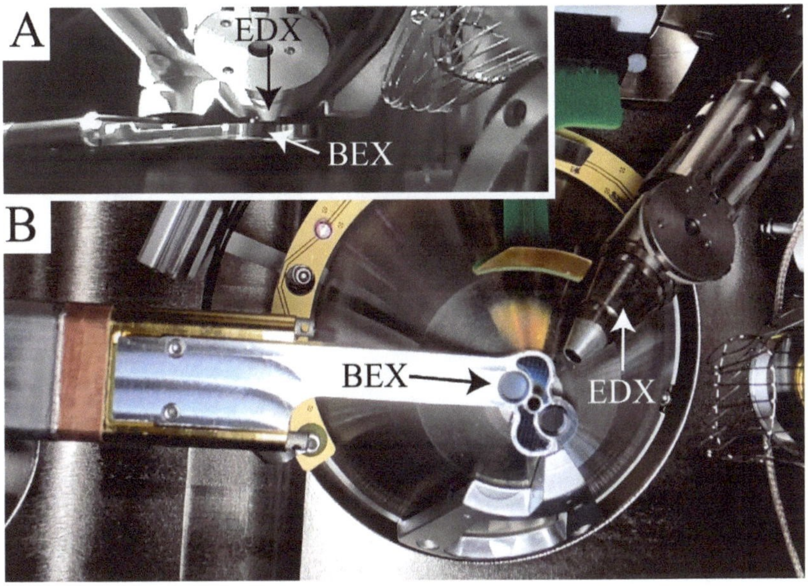

Figure 3. (**A**) Side view of relative position of the Unity BEX (BEX) detector and the X-Max^N 150 mm detector (EDX) within the SEM chamber. (**B**) View of the underside of the SEM polepiece within the chamber, showing both BEX and EDX detectors.

3. Results

Both the polished block and stub were successfully scanned in a circular-shaped grid pattern (Figure 4A,B). In both cases, a range of grain types can be seen, illustrated by variations in the false-colour layered images produced by overlaying elemental data and backscattered (BSE) images (Figure 4). In addition, many grains are composite in nature (Figure 4C,D), and some particles have a complex interlocking texture (Figure 4D). Examples of elemental composition maps (Figure 5) collected from the polished block are shown for specific elements. From these, zircon grains can easily be identified from maps for Zr (Figure 5A). Identifying the occurrence and distribution of other indicative HM phases is also possible, including with titanite and rutile using iron and titanium maps (Figure 5B,C), for minerals high in calcium (Figure 5D) such as apatite and calcite, and for other silicate minerals using silica and aluminium maps (Figure 5E,F).

The collected elemental maps were imported into AZtecLAM (large area maps), where they were successfully differentiated into six separate phases (Figure 6A–F). These phases were further investigated with AZtecMatch. This uses a traffic light system to indicate the degree of confidence for any given match, with green representing a good match, yellow a more tentative suggestion, and red a low probability. Zircon, titanite, and apatite were automatically identified, showing good matches (green traffic light) with individually tested particles (Figure 7A–C). However, it should be noted that some particles in the third phase (Figure 6C) could also be calcite (green traffic light). The fourth phase (Figures 6D and 7D) was identified as epidote, but with less confidence (yellow traffic light), with other alternatives including andradite, grossular garnet, bytownite, lawsonite, anorthite, and zoisite/clinozoisite. The fifth and sixth phases were far less certain in their identification (red traffic light) with siderite, goethite, magnesium almandine (fifth phase, Figures 6E and 7E), and one possible olivine (fayalite, yellow traffic light), while the sixth phase was potentially identified as chromite (Figure 7F). The percentage coverage as part of the illustrated field of view (Figure 6A–F) was zircon = 0.11%, titanite = 4.73%, apatite = 0.89%, epidote = 15.23%, siderite = 0.55%, and chromite = 0.39%.

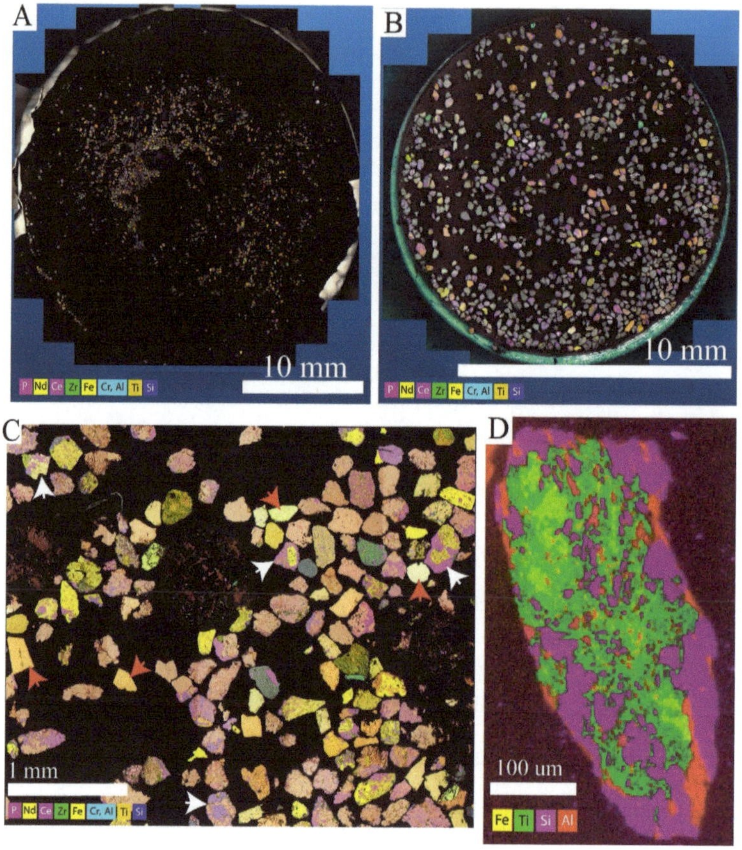

Figure 4. (**A**,**B**) stitched overview layered images of a polished block (**A**) and stub with natural grain surfaces (**B**). Images are a combination of backscattered (BSE) images and elemental maps for P, Nd, Ce, Zr, Fe, Cr, Al, Ti, and Si. (**C**) Zoomed image of area in (**A**), showing details of compositional variation within and between individual grains. Simple grains (red arrows), composite grains (white arrows). Composite grains with multiple phases indicate a degree of textural immaturity. Note that for all three images, colour variation indicates the occurrence of different mineral phases. (**D**) Detail of individual polished grain from (**A**), showing complex elemental distribution. Elemental spectrums can be further extracted from each grain or sub-areas within grains (see Figure 7).

AZtecFeature was successfully tested to extract physical information from a test area on the polished block (Figure 8), with representative numerical data for 20 grains presented here (Table 2). As shown, a range of useful data was extracted, which included the area, aspect ratio, breadth, length, equivalent circular diameter (ECD), perimeter, and shape (Table 2). Data also recorded the mean grey level of each individual particle and, in this case, a redundant measurement of grain orientation (direction, °). From the limited dataset, the aspect ratio and shape indicated a range of morphologies from equant rounded grains to elongated grains, with the size range of the exposed axes of these grains ranging from 24 to 242 μm.

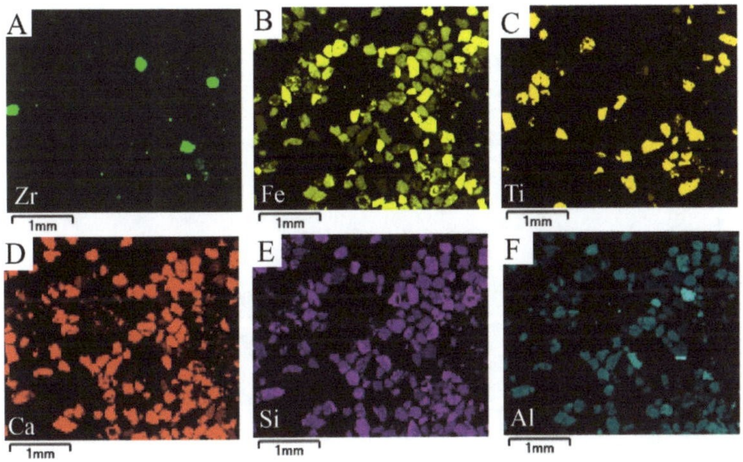

Figure 5. (**A–F**) Individual maps for the elements Zr, Fe, Ti, Ca, Si, and Al, respectively, from the area in Figure 4C. Note that Zr clearly indicates the occurrence of zircon, as does Ti for the presence of Ti-enriched phases, some of which also correlate with Fe bearing phases.

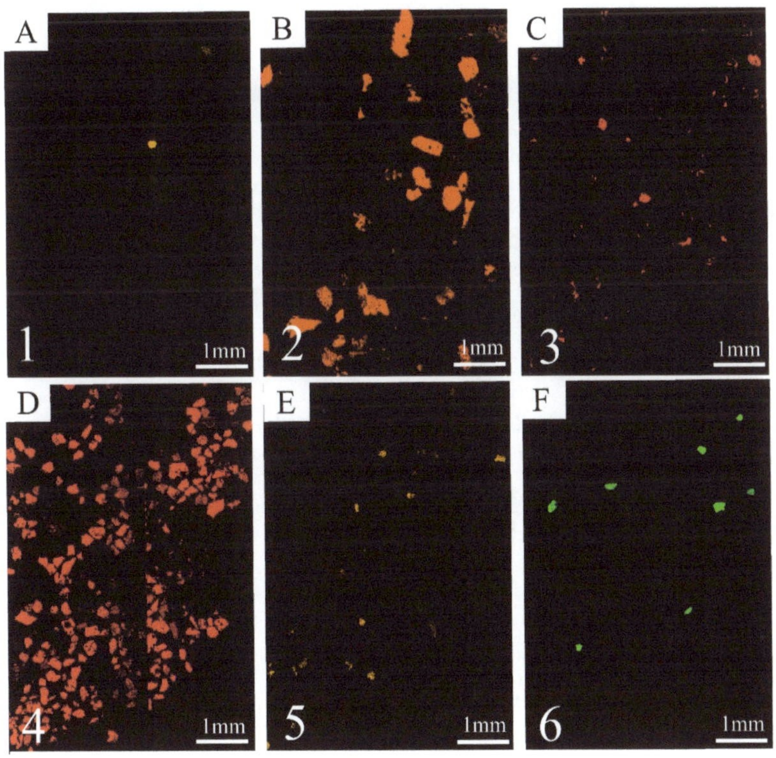

Figure 6. Elemental phase maps of selected area from polished block in Figure 4A. (**A–F**) Maps for six differentiated elemental phases (1–6) made using AZtecLAM (large area maps).

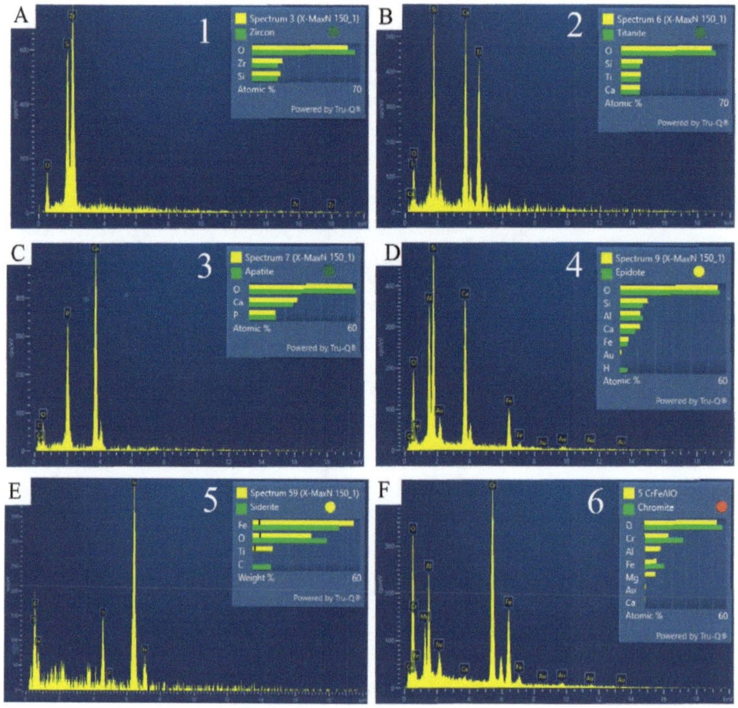

Figure 7. X-ray spectra, and identification of different mineral phases using AZtecMatch, from their corresponding elemental phase maps (Figure 6**A–F**). (**A**) zircon, (**B**) titanite, (**C**) apatite, and potentially (**D**) epidote, (**E**) siderite, and (**F**) chromite. Numbers refer to phase maps in Figure 6.

Table 2. Representative values recorded from 20 heavy mineral particles from the polished block, using AZtecFeature.

Area (μm^2)	Aspect Ratio	Breadth (μm)	Direction (°)	ECD (μm)	Length (μm)	Mean Grey Level	Perimeter (μm)	Shape
1035	4	20	176	36	79	13,809	166	2
245	1	17	35	18	24	9824	64	1
2718	4	31	177	59	132	15,247	276	2
1343	3	31	8	41	80	12,253	190	2
2892	4	39	3	61	152	14,422	326	3
316	2	17	162	20	27	12,874	66	1
1027	2	31	158	36	61	14,641	136	1
1533	2	34	166	44	58	13,776	144	1
379	1	22	135	22	28	12,616	75	1
743	3	28	139	31	70	14,185	164	3
1446	1	39	38	43	54	11,803	143	1
6186	3	73	3	89	208	14,042	553	4
490	9	10	42	25	88	14,088	171	5
948	4	22	34	35	95	15,642	202	3
837	3	20	51	33	62	14,931	135	2
1991	2	42	121	50	66	12,733	167	1
14,316	3	96	3	135	242	17,370	572	2
1296	1	41	45	41	60	12,621	151	1
13,202	2	101	11	130	197	12,396	486	1
158	4	11	143	14	42	16,341	85	4

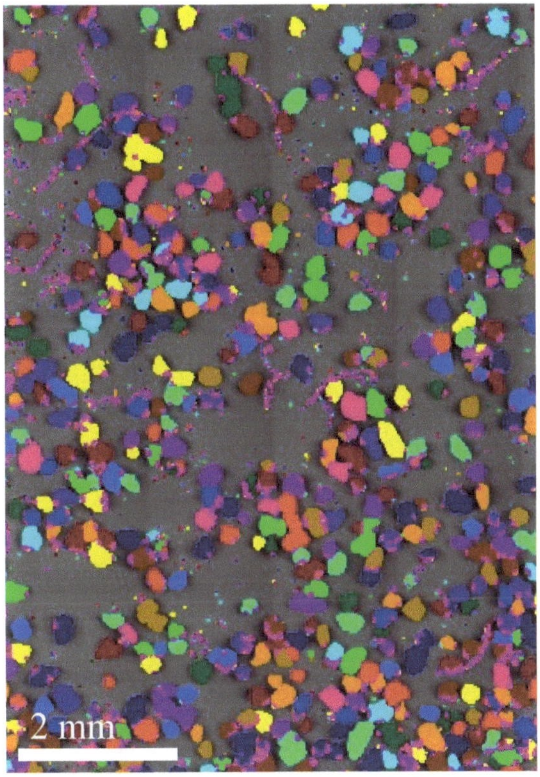

Figure 8. Selected area from stub sample, with grains highlighted to show separate features (particles) using AZtecFeature. The physical data (size, shape, aspect ratio, etc.) derived from these particles are shown in Table 2.

4. Discussion

The Unity BEX system can be up to 100 times faster than existing EDX systems. In this study, the Unity BEX system was directly monitored, registering as 18 times faster than the installed X-Max^N 150 mm EDX detector and recording an output rate in the region of 900 K cps. Other modern EDX detector systems may approach this improved analytical speed, especially in the case of multi-detector installations. However, improvements in the Unity BEX system are not limited to speed but also include the lack of shadowing artefacts in 3D grain mounts (Figure 9); significantly, it can be operated at a broader range of operational working distances (WD) and is not restricted to a single useable analytical WD.

The workflow shown here is ideal for the rapid location and distribution mapping of important mineral species such as zircon, apatite, titanite, and other heavy mineral phases. Zircon can be easily identified through maps of the element Zr, allowing for the possibility of later relocation for cathodoluminescence imaging (Figure 10), which is important for the recognition, imaging, and eventual dating of different growth zones [30]. Furthermore, if the resin mounting and machine polishing workflow (Figure 2) is followed, then this also allows for the in situ U-Pb geochronology of key minerals using LA-ICP-MS (e.g., zircon, apatite, titanite).

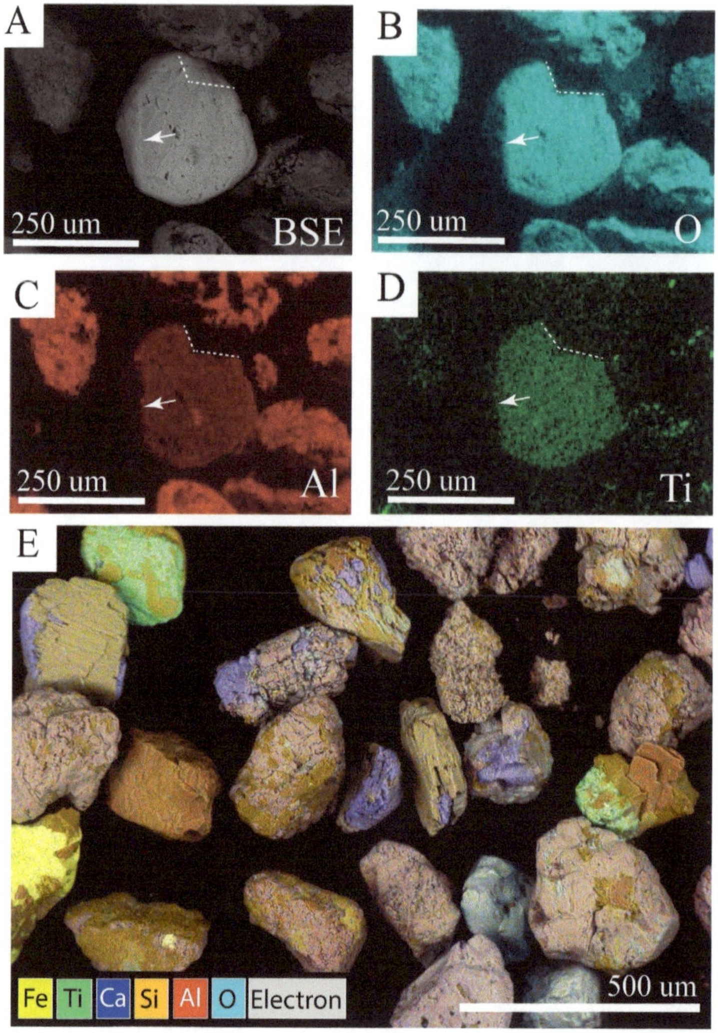

Figure 9. (**A**) BSE SEM image of grains on stub. (**B–D**) elemental maps for O, Al, and Ti over the same area as in (**A**). Note that (**B–D**) were taken using traditionally mounted X-Max^N 150 mm EDX detector and, hence, show shadowing effects (dashed line and arrow) resulting in areas that have no X-ray signal. (**E**) Similar composite BSE/elemental image of grains on stub, but taken with Unity BEX detector, which does not exhibit shadowing artefacts due to the dual X-ray detectors placed at 180° to each other directly under the polepiece.

The combined backscattered electron and X-ray data produced by the Unity BEX also allow for the rapid identification of interlocking and complex chemical (mineralogical) textural relationships within single grains (Figure 4C,D). These complex chemical–textural relationships cannot be deconvolved using other mineral identification techniques (e.g., Raman spectroscopy [23]) and can be important for the interpretation of sediment maturity, which is dependent on the distance of transportation and the degree of weathering of HM suites or may indicate sedimentary recycling (e.g., lithic grains, surface corrosion, etc.) [3,5,6].

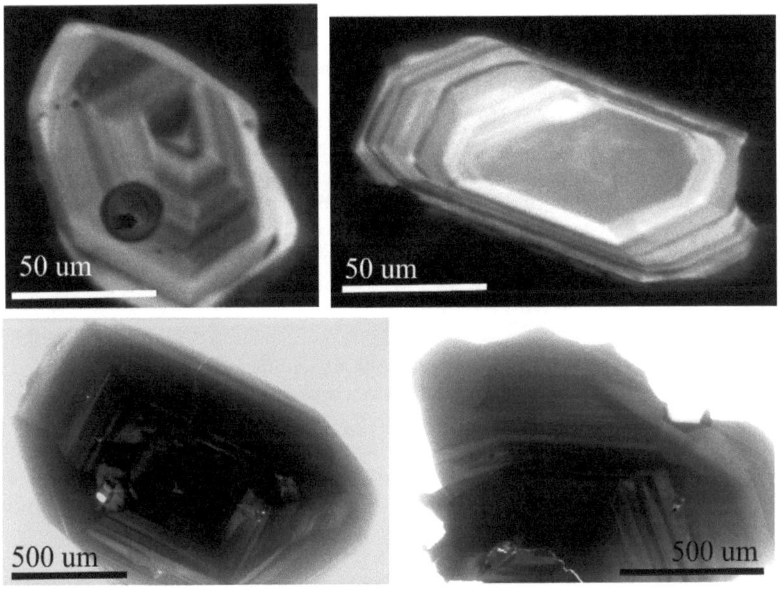

Figure 10. Typical SEM CL images of zircon, illustrating zones within the grains. Fast elemental mapping can be used to map out the locations of zircons (and other zoned mineral phases), allowing for further in situ analysis such as CL or U-Pb geochronology using LA-ICP-MS.

Despite these advantages in the rapid mapping and identification of heavy mineral phases, the Unity BEX system possess areas that require future optimisation. For example, the system cannot detect elements lighter than Na and can only produce qualitative results for the other elements. Light element analysis and full quantification can be achieved through integration with pre-existing standard EDX-type detectors, but this requires analysis to take place at a static working distance set for the EDX detector (e.g., 10 mm). In addition, results will vary depending on the type and age of the EDX detector partnered to the Unity BEX system. Newer and faster large-windowed EDX detectors will improve quantitative results, especially for lower atomic number elements (e.g., carbon and oxygen), and crucially allow for the reduction in scan times, increasing sample throughput without compromising on quality.

In addition, neither AZtecFeature nor AZtecMatch can currently take advantage of data derived through BEX and are restricted to utilising EDX spectral data, in this case from an X-Max^N 150 mm detector system. For example, in AZtecFeature, although it is possible to threshold the BSE images obtained through BEX and, therefore, extract data on particle physical parameters (Table 2), it is not currently possible to utilise either BEX or EDX spectral data within the programme. In the case of AZtecMatch, it is possible to utilise the EDX data. However, the superior BEX spectra with higher count rates cannot be used for mineral matching.

The optimisation and integration of BEX spectra with AZtecMatch will be the key to developing automated heavy mineral identification using SEM-Unity BEX into a standard sedimentary provenance technique. Current quantitative techniques for automated heavy mineral identification such as Raman spectroscopy have been in use for a longer period and, as such, have well-developed and tested neural networks in place to identify a wider range of heavy minerals with high accuracy (e.g., [31]). A separate aliquot of the sample tested in this present study (MCM-18-004) has also been analysed using a Horiba XploRA Raman 532 ųm laser coupled to an Olympus polarising microscope with a lens set at ×50 objective [32]. The heavy minerals assemblage identified using Raman spectroscopy shows a dominance of epidote (58.4%) and garnet (13.6%). There are minor amounts of titanite (8.8%), amphibole (7.4%), apatite (3.9%), zircon (1.6%), and rutile (1.1%), along with a lower content of tourmaline (0.8%), chloritoid (0.6%), anatase (0.4%), chlorite (0.3%), and spinel (0.1%), with "Other" heavy minerals representing 3%.

The Unity BEX and AZtecMatch data and mineral identification presented here correlate well with this (Table 3), except for the abundant garnet signature (although it is noted that the abundant epidote can also be interpreted as magnesium almandine based on AZtecMatch data). Some of the differences may be due to the small sample size used for AZtecMatch. However, the complex interlocking nature of many of the grains, as well as the limitations of the factory library, have also influenced the results. It is worth noting that chromite from this present study may in part equate to spinel sensu stricto, as it also contains Mg and Al (Figure 7F). Separating forms of "spinel" is not easy, even with other techniques such as Raman spectroscopy (see [33]). Some material identified as chromite may, in part, also be garnet. In addition, the phase identified in AZtecMatch as siderite (yellow traffic light) may at least in part contain a proportion of a Ti-bearing mineral such as titanite, rutile, or anatase, as some EDX spectra contain up to 7 atomic% of Ti (Figure 7E). SEM-BEX is a powerful tool for rapid in situ mineral location and identification and provides crucial textural data that other techniques, such as Raman or QEMSCAN, cannot. However, the AZtecMatch database requires additional building and training to deconvolve complex heavy mineral assemblages fully.

Table 3. Comparison of heavy minerals observed by Raman spectroscopy and SEM-EDX for the same sample.

Mineral	Raman (%)	SEM-EDX * (%)
Epidote	58.4	69.60
Garnet	13.6	
Titanite	8.8	21.62
Amphibole	7.4	
Apatite	3.9	4.07
Zircon	1.6	0.5
Rutile	1.1	2.51 †
Tourmaline	0.8	
Chloritoid	0.6	
Anatase	0.4	
Chlorite	0.3	
Spinel	0.1	1.78 ‡
Other HM	3	

* note values from phases 1–6 normalised to 100%. † identified as siderite by AZtecMatch, but also contains up to 7 atomic% Ti. ‡ "Chromite" but with Mg and Al (spinel).

5. Future Workflow Optimisation

Based on the initial workflow development and testing of heavy mineral population identification using multiple analytical techniques (SEM-BEX and Raman spectroscopy), some recommendations can be made to improve future analysis.

The improved sample preparation and maximisation of particle concentration through the placement of multiple samples onto densely packed and machine-polished thin sections will increase productivity and data quality (Figure 11).

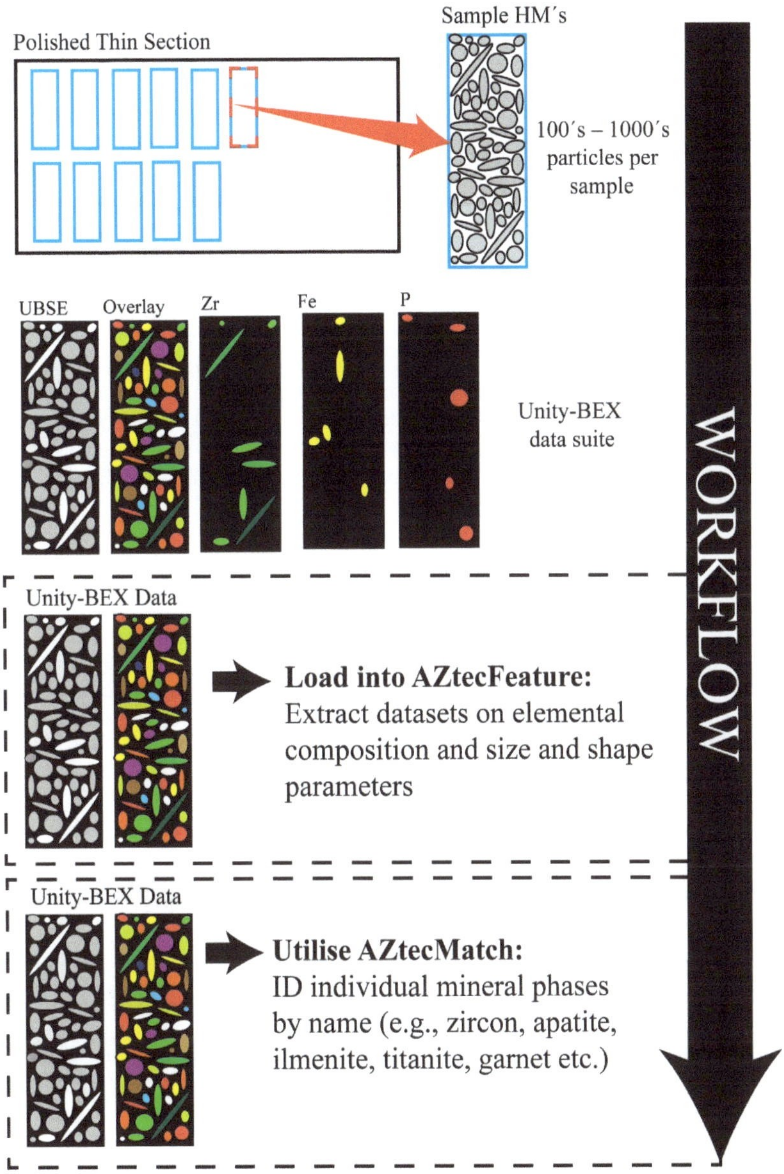

Figure 11. Future idealised workflow using subsets of heavy minerals on polished thin sections to increase efficiency in acquisition and implementation of phase identification. Note that the same workflow can equally be applied to grain mounts (unpolished samples). The workflow can also include other techniques such as Raman spectroscopy.

Fuller integration between the AZtec software options (i.e., BEX, AZtecFeature, AZtec-Match) and the integration of phase analysis from AZtecLAM into BEX would also greatly improve and optimise the workflows on heavy mineral suites. Therefore, some software development from Oxford Instruments would greatly improve usability and enhance semi-automated or automated interaction between the AZtec options (Figure 11). In the case of AZtecFeature, this would allow for the measurement of a range of shape and size parameters (see example in Table 2), as well as the elemental spectral characterisation of each particle from Unity BEX maps. Such compatibility is possible between AZtecLAM and AZtecFeature but is not currently available under the Unity BEX analysis package. The combination of fast elemental mapping through Unity BEX, including on grain mounts, with comprehensive particle analysis would be particularly useful in the study of heavy minerals, as well as in other fields of investigation. The partial use of AZtecMatch herein, which could only utilise X-rays captured by the X-Max^N 150 mm EDX detector, illustrates the potential to leverage the benefits of faster, more in-depth elemental data from Unity BEX. This would greatly benefit from full integration with AZtecMatch, allowing the construction of fast maps with particle analysis based on named mineral distributions.

AZtecMatch can utilise the user-defined libraries of mineral X-ray spectra. These can be collected under specific SEM conditions optimised for data collection (i.e., high-vacuum gold-coated, low-vacuum uncoated, High kV, low kV, etc.). The future development of such data libraries, specifically for heavy mineral suites, will help improve accuracy and allow better speciation of heavy mineral assemblages.

Recent developments in scanning electron microscopy (SEM) now allow for the integration of Raman spectroscopy within SEM. We highly recommend that this current proposed workflow (Figure 11) be modified to include an additional Raman spectroscopy step where appropriate.

6. Conclusions

This proposed workflow demonstrates that automated HM analysis using fast elemental X-ray mapping in conjunction with BSE imaging (Unity BEX) is practical and has the potential for the direct identification of mineral phases as well as the characterisation of several other physical parameters (size, shape, etc.) that are important for comprehensive provenance analysis. Further automation and integration of options within AZtec software (e.g., AZtecMatch) can only improve these possibilities. Such workflows are faster than traditional techniques that use optical microscopy or Raman spectroscopy, have the potential to be fully automated, and do not rely on the availability of personnel with specialist knowledge in heavy mineral grain identification.

Author Contributions: Conceptualization, J.B.; methodology, J.B.; formal analysis, J.B., A.G. and M.W.; resources, A.G., M.W. and J.B.; data curation, J.B. and A.G.; writing—original draft preparation, J.B., A.G. and M.W.; writing—review and editing, J.B., A.G. and M.W.; visualization, J.B. and M.W. All authors have read and agreed to the published version of the manuscript.

Funding: This research received no external funding.

Data Availability Statement: No data is available.

Acknowledgments: Royal Holloway University of London (RHUL) for use of heavy mineral separation facilities. Conor McMillan is thanked for collecting the sample used in this study. Scanning electron microscopy work was carried out at the Centre for Environmental Scanning Electron Microscopy (CESEM) at Heriot-Watt University (HWU).

Conflicts of Interest: The authors declare no conflicts of interest.

References

1. Morton, A.C. Heavy minerals. In *Sedimentology Encyclopedia of Earth Science*; Springer: Berlin/Heidelberg, Germany, 1978. [CrossRef]
2. Morton, A.C. Heavy Minerals in Provenance Studies. In *Provenance of Arenites*; NATO ASI Series 148; Springer: Dordrecht, The Netherlands, 1985. [CrossRef]
3. Coll, X.; Gómez-Gras, D.; Roigé, M.; Teixell, A.; Boya, S.; Mestres, N. Heavy-mineral provenance signatures during the infill and uplift of a foreland basin: An example from the Jaca basin (Southern Pyrenees, Spain). *J. Sediment. Res.* **2021**, *90*, 1747–1769. [CrossRef]
4. Joshi, K.B.; Banerji, U.S.; Dubey, C.P.; Oliveira, E.P. Heavy minerals in provenance studies: An overview. *Arab. J. Geosci.* **2021**, *14*, 1330. [CrossRef]

5. Morton, A.C.; Smale, D. The effects of transport and weathering on heavy minerals from the Cascade River, New Zealand. *Sediment. Geol.* **1990**, *68*, 117–123. [CrossRef]
6. Sevastjanova, I.; Hall, R.; Alderton, D. A detrital heavy mineral viewpoint on sediment provenance and tropical weathering in SE Asia. *Sediment. Geol.* **2012**, *280*, 179–194. [CrossRef]
7. Morton, A.C.; Hallsworth, C.R. Identifying provenance-specific features of detrital heavy mineral assemblages in sandstones. *Sediment. Geol.* **1994**, *90*, 241–256. [CrossRef]
8. Morton, A.C.; Whitham, A.G.; Fanning, C.M. Provenance of Late Cretaceous to Paleocene submarine fan sandstones in the Norwegian Sea: Integration of heavy mineral, mineral chemical and zircon age data. *Sediment. Geol.* **2005**, *182*, 3–28. [CrossRef]
9. McNeil, J.D.; Gough, A.; Hall, R.; Lünsdorf, N.K.; Webb, M.; Feil, S. A multi-proxy provenance study of Eocene to Oligocene sandstones in the Salin Sub-basin, Myanmar. *J. Asian Earth Sci.* **2021**, *216*, 104825. [CrossRef]
10. Pickering, K.T.; Carter, A.; Andò, S.; Garzanti, E.; Limonta, M.; Vezzoli, G.; Milliken, K.L. Deciphering relationships between the Nicobar and Bengal submarine fans, Indian Ocean. *Earth Planet. Sci. Lett.* **2020**, *544*, 116329. [CrossRef]
11. Webb, M.; Gough, A.; Endinanda, F. Depositional environments and sedimentary provenance of the Cenozoic deposits of Natuna Island, Indonesia: Implications for basin evolution in central Sundaland. *Gondwana Res.* **2024**, *134*, 298–325. [CrossRef]
12. Wagner, G.A. Fission track dating of apatites. *Earth Planet. Sci. Lett.* **1968**, *4*, 411–415. [CrossRef]
13. Naeser, C.W.; Zimmermann, R.A.; Cebula, G.T. Fission-track dating of apatite and zircon: An interlaboratory comparison. *Nucl. Tracks* **1981**, *5*, 65–72. [CrossRef]
14. Lee, T.-H. SHRIMP U-Pb ages of the detrital zircons from the Cretaceous Ullyeonsan Formation, SE Korea: A small piece of stratigraphic correlation in eastern Asia. *Cretac. Res.* **2023**, *151*, 105626. [CrossRef]
15. Dung, N.T.; Bac, B.H.; An, D.M.; Anh, T.T.V. Distribution and Reserve Potential of Titannium-Zirconium Heavy Minerals in Quang an Area, Thua Thien Hue Province, Vietnam. In *Advances and Applications in Geospatial Technology and Earth Resources*; Bui, D.T., Do, A.N., Bui, H.-B., Hoang, N.-D., Eds.; Springer International Publishing: New York, NY, USA, 2018. [CrossRef]
16. Nzeh, N.S.; Popoola, P.; Okanigbe, D.; Adeosun, S.; Adeleke, A. Physical beneficiation of heavy minerals—Part 1: A state of the art literature review on gravity concentration techniques. *Heliyan* **2023**, *9*, e18919. [CrossRef] [PubMed]
17. Commeau, J.A.; Poppe, L.J.; Commeau, R.F. *Separation and Identification of the Silt-Sized Heavy-Mineral Fraction in Sediments*; U.S. Geological Survey Circular: Reston, VA, USA, 1991; Volume 1071.
18. Rahman, M.A.; Zaman, M.N.; Biswas, P.K.; Nandy, P.K. Physical separation for upgradation of valuable minerals: A study of sands of the Someswari river. *Bangladesh J. Sci. Ind. Res.* **2015**, *50*, 53–58. [CrossRef]
19. Andò, S. Gravimetric separation of Heavy Minerals in Sediments and Rocks. *Minerals* **2020**, *10*, 273. [CrossRef]
20. Andò, S.; Garzanti, E.; Padoan, M.; Limonta, M. Corrosion of heavy minerals during weathering and diagenesis: A catlogue for optical analysis. *Sediment. Geol.* **2012**, *280*, 165–178. [CrossRef]
21. Mounteney, I.; Burton, A.K.; Farrant, A.R.; Watts, M.J.; Kemp, S.J.; Cook, J.M. Heavy mineral analysis by ICP-AES a tool to aid sediment provenancing. *J. Geochem. Explor.* **2018**, *184*, 1–10. [CrossRef]
22. Le Pera, E.; Tangari, A.C.; Marinangeli, L.; Morrone, C.; Riber, R.; Andò, S. Provenance of modern sands from Baja California rivers (Mexico): Petrographic constraints from light and heavy minerals. *J. Sediment. Res.* **2023**, *93*, 617–641. [CrossRef]
23. Dunkl, I.; von Eynatten, H.; Andò, S.; Lünsdorf, K.; Morton, A.; Alexander, B.; Yoshida, K. Comparability of heavy mineral data–The first interlaboratory round robin test. *Earth Sci. Rev.* **2020**, *211*, 103210. [CrossRef]
24. Nie, J.S.; Peng, W.B. Automated SEM-EDS heavy mineral analysis reveals no provenance shift between glacial loess and interglacial paleosol on the Chinese Loess Plateau. *Aeolian Res.* **2014**, *13*, 71–75. [CrossRef]
25. Hao, H.; Guo, R.; Gu, Q.; Hu, X. Machine learning application to automatically classify heavy minerals in river sand by using SEM/EDS data. *Miner. Eng.* **2019**, *143*, 105899. [CrossRef]
26. Andò, S.; Garzanti, E. Raman spectroscopy in heavy-mineral studies. *Geol. Soc. Lond. Spec. Public.* **2014**, *386*, 395–412. [CrossRef]
27. Tangari, A.C.; Cirillo, D.; De Luca, R.; Miriello, D.; Pugliese, E.; Le Pera, E. Heavy Minerals Distribution and Provenance in Modern Beach and Fluvial Sands of the Betic Cordillera, Southern Spain. *Geosciences* **2024**, *14*, 208. [CrossRef]
28. Mange, M.A.; Maurer, H.F.W. *Heavy Minerals in Colour*; Springer Science & Business Media: Berlin/Heidelberg, Germany, 2012; p. 147. [CrossRef]
29. Gough, A.; Hall, R.; BouDagher-Fadel, M.K. Mid-Cenozoic fluvio-deltaic to marine environments of the Salin Sub-basin, Central Myanmar. *J. Asian Earth Sci.* **2020**, *190*, 104143. [CrossRef]
30. Corfu, F.; Hanchar, J.M.; Hoskin, P.W.; Kinny, P. Atlas of zircon textures. *Rev. Mineral. Geochem.* **2003**, *53*, 469–500. [CrossRef]
31. Lünsdorf, N.K.; Kalies, J.; Ahlers, P.; Dunkl, I.; von Eynatten, H. Semi-automated heavy-mineral analysis by Raman spectroscopy. *Minerals* **2019**, *9*, 385. [CrossRef]
32. McMillan, C. Sedimentary Provenance and Depositional Environments from the Oligocene Formations in the North of the Salin Sub-basin, Onshore Myanmar. Master's Thesis, Royal Holloway University of London, London, UK, 2019.
33. Zhang, L.; He, K.; Guo, Q. Spectroscopic Characteristics and Coloring Mechanisms of Different Colored Spinels from Myanmar. *Crystals* **2023**, *13*, 575. [CrossRef]

minerals

MDPI

Article

Adsorption of Ciprofloxacin and Lidocaine by Non-Fibrous Raw Mg-Clays: The Role of Composition and Texture

Maria Eugenia Roca-Jalil [1,2,*], Telma Musso [2,3], Vanina Rodriguez-Ameijide [1,2], Micaela Sanchez [1,2], Andrea Maggio [1,2], Miria Teresita Baschini [1,2], Gisela Pettinari [2,3], Luis Villa [4], Manuel Pozo [5,*] and Alejandro Pérez-Abad [4]

1 Institute of Research and Development in Process Engineering, Biotechnology, and Alternative Energies (PROBIEN, CONICET-UNCo), Neuquén 8300, Argentina; vanina.rodriguez@fain.uncoma.edu.ar (V.R.-A.); micaela.sanchez@fain.uncoma.edu.ar (M.S.); andrea.maggio@probien.gob.ar (A.M.); miria.baschini@fain.uncoma.edu.ar (M.T.B.)

2 Department of Chemistry, Faculty of Engineering, National University of Comahue, Neuquén 8300, Argentina; telma.musso@probien.gob.ar (T.M.); gisela.pettinari@probien.gob.ar (G.P.)

3 Department of Geology and Petroleum, Faculty of Engineering, National University of Comahue, Neuquén 8300, Argentina

4 Minersa Group, 48992 Getxo, Spain; luis.villa@minersa.com (L.V.); alejandro.perez@minersa.com (A.P.-A.)

5 Department of Geology and Geochemistry, Faculty of Sciences, Cantoblanco Campus, Autonomous University of Madrid, 28049 Madrid, Spain

* Correspondence: eugenia.rocajalil@probien.gob.ar (M.E.R.-J.); manuel.pozo@uam.es (M.P.); Tel./Fax: +54-2994490688 (M.E.R.-J.); +34-660050721 (M.P.)

Abstract: This study evaluated non-fibrous Mg-clays as potential adsorbents of emerging contaminants (ECs) from water. The materials were characterized, and their textural and structural properties were related to their ability to remove two model EC molecules: ciprofloxacin (CPX) and lidocaine (LID). The results showed that Ad-6 and Ad-7 are mixed-layer kerolite/stevensite, while Ad-5 and Ad-8 are mainly composed of smectite minerals like stevensite and saponite, respectively. Ad-8 exhibited the highest CPX-adsorption capacity (0.91 mmol·g^{-1} clay), likely due to its saponite content. Mixed-layer materials also performed well, with Ad-6 and Ad-7 achieving an adsorption capacity of 0.8 and 0.55 mmol·g^{-1} clay, respectively. Adsorption studies suggested that CPX is adsorbed through ion exchange in materials with high smectite content (Ad-8 and Ad-5), while interstratified materials showed enhanced retention due to kerolite presence, which improves their porous structures. Similar findings were observed for LID, indicating a cationic-exchange mechanism for LID adsorption in all the materials and suggesting that the molecular size of the EC could regulate the removal capacity of these materials. This work showed that the studied Mg-clays could be effectively used for the removal of pharmaceutical pollutants, expanding their commercial possibilities.

Keywords: Mg-clays; emerging contaminants; removal from water; adsorption

1. Introduction

The large production of medicines worldwide (around 4000 types in the order of 200,000 tons per year) is a major cause of the progressive increase in pharmaceutical contaminants that are released to the environment with undervalued consequences on ecosystems and water quality [1–6]. The greatest concern of releasing antibiotics to aquatic environments, either directly or indirectly by surface runoff or soil leaching, is related to the evolution of antimicrobial resistance genes and antimicrobial-resistant bacteria, which reduces the therapeutic potential against human and animal pathogens [7–10].

Ciprofloxacin (CPX) and lidocaine (LID) are considered contaminants of emerging concern due to their presence in wastewater effluents, surface waters, and groundwaters [6,8,11]. CPX is one of the most widely used antibiotics in human health [12,13]. This compound is highly soluble in water (about 1.35 mg·mL^{-1}) under different pH conditions and highly stable in wastewater and soil [14]. Meanwhile, lidocaine is a topical local anesthetic of the amide type with immediate onset of action. Both humans and animals metabolize LID to various metabolites. Its n-octanol/water partition coefficients indicate a tendency to remain in the water phase instead of accumulating in sewage sludge or in aquatic organisms [1,11,15].

Several methods have been put in place for the sequestration of pharmaceutical pollutants from their aqueous environment. Conventional methods like lime softening, reverse osmosis, distillation, and sedimentation were determined to be not efficient in removing antibiotics from wastewater completely [16–18]. Other methods different from conventional procedures, like membrane separation, electrodialysis, ozonation, and advanced oxidation processes, were also evaluated [19]. However, some of these methods produce toxic by-products, which are even more harmful than the original contaminants [20]. Compared to other remediation methods, adsorption has several benefits, such as simplicity, reliability, low energy requirement, and easy recovery of the adsorbent [21]. The two types of adsorptive processes used for the removal of antibiotics from water or wastewater are the Batch process and the Continuous process, either of which can be utilized using different adsorbents like Activated Carbons, Biochars, clay minerals, Carbon Nanotubes, ion-exchange resins, and Zeolites [20].

Clay minerals occur as large deposits worldwide, which guarantees their continuing abundance and availability for numerous applications due to their relatively low cost. These phyllosilicates differ in the arrangement of the tetrahedral and octahedral sheets and the distribution of surface electric charge. They exhibit special properties such as high adsorption, ion-exchange capability, or swelling behavior, resulting from their nonpareil structure, the presence of surface OH groups, and weak electrostatic interactions between layers and/or sheets and the exchangeable cations [22,23].

Several types of clay minerals, such as montmorillonite [24], halloysite [25], kaolinites [26], sepiolite [27], and palygorskite [28], have been investigated for the removal of pollutants in aqueous media. Particularly, montmorillonite-rich bentonites have proven to be effective adsorbents for a variety of drugs, such as antibiotics, anti-inflammatories, and anesthetics, among others [29–31]. Also, in the search for adsorbent materials that are both natural and versatile in retaining such molecules or their analytes, alternative clay minerals have gained attention [27,32–39].

The magnesian clays consist of mostly 2:1 clay minerals that are differentiated as non-fibrous clay minerals (kerolite, stevensite, and saponite) and fibrous clay minerals (sepiolite and palygorskite). Saponite and palygorskite contain more aluminum than stevensite, kerolite, and sepiolite. Kerolite (turbostratic talc for the IMA), Mg-smectite, and sepiolite form a clay assemblage found in Neogene continental deposits of the Madrid Basin (Spain) that originated in saline-alkaline lacustrine–palustrine environments [40,41]. Kerolite-Mg-smectite clays from the Madrid Basin deposits show exceptional suitability as adsorbent material in biotechnology processes, including as adsorbent for mycotoxins [42,43]. Moreover, pesticide removal from water has also been observed [44]. Bentonites from the Madrid Basin are characterized by the fact that they consist of magnesian smectites, mainly saponite. These bentonites have a large number of applications, which mainly include pet litter, drilling muds, and foundry sands, but also as oil and grease absorbents, agricultural pesticide carriers, and clarifying and decolorizing agents [45].

Adsorption studies on stevensite for the removal of different types of contaminants from water, such as heavy metals, methyl violet, and tetracycline, have been reported [39,46,47]. Kerolite has been tested for removing pesticides and herbicides from an aqueous solution [44,48,49]. Saponite has been evaluated in photocatalytic and adsorption processes for water treatment [38].

Several references can be found on CPX adsorption on natural and modified clays such as bentonite, kaolinite, or pillared clay [14,29,30,50–52]. However, neither CPX nor LID adsorption on Mg-clays has been documented. Unlike smectite, the surface characteristics of non-fibrous Mg-clays remain underexplored for its potential application in environmental remediation. Thus, the aim of this work was to evaluate four non-fibrous Mg-clays as novel materials for the removal of ciprofloxacin and lidocaine compounds, which are considered emerging contaminants. Their adsorptive behavior was explained on the basis of their composition and textural properties. In this way, promising clay materials are presented for the removal of pharmaceutical pollutants.

2. Materials and Methods

2.1. *Raw Mg-Clays*

Four samples of raw Mg-clays (Ad-5, Ad-6, Ad-7, and Ad-8) of industrial interest were selected from the Esquivias–Batallones quarries located in the Madrid basin Neogene deposits [53]. Hand specimen observations are described below. Ad-5: compacted claystone, with intraclastic morphologies of different sizes enclosed by cementing clay. It shows some morphologies with concentric structures and burrower fillings. The large number of small circular pores (<0.2 mm) is noteworthy. Predominantly greenish in color, although locally pinkish tones are observed (5GY 6/1). Ad-6: clay-clast intrarenite of moderate compactness, massive, and with evidence of burrowing activity and small perforations with fillings. Staining of black Mn-Fe oxides and waxy films on surfaces. Pinkish gray in color (5YR 8/1). Ad-7: clay-clast intrarenite, consisting of crumbly granular morphologies with local cementing clayey material and abundant pore roots. Pinkish gray in color (5YR 8/1). Ad-8: compacted and massive claystone, showing slickensides with striations and staining of black Mn-Fe oxides. It presents a yellowish-green color (5Y 7/2).

Structural and Physicochemical Characterization of Raw Mg-Clays

The four samples were subjected to mineralogical analysis by means of X-ray diffraction (XRD) in a Bruker D8 with Lynxeye XE-T detector equipment (Bruker, Billerica, MA, USA) using CuKα radiation (40 kV, 20 mA) and a scanning speed of 1° 2θ/min. The powdered whole-rock samples were used to determine bulk mineralogy. After clay fraction separation (<2 µm), sample mounts were prepared from suspensions oriented on glass slides. The identification of clay minerals was carried out on the samples air-dried (AD), with ethylene glycol solva-tion (EG), and heated at 550 °C. The mineral intensity factors (MIF) method was applied to XRD reflection intensity ratios normalized to 100% with calibration constants for the quantitative estimation of mineral content [54]. In the mixed-layer samples formed by mixed-layers, X-ray diagrams modeled using the NEWMOD program were used as a reference [40,41].

The thermal analyses (DSC/DTA/TGA Q600 of TA Instruments) (TA Instruments, Inc., New Castle, DE, USA) of the samples were performed using 10 mg of powdered clay sample in a Pt sample holder at an average heating rate of 10 °C/min with an alumina reference. The infrared spectroscopy (FTIR) (Bruker IFS 66v standard) (Bruker, Billerica, MA, USA) spectra were recorded with a spectral resolution of 2 cm^{-1} in transmission mode in the region of 400–4000 cm^{-1}. The samples were examined in KBr pellets (3 mg·300 mg^{-1} KBr). Samples were examined under scanning electron microscopy (SEM) using an S-3000N microscope with an ESED detector (Hitachi, Chiyoda, Japan) attached to an Oxford Instruments EDX analyzer (Oxford Instruments, Abingdon, UK), model ICAx-sight. The surface of the fractured sample was coated with a 10 nm gold film.

Chemical analyses were carried out using sample fusion and inductively coupled plasma atomic emission spectroscopy (ICP-AES) at the ACTLABS laboratory (Canada). The detection limits for analyses were between 0.01 and 0.001 wt.% for major elements.

Cation-exchange capacity (CEC) was measured using the copper triethylenetetramine $[Cu(trien)]^{2+}$ method [55]. Textural properties were studied by nitrogen adsorption–desorption isotherms at −196.15 °C. These were measured using a Micromeritics ASAP2020 manometer (Micromeritics Instrument., Norcross, GA, USA) on degassed samples at 200 °C for 12 h. The apparent specific surface area was calculated by the Brunauer, Emmet, and Teller (S_{BET}) method. Micropore volume ($V_{\mu p}$) was calculated by applying the α-plot method [56]. The total pore volume (V_T) was calculated by applying Gurvich's rule at a relative pressure of 0.98, and the mesopore volume (V_{mp}) was calculated by using of

BJH method [57]. Pore classification used was the one proposed by IUPAC [57], which states that pores whose size exceeds 50 nm are macropores, pores whose size is between 2 and 50 nm are considered mesopores, and pores whose size is less than 2 nm are called micropores. The water vapor isotherms were obtained at 20 °C using the technique described by [30]. Hydrophilicity and water adsorption capacity were estimated from these isotherms. Natural humidity content (NH) was obtained by calculating the difference between the initial mass of the natural sample and its mass at 0% relative humidity. The amount of water adsorbed at relative humidity (RH) values (15%–20%) was also calculated. This value correlates well with the monolayer coverage on the surface of clay minerals and allows determining the amount of adsorbed water measured in mmol·g^{-1} (q_{ads}) and the monolayer surface area (*S*) in samples at 16.3% RH samples to be determined [58,59].

2.2. Adsorbates

The organic molecules selected as models were an anesthetic (LID) and an antibiotic (CPX), two pharmaceutical compounds with a high environmental impact. The CPX is usually used as ciprofloxacin hydrochloride (CPX.HCl), but its molecule exhibits a nearly planar structure, lacks chiral carbons, and, therefore, does not display optical isomerism. Nevertheless, the presence of protonable carboxyl, carbonyl, and amino groups in its structure imparts amphoteric behavior, leading to the formation of different species depending on the physical state or pH of the solution in which it is present [60]. Physically, it is a white crystalline solid with a melting point of 75 °C, exhibiting high stability under various pH values and temperature conditions [61]. LID can exist as a lidocaine base, meaning it is neutral and therefore insoluble in water, or its more soluble form as lidocaine monohydrate hydrochloride (LID.HCl.H_2O) [62]. Chemically, the LID molecule features an amide group and an amino group; the latter can be protonated–deprotonated, depending on the pH of the contacting medium, leading to the generation of different species in solution [63]. The structural and chemical properties of CPX and LID are shown in Table 1.

Table 1. General characteristics of organic molecules studied.

	Ciprofloxacin	Lidocaine
Structure	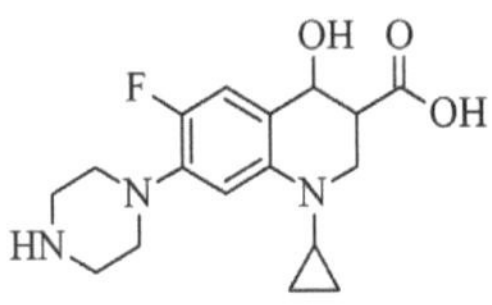	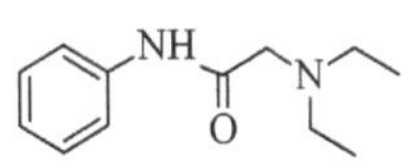
Chemical formula	$C_{17}H_{18}FN_3O_3$	$C_{14}H_{22}N_2O$
Molecular weight (g.mol^{-1})	331.45	288.80
Molecular dimensions (nm)	1.30 × 0.30 × 0.70 [60]	1.04 × 0.60 × 0.59 *
pKa	pKa_1 = 5.90; pKa_2 = 8.90	pKa = 7.92

* The dimensions of the molecule were calculated by using Avogadro software (1.1.1 version, 2013).

2.3. Adsorption Experimental Procedure

CPX and LID were employed in their hydrochloride forms as adsorbates. CPX.HCl, sourced from Magel S.A., and LID.HCl.H_2O, obtained from Parafarm (Parapharm Pharmaceuticals Group, Athens, Greece), were used, both with a purity of 99.9%. Adsorption studies were conducted under conditions previously reported by the research group for other materials [30,50,51]. All adsorption tests were carried out using 20 mg of adsorbent in contact with a drug solution at a ratio of 2.5 g·L^{-1}. These studies were conducted at 20 °C, without initial pH adjustment (pH values of around 5–6), and in all cases, separation was achieved through centrifugation at 8000 rpm for 20 min using a SOVALL ultracentrifuge (Thermo Fisher Scientific, Waltham, MA, USA).

Initial concentrations of 0.3 mM for CPX and 1.69 mM for LID were used for kinetic studies. Adsorption isotherms were obtained with concentrations ranging from 0.15 to 3 mM for CPX and 0.2 to 5 mM for LID, allowing the system to reach equilibrium over a 24-h contact period, both in accordance with previous works [30,51].

CPX and LID concentrations were quantified using a UV-Vis T60 spectrophotometer at their respective maximum wavelengths, and the adsorbed quantity was calculated by Equation (1).

$$q = \frac{(C_i - C_e) * V}{w} \tag{1}$$

where C_i and C_e are the initial and equilibrium concentrations (mM), V is the solution volume (L), and w is the mass of adsorbent (g).

The experimental data were fitted to the widely used theoretical models of Langmuir, Freundlich, and Sips in order to calculate the maximum adsorption capacity of the samples under the studied conditions.

3. Results and Discussion

3.1. Characterization of Mg-Clays

XRD analyses indicated that samples Ad-6 and Ad-7 showed a high phyllosilicate content (≥98%) with traces (<5%) of quartz and potassic feldspars (Figure 1A, Table 2). In the bulk sample, the powder patterns showed a band between 2 and 10° 2θ, characteristic of an irregular kerolite/Mg-smectite mixed layer [40,64], which was corroborated by the existence of reflections typical of this material. In particular, the d_{060} reflection at approximately 1.527 Å supports the trioctahedral character of the constituent clay minerals. The study of the clay fraction with different treatments indicated the presence of both a swelling and a non-swelling phase in Ad-6, but almost exclusively non-swelling in Ad-7 (Figure 1B). The use of the NEWMOD program to model the interstratification showed in sample Ad-6 a predominance of Mg-smectite (80%–90%), while in Ad-7, a predominance of kerolite (80%) was observed. These results were coherent with those reported in other papers [40,41].

In sample Ad-5, phyllosilicates are the main constituents (99%), with minor amounts of quartz and potassic feldspar (<5%). In the bulk sample, two reflections at 15 Å and at around 10 Å can be observed. Also, the d_{060} value indicated the presence of trioctahedral clay minerals. Oriented samples showed the presence of a swelling phase (87%) and illite (13%) (Figure 1B).

Sample Ad-8 was also predominantly composed of phyllosilicates at 14–15 Å (96%), with traces (<5%) of quartz and feldspars (both potassic and plagioclase). The d_{060} spacing was predominantly trioctahedral (1.533 Å), although a minor reflection at 1.502 Å was also identified, indicating the presence of dioctahedral phases (Figure 1A). In the clay fraction, predominantly smectite (97%) and a low content of illite (3%) were identified. This material can be considered as a bentonite.

Table 3 shows the chemical composition of the raw Mg-clays. The results were consistent with the mineralogy of the samples. Thus, the kerolite/Mg-smectite mixed layer samples exhibit the highest percentages of MgO and the lowest percentages of Al_2O_3. The SiO_2/MgO molar ratio is the lowest, between 1.25 and 1.46. MgO content in Ad-5 is also significant, although this clay has a higher Al_2O_3 content than the kerolite/Mg-smectite mixed layer samples. Bentonite has the highest Al_2O_3 content but the lowest MgO content, resulting in a SiO_2/MgO molar ratio of 2.25. The levels of aluminum and titanium in Ad-5 and Ad-8 correspond to the illite phase, which is absent in the other samples.

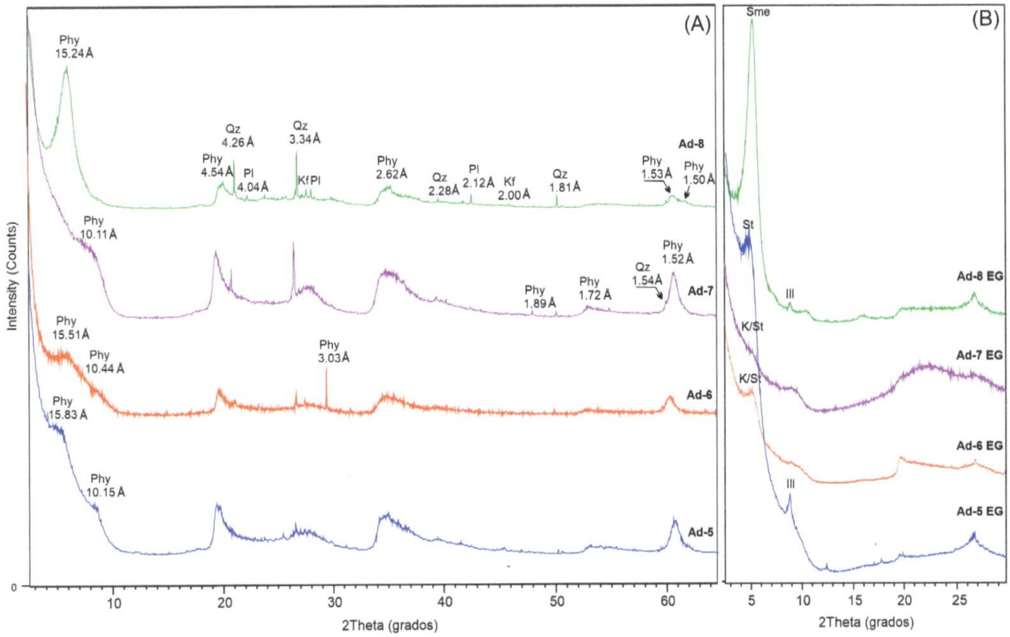

Figure 1. Powder X-ray patterns of the Mg-clays: (**A**) X-ray patterns of <2 μm grain size fractions, oriented ethylene glycol solvated samples (**B**) Phy: phyllosilicates; Sme: smectite; Qz: quartz; Kf: potassium feldspar; Pl: plagioclase; K/St: kerolite–stevensite mixed layer. St: stevensite; Ill: illite.

Table 2. Bulk (% w/w) and clay mineralogy (% w/w) of the Mg-clays determined by XRD. Phy: phyllosilicates, Qz; quartz; Kf: potassium feldspar; Pl: plagioclase; Sme: smectite; K/St kerolite–stevensite mixed layer; Ill: illite. Tr: trace.

Samples	Bulk Mineralogy				Clay Fraction		
	Phy	Qz	Kf	Pl	Sme	Ill	K/St
Ad-5	99	<1	<1	-	87	13	-
Ad-6	99	1	Tr	-	-	-	$K_{20}St_{80}$
Ad-7	98	2	-	-	-	-	$K_{80}St_{20}$
Ad-8	96	2	1	1	97	3	-

Table 3. Major element chemistry of studied non-fibrous Mg-clays.

Chemical Analysis (%)	Ad-5	Ad-6	Ad-7	Ad-8
SiO_2	50.15	51.43	53.17	48.23
Al_2O_3	4.96	2.54	1.27	9.37
Fe_2O_3	1.92	0.63	0.46	3.03
MnO	0.025	0.013	0.006	0.045
MgO	23.09	24.79	27.09	14.37
CaO	0.54	0.96	0.38	0.87
Na_2O	0.09	0.08	0.06	0.22
K_2O	0.91	0.6	0.34	1.11
TiO_2	0.233	0.115	0.058	0.363
P_2O_5	0.08	0.03	<0.01	0.07
LOI	18.15	17.34	16.09	20.57

These results, together with the test conducted [65] to differentiate between the different magnesian smectites, allowed us to identify a predominance of stevensite in sample

Ad-5 and in the kerolite/Mg-smectite mixed layer, while the composition of the Mg-smectite present in the bentonite is a saponite.

The infrared spectra of the samples formed by stevensite and kerolite/stevensite mixed layer encompassed a broad band in the range of 3000–3400 cm^{-1} with maximum absorption at 3451 cm^{-1}, presenting a small shoulder at a somewhat higher frequency corresponding to the OH-stretching bands of adsorbed and interlaminar water, respectively (Figure 2). The presence of an absorption band at 3677 cm^{-1} is characteristic of trioctahedral minerals and corresponds to the OH-Mg-Mg stretching band [66]. At lower frequencies, an intense absorption band at 1019 cm^{-1} is typical of silicates (Si-O stretching band), while the band at 667 cm^{-1} confirms the presence of trioctahedral minerals, as it corresponds to the Si-O-Mg deformation band (Figure 2). The presence of quartz was observed in all samples (around 784 cm^{-1}), which is in line with the XRD results.

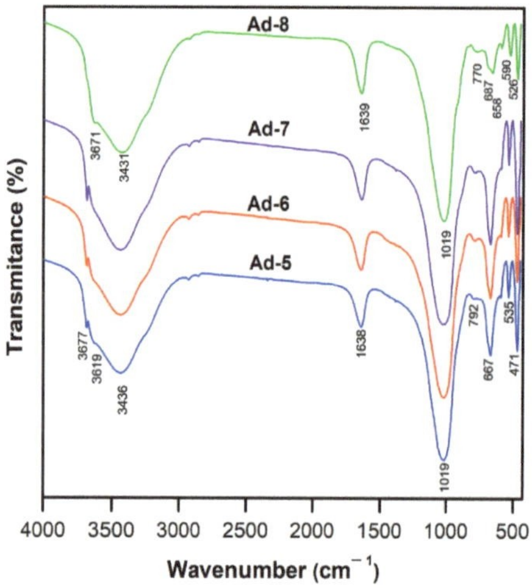

Figure 2. FT-IR spectra of the bulk samples of Mg-clays. Wavenumber peaks are indicated.

The infrared spectrum of the bentonite sample was very similar to that of the stevensite and kerolite/stevensite mixed-layer samples. Between 2000 and 4000 cm^{-1}, the large absorption band at 3428 cm^{-1} in Ad-8 coincides with that present between 3431 and 3434 cm^{-1} in the other samples (Figure 2) [66]. The presence of the effect at 3617 cm^{-1} was indicative of the presence of Al in the octahedral layer. Between 2000 and 450 cm^{-1}, the absorption bands around 1636 cm^{-1} and the intense one at 1017 cm^{-1} coincide, as well as those at 667 cm^{-1} and 469–470 cm^{-1}. A small absorption peak at 590 cm^{-1} and the existence of a small band at 658 cm^{-1} with a shoulder at 687 cm^{-1} stood out in Ad-8, which contrasts with a better-defined peak at 667 cm^{-1} in the other samples (Figure 2).

From the point of view of thermal behavior (Figure 3), the samples formed by the stevensite and kerolite/stevensite mixed layer behaved similarly. The first endothermic peak was observed at low temperatures (66–74 °C), which is attributed to the loss of zeolitic water: water adsorbed on the surface and hydration of the exchange cations [67]. The second endothermic peak at high temperatures (803–810 °C) can be attributed to dehydroxylation followed by an exothermic peak (828–830 °C) due to the formation of a high-temperature phase. The total weight loss of the samples had a variation between 15 and 20%, with the sample with the highest loss featuring a mineralogy with a predominance of stevensite (Ad-5).

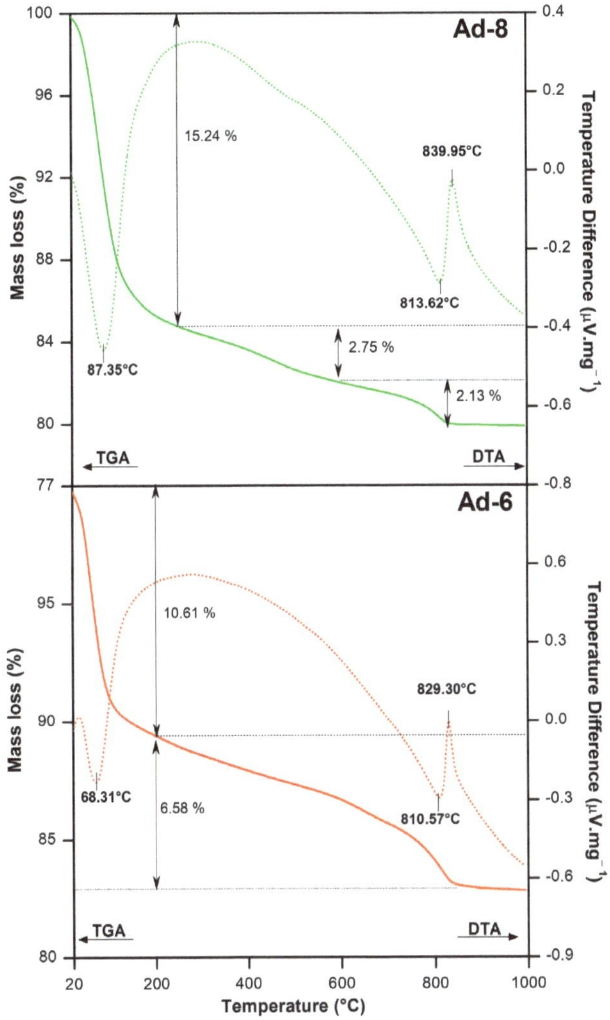

Figure 3. DTA (dotted lines) and TG (full lines) curves of the bentonite (Ad 8) and one representative kerolite/stevensite mixed layer (Ad 6).

The bentonite sample showed three endothermic peaks. One was at a low temperature (<90 °C), which is attributed to dehydration, loss of adsorbed water on external surfaces, and hydration of the exchange cations [67]. Another was at medium temperature (477 °C), assigned to dehydroxylation of the remaining coordination water and hydroxyls or structural water [68]. Finally, at high temperatures (813 °C), a third possibly corresponds to the residual dehydroxylation of the silanol groups [69]. The exothermic peak (828 °C) represents the destruction of the structure and formation of a high-temperature phase [69]. The total weight loss value of the sample is 20%.

The microfabric characteristics of the four Mg-clays studied are described below. The terminology for the microfabric type is that proposed by [70].

Ad-5. Matrix-skeletal microfabric with complex distribution of constituents and different types of porosity. Some morphologies with coatings and scattered mineral grains (Figure 4a). In wrinkled morphologies, there are abundant pores with diameters generally less than 10 μm. The particle size is small (<0.5 μm), and morphological features are characteristic of smectites.

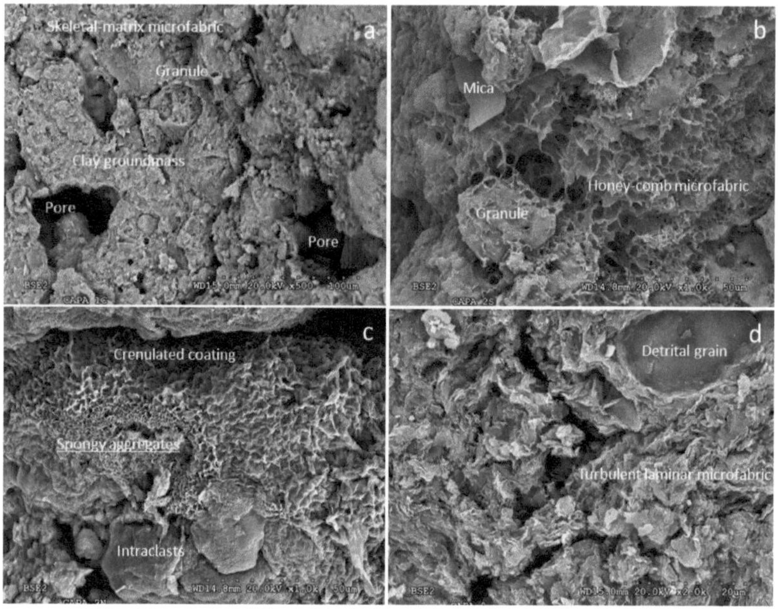

Figure 4. (**a**) Ad-5. Matrix-skeletal microfabric showing the presence of granules embedded in a clayey mass (matrix) with small pore sizes (up to 80 μm) related to bioturbation. (BSE); (**b**) Ad-6. Porous coatings on the clayey mass and granules showing a honeycomb microfabric, with some mica grains included. (BSE); (**c**) Ad-7. Spongy aggregates have been observed coating porosities or enveloping morphologies (granules and intraclasts) with wrinkled shapes on the surface. (BSE); (**d**) Ad-8. The clay has a turbulent laminar micro-fabric, where the micro-laminae are less than 5 μm thick and may envelop detrital grains. (BSE).

Ad-6. Matrix-skeletal-type microfabric, where granular morphologies of variable size and porous coatings on surfaces can be recognized. Different textural types are identified in the clayey material: one is massive with a clastic appearance, another has a large wrinkled appearance, and a third has smaller wrinkles made up of a very porous microfabric with face-edge and edge-edge (honeycomb) arrangements (Figure 4b). This sample presents laminar morphologies characteristic of smectite with sizes smaller than 2 μm that may be accompanied by sepiolite fibers.

Ad-7. Matrix-skeletal to sponge-like microfacies in areas where dense morphologies, detrital grains, and spongy clay films are recognized. The spongy aggregates show a high porosity of small size. Locally, there are thin clayey films formed by fibrous aggregates coating dense morphologies and spongy aggregates (Figure 4c). The sample presents very small morphologies (<0.5 μm) together with other more wrinkled morphologies of larger size.

Ad-8. A matrix-type microfabric, although locally, it may show a turbulent laminar microfabric formed by dense aggregates of lamellar particles that may show incipient fiber growth (Figure 4d). There are laminar morphologies of small size (0.5 μm) forming aggregates, characteristic of smectite. Locally incipient sepiolite fibers are observed.

CEC values were related to the expansible clay mineral content of the analyzed samples. The highest value was found in the bentonite sample (saponite, Ad-8), followed by the sample with a predominance of stevensite (Ad-5) (Table 4). In these two smectitic samples, CEC behavior was in agreement with saponite's higher layer charge with regard to stevensite. The kerolite/stevensite mixed layer showed values of 38.4 cmol(+)·Kg^{-1} (Ad-7), 48.6 cmol(+)·Kg^{-1} (Ad-6), evidencing a proportional relationship between the stevensite content and the increment in cation-exchange capacity.

Table 4. Physicochemical and textural properties of Mg-clay.

	Ad-5	Ad-6	Ad-7	Ad-8
CEC (cmol(+).Kg^{-1})	53	48.6	38.4	90.86
S_{BET} ($m^2.g^{-1}$)	240	244	280	203
S_{ext} ($m^2.g^{-1}$)	195	201	245	116
V_T ($cm^3.g^{-1}$)	0.24	0.31	0.35	0.17
$V_{\mu p}$ ($cm^3.g^{-1}$)	0.02	0.02	0.01	0.03
V_{mp} ($cm^3.g^{-1}$)	0.18	0.24	0.31	0.10

Figure 5 shows the N_2 adsorption–desorption isotherms at −196.15 °C. The isotherms were classified as type II according to the IUPAC classification, although the hysteresis cycles exhibit different classifications. Mg-clays Ad-5 and Ad-8 exhibit a type H4 cycle, which is associated with the presence of lamellar particle aggregates, common in smectite clays, whereas Ad-6 and Ad-7 display type II b cycles, characteristic of structures with interconnected pores, some of which are blocked during desorption [57]. This difference can be related to the presence of kerolite, given that its content in the mixed layer structure increased from 20% to 80%. The desorption behavior in samples with kerolite suggests greater pore interconnection. This observation is in agreement with the textural properties shown in Table 4, where these samples are predominantly mesoporous.

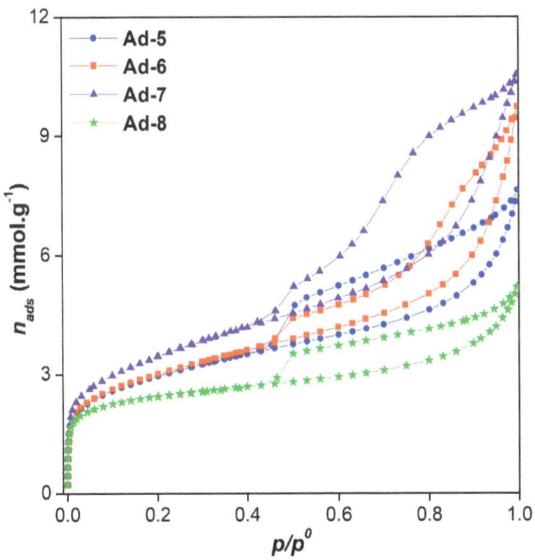

Figure 5. N_2 adsorption–desorption isotherms at −196.15 °C of Mg-clays.

The presence of kerolite appeared to enhance the specific surface area (S_{BET}) and mesoporosity, the latter closely related to the total pore volume. Moreover, the increasing amount of kerolite correlated well with changes in hysteresis loop patterns, suggesting that kerolite promotes interconnectivity within the mesoporous structure of the Mg-clays. The bentonite Ad-8, predominantly made up of trioctahedral smectite of saponite composition, exhibited textural properties typical of such minerals, with hysteresis cycles indicating mesopores accounting for 80% of the total pore volume.

When comparing these results across the four Mg-clays, the specific surface area (S_{BET}) values are notably higher for the kerolite/stevensite mixed layer compared to bentonite. Stevensite sample Ad-5 had an S_{BET} value higher than the saponite but lower than the kerolite/stevensite mixed-layer samples. This difference underscores how the presence of kerolite enhances porosity. The results obtained for S_{BET} values are in line with those previously reported for similar samples [46,71,72].

The water vapor adsorption isotherms are shown in Figure 6, with corresponding parameters detailed in Table 5. All Mg-clays display a water vapor adsorption behavior similar to that observed in swelling clay minerals. At low relative humidity (RH), Ad-7, which contains the highest proportion of kerolite, exhibited the lowest water adsorption. This can be attributed to the limited capacity of kerolite to retain water within its structure, unlike conventional clay minerals.

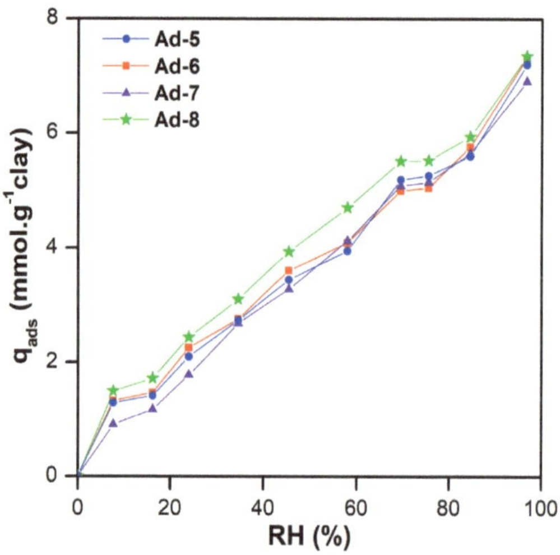

Figure 6. Water adsorption isotherms at 20 °C of Mg-clays.

Table 5. Parameters calculated from water adsorption isotherms for the four samples.

	Ad-5	Ad-6	Ad-7	Ad-8
NH (%)	5.10	3.56	3.05	3.58
q_{ads} ($mmol.g^{-1}$)	1.41	1.47	1.17	1.72
S ($m^2.g^{-1}$)	90.2	93.6	74.8	109.5

However, the bentonite Ad-8 displayed the highest water adsorption across the entire range studied, owing to its elevated smectite content of saponite composition [59]. An analysis of the parameters derived from the water adsorption isotherms revealed a direct relationship between the natural humidity content (NH) of kerolite/stevensite clays and the proportion of stevensite present. NH values had a high correlation with S_{BET}, suggesting that water vapor adsorbs not only to the interlayer space but also to the mesoporous structure. The amount of water adsorption and the surface area calculated at 16.3% RH demonstrated that Mg-clays containing equivalent amounts of magnesian smectite (stevensite) exhibit similar characteristics.

Bentonite Ad-8 showed the highest total water adsorption, although its HN value was not the highest due to its swelling nature, which causes a significant portion of the total water to be interlayered (absorbed). Additionally, the monolayer surface areas of all materials correlated with the amount of smectite present: Ad-8 exhibited the highest surface area, followed by Ad-5 and Ad-6 (both containing similar amounts of stevensite), with Ad-7 displaying the lowest.

This reverse order observed in the nitrogen adsorption–desorption isotherms indicates that while the presence of kerolite impacts the porous structure of the Mg-clays, it decreases their affinity for polar molecules such as water.

3.2. Adsorption Studies

The kinetic studies for both CPX and LID on Mg-clays are shown in Figure 7. The adsorption processes quickly achieved equilibrium removal at 120 min in all analyzed systems.

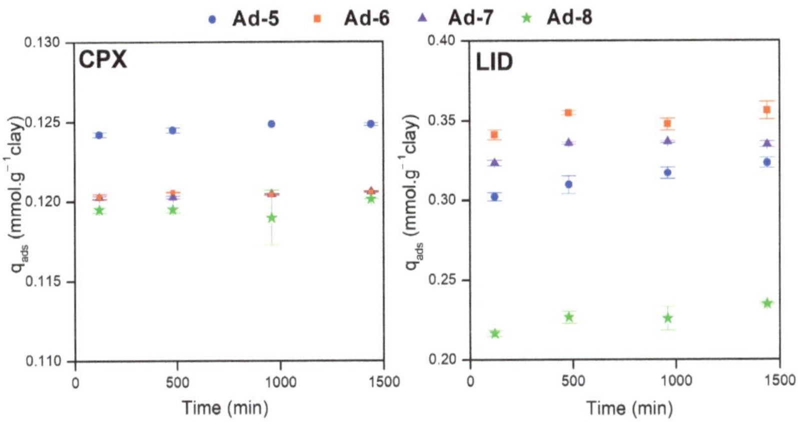

Figure 7. Kinetic studies for CPX and LID on Mg-clays at 20 °C.

The results showed that for CPX, the adsorbed amounts were similar in the four Mg-clays. Conversely, the amount of LID adsorption varied among the analyzed Mg-clays, the bentonite being the sample with the lowest values. This could suggest an influence of porous structure on LID molecule adsorption. This information indicated that all adsorption data obtained in this study can be considered obtained under equilibrium conditions.

In order to explain the adsorption results, it is important to consider that both CPX and LID are molecules with protonable groups and are pH-dependent in a solution [60,63]. In the studied conditions, the solutions were maintained at pH values close to 5, causing the predominant species of CPX to be the cationic form (CPX^{+}), with a minor fraction existing as the zwitterion ($CPX^{\pm}$), whereas for LID, the predominant species was its cationic form (LID^{+}).

Figure 8 presents the experimental results and fitted isotherms for CPX and LID adsorbed on the Mg-clays. It was observed that CPX adsorption isotherms were L-type (Langmuir) for Ad-5 and H-type (high-affinity) for the other materials [73]. The L-type classification for Ad-5 suggests that the predominant species of CPX present in solution were adsorbed onto surface sites until saturation. On the other hand, H-type isotherms indicate that the predominant CPX species in solution exhibits a strong affinity toward the surface, evidenced by a much higher slope at initial concentrations. Both L- and H-type isotherms are associated with the adsorption of ionic solutes on materials and suggest that there is no significant competition among CPX species and the solvent for the available sites on the surface.

The Langmuir, Freundlich, and Sips isotherms were used to model the experimental data (Table 6). The Freundlich model correlated better with the experimental data of Ad-5 and Ad-7 samples, while the Sips model was the best fit for the experimental data of Ad-6 and Ad-8 samples. In any case, these models suggested that adsorption occurred in heterogeneous systems.

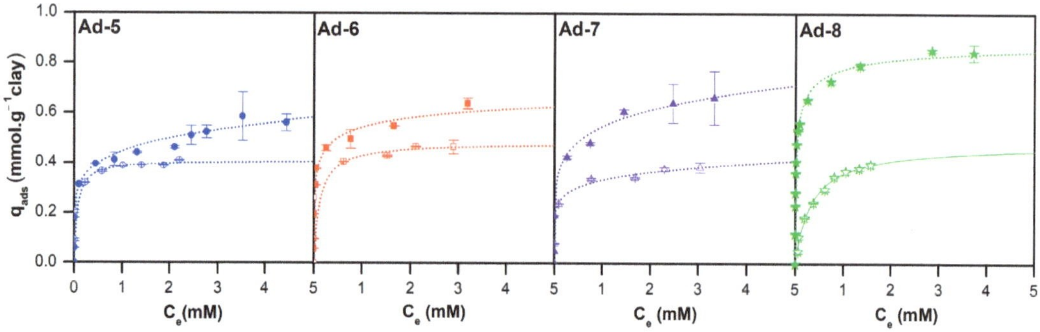

Figure 8. Adsorption isotherms of CPX (full points) and LID (empty points) at 20 °C on Mg-clays and their best fits.

Table 6. Langmuir, Freundlich, and Sips parameters for CPX and LID adsorption on Mg-clays.

		Ad-5		Ad-6		Ad-7		Ad-8	
		CPX	LID	CPX	LID	CPX	LID	CPX	LID
Langmuir	q_m (mmol g^{-1})	0.52	0.41	0.56	0.48	0.65	0.37	0.76	0.45
	k (L $mmol^{-1}$)	13.5	16.4	47.9	8.22	11.9	19.3	75.4	3.00
	R^2	0.79	0.87	0.80	0.87	0.83	0.86	0.92	**0.99**
Freundlich	k_f (L g^{-1})	**0.45**	0.38	0.53	0.41	**0.55**	**0.33**	0.74	0.35
	n	**6.34**	6.92	6.18	5.93	**6.11**	**7.63**	6.01	2.34
	R^2	**0.99**	0.89	0.92	0.90	**0.99**	**0.88**	0.92	0.95
Sips	q_m (mmol g^{-1})	1.81	**0.41**	**0.80**	**0.48**	1.26	0.57	**0.91**	0.51
	b (L $mmol^{-1})^{1/n}$	0.01	**16.4**	**10.2**	**8.22**	0.34	3.96	**32.1**	2.57
	n	4.39	**1.00**	**3.01**	**1.00**	4.07	3.40	**1.95**	1.09
	R^2	0.96	**0.99**	**0.96**	**0.99**	0.98	0.84	**0.99**	0.98

The Ad-8 bentonite exhibited the highest CPX-adsorption capacity, at 0.911 mmol·g^{-1}, equivalent to its CEC value (0.909 mmol·g^{-1}). In coherence with this, it also exhibited the highest smectite content of saponite composition, which may have favored CPX^+ adsorption mainly through cationic exchange, which is the main adsorption mechanism reported for other smectite clay minerals [74].

For the kerolite/stevensite mixed layer, CPX adsorption was higher for Ad-6 than Ad-7, exhibiting adsorbed amounts exceeding their CEC values (Table 6). This would evidence that kerolite content favored CPX removal from the solution, which could be explained by two interrelated effects. As previously observed, kerolite/stevensite clays developed mesoporosity in their structure, potentially increasing the availability of surface sites such as broken edge sites for adsorbing the zwitterionic species of CPX in solution. The $CPX^{\pm}$ species presented a protonated amino group in its structure, while the carboxyl group was deprotonated, resulting in a neutral total charge that decreased its adsorption on the negative structure. The presence of the deprotonated carboxylate group could favor CPX interaction with the broken edge groups on the surface to form inner-sphere complexes, increasing the total amount adsorbed for these Mg-clays [39,52]. This was also observed by other authors in similar systems using pillared clays as adsorbents, which are characterized by a micro-mesoporous structure that favors CPX adsorption [30,50,51]. This suggests that in these non-swelling clay minerals, CPX could be adsorbed both as cation species on the negative charged surface and in zwitterion form on reactive sites located on the broken edge of the 2:1 structure. However, more specific determinations, such as the effect of pH and ionic strength on adsorption, are needed to elucidate the adsorption mechanisms of these clays.

Stevensite sample Ad-5 presented the lowest adsorption capacity. This could be explained by a much lower CEC than the bentonite Ad-8 and the absence of a kerolite phase that could have enhanced the adsorption, as explained above.

Adsorption isotherms for LID adsorption on Mg-clays were L-type (Langmuir) in Ad-5, Ad-6, and Ad-8 samples, while in the Ad-7 sample, these were H-type according to the classification proposed by Giles et al. [73]. Again, this would be evidence that ionic species adsorption, in this case, LID^+, occurred without competing with water for the solid surface sites.

The adsorption parameters of Freundlich, Langmuir, and Sips models are presented in Table 6. The quadratic correlation values indicated in the good adjustment of samples Ad-5 and Ad-8 point to the Langmuir model and, in the case of Ad-6 and Ad-7, to the Freundlich model. These results suggest that in the Mg-clays with a higher smectite content, LID^+ adsorption took place in reactive sites of the same nature, while in the clays with kerolite content, adsorption occurred on reactive sites of a different nature.

The maximum adsorption capacity (q_m) for LID^+ on the analyzed clays, and therefore their removal efficiency, followed the order Ad-8 > Ad-6 > Ad-5 > Ad-7. Except for the bentonite Ad-8, the adsorbed amounts q_m obtained were equivalent to their CEC, suggesting that cation exchange is the main adsorption mechanism. In the case of the bentonite Ad-8 (97% saponite), the q_m value was lower than its CEC, suggesting that the LID molecule could occupy more than one available site in the saponite interlayer space, thus decreasing the number of sites. However, additional studies of Ad-8-LID adsorption complexes are needed in order to corroborate this.

All the samples showed a higher affinity and adsorption capacity for CPX than for LID in the studied conditions. This can be explained by considering the morphology and the size of the analyzed molecules as well as the Mg-clays properties. In the first place, CPX is a flat-shaped molecule with a lower thickness than LID. In addition, the positive charge of the CPX is located in its amine group situated at one end of the molecule, which would facilitate the interaction with the solids' surface. Meanwhile, in the LID molecule, voluminous carbon chains surround the protonated group, hindering interaction with the solid surface. Additionally, at the analyzed pH, the zwitterionic form of CPX is also present and therefore can be adsorbed on different surface reactive sites.

Taking all of this into account, bentonite Ad-8 exhibited the highest adsorption capacity for both CPX and LID under the studied conditions, which may be attributed to its higher smectite content of saponite composition and CEC value. This suggests that the primary mechanism of CPX and LID adsorption on the studied Mg-clays was cation exchange. This adsorption is likely facilitated by electrostatic forces between the positively charged CPX and LID molecules and the negative charge of the smectite component in the clays (saponite and stevensite).

However, in the case of CPX adsorption, the kerolite/stevensite mixed layer (Ad 6 and Ad-7) showed q_m values higher than their CEC values. In this case, the results suggested that $CPX^{\pm}$ present could be adsorbed through another type of interaction, such as inner-sphere complexes or hydrogen bonds involving its deprotonated carboxyl group and -Si-OH or -Mg-OH groups exposed in broken edge sites. Such sites are associated with the porosity generated by the presence of kerolite in those samples, thereby increasing their adsorption capacity relative to their CEC values.

Finally, for a comparative performance assessment of the studied materials for CPX adsorption from aqueous media, the reported adsorption capacities for other natural clays are summarized and presented in Table 7. According to other authors, kaolinite clays feature the lowest adsorption capacities, followed by montmorillonite clay minerals (Na and Ca type). The studied Mg-clays in samples Ad-5 and Ad-7 presented values similar to montmorillonite types, but the values obtained in samples Ad-6 and Ad-8 surpassed them. This suggests that the composition of the mixed layer K_{20}/St_{80} of Ad-6 adsorbed as much as a bentonite clay (Ad-8) with high purity (97% saponite).

Table 7. Adsorbed reports for CPX on natural clays.

Adsorbent	q_m CPX (pH) (mmol.g^{-1})	Adsorbent dosage (g.L^{-1})	Reference
kaolinite	0.02 (5–6)	0.2	[75]
Na-Bentonite from Turkey	0.44 (4.5)	2.5	[76]
Na-Montmorillonite from Argentine	0.42 (6)	2.5	[60]
Ca-bentonite	0.49 (6.5)	0.2	[12]
Clay 4 (Na-montmorillonite type)	0.58 (6–7.6)	1.0	[77]
Clay 2 (Kaolin type)	0.09 (6–7.6)		
Ad-5	0.44	2.5	This work
Ad-6	0.80		
Ad-7	0.55		
Ad-8	0.91		

The LID adsorption capacities were not included because no values have been previously reported for this molecule.

4. Conclusions

In this study, four Mg-clays were evaluated as potential adsorbents for the removal of two pharmaceutical pollutants: CPX and LID. Mineralogical analysis revealed that two of the clays are composed of a kerolite/stevensite mixed layer, the third is predominantly stevensite, and the fourth is classified as a bentonite with 97% saponite as its primary mineral component. Physicochemical characterization showed a strong correlation between the percentage of swelling clay minerals and the cation-exchange capacity (CEC) values. Textural properties indicated a significant influence of kerolite content, with a close relationship observed between the amount of kerolite and the percentage of mesoporosity.

Batch adsorption tests demonstrated that all materials are suitable for CPX and LID removal from water under the studied conditions. The results showed that for smectitic samples (saponite and stevensite), CPX adsorption primarily occurs through ionic exchange. Conversely, for samples containing kerolite, porosity plays a critical role by providing additional reactive sites within the porous structure and at the broken edges of clay particles. LID adsorption also appears to occur via ionic exchange, but due to the molecular shape of LID, this EC is more effectively accommodated in samples containing kerolite due to its higher pore volume.

This study not only underscores the potential of non-fibrous Mg-clays as adsorbents for emerging contaminants but also establishes a link between their composition, textural properties, and contaminant retention capacity. This understanding is essential for optimizing their application in various industrial contexts.

Author Contributions: Conceptualization: M.E.R.-J., M.P., T.M., M.T.B., G.P., L.V. and A.P.-A.; methodology: M.E.R.-J., V.R.-A., M.S. and A.M.; formal analysis: V.R.-A., M.S., A.M., M.E.R.-J., T.M., G.P. and M.P; investigation: M.P., T.M., G.P., M.T.B. and M.E.R.-J.; resource: M.E.R.-J. and M.P.; data curation: V.R.-A., M.S., A.M., T.M., G.P. and M.P. writing—original draft preparation: M.E.R.-J., T.M. and M.P.; writing—review and editing: M.E.R.-J., T.M., M.T.B. and M.P.; project administration: M.E.R.-J. and M.P.; funding acquisition: M.P., L.V., M.T.B. and G.P. All authors have read and agreed to the published version of the manuscript.

Funding: This research was support from supported by the Autonomous University of Madrid (FUAM.2023/006) and the National University of Comahue (Argentina) under the auspices of Project I04-249, titled "Natural and Modified Clay Minerals for Applications in Health and Environment", within the Faculty of Engineering.

Data Availability Statement: Data will be made available on request.

Acknowledgments: The authors gratefully acknowledge Minersa Group, the National University of Comahue (UNCo), PROBIEN-CONICET-UNCo, and the Autonomous University of Madrid for providing technology and financial support.

Conflicts of Interest: The authors declare no conflicts of interest.

References

1. Nikolenko, O.; Pujades, E.; Teixidó, M.; Sáez, C.; Jurado, A. Contaminants of emerging concern in the urban aquifers of Barcelona: Do they hamper the use of groundwater? *Chemosphere* **2023**, *341*, 140023. [CrossRef] [PubMed]
2. Gil, A.; Santamaría, L.; Korili, S.A.; Vicente, M.A.; Barbosa, L.V.; de Souza, S.D.; Marçal, L.; de Faria, E.H.; Ciuffi, K.J. A review of organic-inorganic hybrid clay based adsorbents for contaminants removal: Synthesis, perspectives and applications. *J. Environ. Chem. Eng.* **2021**, *9*, 105808. [CrossRef]
3. Gould, S.L.; Winter, M.J.; Norton, W.H.J.; Tyler, C.R. The potential for adverse effects in fish exposed to antidepressants in the aquatic environment. *Environ. Sci. Technol.* **2021**, *55*, 16299–16312. [CrossRef] [PubMed]
4. Reichert, G.; Hilgert, S.; Fuchs, S.; Azevedo, J.C.R. Emerging contaminants and antibiotic resistance in the different environmental matrices of Latin America. *Environ. Pollut.* **2019**, *255*, 113140. [CrossRef]
5. Kakakhel, M.A.; Nishita, N.; Navish, K.; Seyed Ali, J.; Syed Zaheer Ud, D.; Jiang, Z.; Khoo, K.S.; Shi, X. Deciphering the dysbiosis caused in the fish microbiota by emerging contaminants and its mitigation strategies—A review. *Environ. Res.* **2023**, *237*, 117002. [CrossRef]
6. Chaturvedi, P.; Shukla, P.; Giri, B.S.; Chowdhary, P.; Chandra, R.; Gupta, P.; Pandey, A. Prevalence and hazardous impact of pharmaceutical and personal care products and antibiotics in environment: A review on emerging contaminants. *Environ. Res.* **2021**, *194*, 110664. [CrossRef]
7. Lofrano, G.; Sacco, O.; Venditto, V.; Carotenuto, M.; Libralato, G.; Guida, M.; Meric, S.; Vaiano, V. Occurrence and potential risks of emerging contaminants in water. In *Visible Light Active Structured Photocatalysts for the Removal of Emerging Contaminants: Science and Engineering*; Elsevier: Amsterdam, The Netherlands, 2020; pp. 1–25.
8. Pulingam, T.; Parumasivam, T.; Gazzali, A.M.; Sulaiman, A.M.; Chee, J.Y.; Lakshmanan, M.; Chin, C.F.; Sudesh, K. Antimicrobial resistance: Prevalence, economic burden, mechanisms of resistance and strategies to overcome. *Eur. J. Pharm. Sci.* **2021**, *170*, 106103. [CrossRef]
9. Polianciuc, S.I.; Gurzău, A.E.; Kiss, B.; Georgia Ștefan, M.; Loghin, F. Antibiotics in the environment: Causes and consequences. *Med. Pharm. Rep.* **2020**, *93*, 231–240. [CrossRef]
10. Tran, N.H.; Hoang, L.; Nghiem, L.D.; Nguyen NM, H.; Ngo, H.H.; Guo, W.; Trinh, Q.T.; Mai, N.H.; Chen, H.; Nguyen, D.D.; et al. Occurrence and risk assessment of multiple classes of antibiotics in urban canals and lakes in Hanoi. *Vietnam. Sci. Total Environ.* **2019**, *692*, 157–174. [CrossRef]
11. Rúa-Gómez, P.C.; Püttmann, W. Impact of wastewater treatment plant discharge of lidocaine, tramadol, venlafaxine and their metabolites on the quality of surface waters and groundwater. *J. Environ. Monit.* **2012**, *14*, 1391–1399. [CrossRef]
12. Jara-Cobos, L.; Peñafiel, M.E.; Montero, C.; Menendez, M.; Pinos-Vélez, V. Ciprofloxacin Removal Using Pillared Clays. *Water* **2023**, *15*, 2056. [CrossRef]
13. Jara-Cobos, L.; Abad-Delgado, D.; Ponce-Montalvo, J.; Menendez, M.; Peñafiel, M.E. Removal of ciprofloxacin from an aqueous medium by adsorption on natural and hydrolyzed bentonites. *Front. Environ. Sci.* **2023**, *11*, 1239754. [CrossRef]
14. Igwegbe, C.A.; Oba, S.N.; Aniagor, C.O.; Adeniyi, A.G.; Ighalo, J.O. Adsorption of ciprofloxacin from water: A comprehensive review. *J. Ind. Eng. Chem.* **2021**, *93*, 57–77. [CrossRef]
15. Rúa-Gómez, P.C.; Püttmann, W. Occurrence and removal of lidocaine, tramadol, venlafaxine, and their metabolites in German wastewater treatment plants. *Environ. Sci. Pollut. Res.* **2012**, *19*, 689–699. [CrossRef] [PubMed]
16. Burch, K.D.; Han, B.; Pichtel, J.; Zubkov, T. Removal efficiency of commonly prescribed antibiotics via tertiary wastewater treatment. *Environ. Sci. Pollut. Res.* **2019**, *26*, 6301–6310. [CrossRef] [PubMed]
17. Ebrahimi, S.M.; Reyhani, R.D.; Asghari JafarAbadi, M.; Fathifar, Z. Diversity of antibiotics in hospital and municipal wastewaters and receiving water bodies and removal efficiency by treatment processes: A sys-tematic review protocol. *Environ. Evid.* **2020**, *9*, 19. [CrossRef]
18. Sharma, S.; Bhattacharya, A. Drinking water contamination and treatment techniques. *Appl. Water Sci.* **2017**, *7*, 1043–1067. [CrossRef]
19. Bangia, S.; Bangia, R.; Daverey, A. Pharmaceutically active compounds in aqueous environment: Recent developments in their fate, occurrence and elimination for efficient water purification. *Environ. Monit. Assess.* **2023**, *195*, 1–27. [CrossRef]
20. Mangla, D.; Annu; Sharma, A.; Ikram, S. Critical review on adsorptive removal of antibiotics: Present situation, challenges and future perspective. *J. Hazard. Mater.* **2022**, *425*, 127946. [CrossRef]
21. Choudhry, A.; Sharma, A.; Khan, T.A.; Chaudhry, S.A. Flax seeds based magnetic hybrid nanocomposite: An advance and sustainable material for water cleansing. *J. Water Process Eng.* **2021**, *42*, 102150. [CrossRef]
22. Varsha, M.; Senthil Kumar, P.; Senthil Rathi, B. A review on recent trends in the removal of emerging contaminants from aquatic environment using low-cost adsorbents. *Chemosphere* **2022**, *287*, 132270. [CrossRef] [PubMed]
23. Barakan, S.; Aghazadeh, V. The advantages of clay mineral modification methods for enhancing adsorption efficiency in wastewater treatment: A review. *Environ. Sci. Pollut. Res.* **2021**, *28*, 2572–2599. [CrossRef] [PubMed]
24. França, D.B.; Trigueiro, P.; Silva Filho, E.C.; Fonseca, M.G.; Jaber, M. Monitoring diclofenac adsorption by organophilic alkylpyridinium bentonites. *Chemosphere* **2020**, *242*, 125109. [CrossRef] [PubMed]
25. Tharmavaram, M.; Pandey, G.; Rawtani, D. Surface modified halloysite nanotubes: A flexible interface for biological, environmental and catalytic applications. *Adv. Colloid. Interface Sci.* **2018**, *261*, 82–101. [CrossRef] [PubMed]

26. Zhao, B.; Liu, L.; Cheng, H. Rational design of kaolinite-based photocatalytic materials for environment decontamination. *Appl. Clay Sci.* **2021**, *208*, 106098. [CrossRef]
27. Song, N.; Hursthouse, A.; McLellan, I.; Wang, Z. Treatment of environmental contamination using sepiolite: Current approaches and future potential. *Environ. Geochem. Health* **2021**, *43*, 2679–2697. [CrossRef]
28. Rusmin, R.; Sarkar, B.; Biswas, B.; Churchman, J.; Liu, Y.; Naidu, R. Structural, electrokinetic and surface properties of activated palygorskite for environmental application. *Appl. Clay Sci.* **2016**, *134*, 95–102. [CrossRef]
29. Maged, A.; Kharbish, S.; Ismael, I.S.; Bhatnagar, A. Characterization of activated bentonite clay mineral and the mechanisms underlying its sorption for ciprofloxacin from aqueous solution. *Environ. Sci. Pollut. Res.* **2020**, *27*, 32980–32997. [CrossRef]
30. Maggio, A.A.; Roca-Jalil, M.E.; Villarroel-Rocha, J.; Sapag, K.; Baschini, M.T. Fe- and SiFe-pillared clays from a mineralogical waste as adsorbents of ciprofloxacin from water. *Appl. Clay Sci.* **2022**, *220*, 106458. [CrossRef]
31. Meziti, C.; Zaibet, W.; Benaidj, Y. Removal of ibuprofen from water using recycled clay waste. *Tob. Regul. Sci. (TRS)* **2023**, *9*, 4793–4804.
32. Akyuz, S.; Akyuz, T.; Akalin, E. Adsorption of isoniazid onto sepiolite-palygorskite group of clays: An IR study. *Spectrochim. Acta A Mol. Biomol. Spectrosc.* **2010**, *75*, 1304–1307. [CrossRef] [PubMed]
33. Cuevas, J.; Dirocie, N.; Yunta, F.; Delgado, C.G.; Santamaría DE, G.; Ruiz, A.I.; Férnández, R.; Eymar, E. Evaluation of the sorption potential of mineral materials using tetracycline as a model pollutant. *Minerals* **2019**, *9*, 453. [CrossRef]
34. Junior, H.B.; da Silva, E.; Saltarelli, M.; Crispim, D.; Nassar, E.J.; Trujillano, R.; Rives, V.; Vicente, M.A.; Gil, A.; Korili, S.A.; et al. Inorganic–organic hybrids based on sepiolite as efficient adsorbents of caffeine and glyphosate pollutants. *Appl. Surf. Sci. Adv.* **2020**, *1*, 100025. [CrossRef]
35. Li, Y.; Tian, G.; Gong, L.; Chen, B.; Kong, L.; Liang, J. Evaluation of natural sepiolite clay as adsorbents for aflatoxin B1: A comparative study. *J. Environ. Chem. Eng.* **2020**, *8*, 104052. [CrossRef]
36. Gómez-Avilés, A.; Sellaoui, L.; Badawi, M.; Bonilla-Petriciolet, A.; Bedia, J.; Belver, C. Simultaneous adsorption of acetaminophen, diclofenac and tetracycline by organo-sepiolite: Experiments and statistical physics modelling. *Chem. Eng. J.* **2021**, *404*, 126601. [CrossRef]
37. Davoud, B.; Mohadeseh, Z.; Shaziya, S. Investigation of Adsorptive Properties of Surfactant Modified Sepiolite for Removal of Ciprofloxacin. *Int. J. Life Sci. Pharma Res.* **2020**, *10*, 12–19.
38. Damacena DH, L.; Trigueiro, P.; Monteiro, V.H.; Honorio LM, C.; Duarte, T.M.; Cunha, R.; Furtini, M.B.; Fonseca, M.G.; da Silva-Filho, E.C.; Osajima, J.A. Saponite-inspired Materials as Remediation Technologies for Water Treatment: An Overview. *Environ. Process.* **2023**, *10*, 15. [CrossRef]
39. Antón-Herrero, R.; García-Delgado, C.; Alonso-Izquierdo, M.; García-Rodríguez, G.; Cuevas, J.; Eymar, E. Comparative adsorption of tetracyclines on biochars and stevensite: Looking for the most effective adsorbent. *Appl. Clay Sci.* **2018**, *160*, 162–172. [CrossRef]
40. De Vidales JL, M.; Pozo, M.; Alia, J.M.; Garcia-Navarro, F.; Rull, F. Kerolite-Stevensite mixed-layers from the Madrid basin, Central Spain. *Clay Miner.* **1991**, *26*, 329–342. [CrossRef]
41. Pozo, M.; Casas, J. Origin of kerolite and associated Mg clays in palustrine-lacustrine environments. The Esquivias deposit (Neogene Madrid Basin, Spain). *Clay Miner.* **1999**, *34*, 395418. [CrossRef]
42. Temme, H.; Sohling, U.; Suck, K.; Ruf, F.; Niemeyer, B. Selective adsorption of aromatic ketones on kerolite clay for separation in biocatalytic applications. *Colloids Surf. A Physicochem. Eng. Asp.* **2011**, *377*, 290–296. [CrossRef]
43. Ulrich Sohling, F.; Haimeirl Agnes, H. Use of Stevensite for Mycotoxin Adsorption. U.S. Patent No. 8,221,807, 17 July 2012.
44. Ureña-Amate, M.D.; Socías-Viciana, M.; González-Pradas, E.; Saifi, M. Effects of ionic strength and temperature on adsorption of atrazine by a heat treated kerolite. *Chemosphere* **2005**, *59*, 69–74. [CrossRef] [PubMed]
45. Santarén, J. Properties and applications of magnesian clays. In *Magnesian Clays: Characterization, Origin and Applications*; Pozo, M., Galán, E., Eds.; AIPEA Educational Series; Pub. N° 2; Digilabs: Bari, Italy, 2015; pp. 123–174. ISBN 978-88-7522-093-8.
46. Elass, K.; Laachach, A.; Alaoui, A.; Azzi, M. Removal of methyl violet from aqueous solution using a stevensite-rich clay from Morocco. *Appl. Clay Sci.* **2011**, *54*, 90–96. [CrossRef]
47. Fernández, R.; Ruiz, A.I.; García-Delgado, C.; González-Santamaría, D.E.; Antón-Herrero, R.; Yunta, F.; Poyo, C.; Hernández, A.; Eymar, E.; Cuevas, J. Stevensite-based geofilter for the retention of tetracycline from water. *Sci. Total Environ.* **2018**, *645*, 146–155. [CrossRef] [PubMed]
48. González-Pradas, E.; Socías-Viciana, M.; Saifi, M.; Ureña-Amate, M.D.; Flores-Céspedes, F.; Fernández-Pérez, M.; Villafranca-Sánchez, M. Adsorption of atrazine from aqueous solution on heat treated kerolites. *Chemosphere* **2003**, *51*, 85–93. [CrossRef]
49. Socías-Viciana, M.M.; Tévar de Fez, J.; Ureña-Amate, M.D.; González-Pradas, E.; Fernández-Pérez, M.; Flores-Céspedes, F. Removal of chloridazon by natural and ammonium kerolite samples. *Appl. Surf. Sci.* **2006**, *252*, 6053–6057. [CrossRef]
50. Roca Jalil, M.E.; Baschini, M.; Sapag, K. Removal of ciprofloxacin from aqueous solutions using pillared clays. *Materials* **2017**, *10*, 1345. [CrossRef]
51. Roca Jalil, M.E.; Toschi, F.; Baschini, M.; Sapag, K. Silica pillared montmorillonites as possible adsorbents of antibiotics from water media. *Appl. Sci.* **2018**, *8*, 1403. [CrossRef]
52. Hu, Y.; Pan, C.; Zheng, X.; Liu, S.; Hu, F.; Xu, L.; Xu, G.; Peng, X. Removal of ciprofloxacin with aluminum-pillared kaolin sodium alginate beads (CA-Al-KABs): Kinetics, isotherms, and BBD model. *Water* **2020**, *12*, 905. [CrossRef]
53. Pozo, M.; Calvo, J.P. An overview of authigenic magnesian clays. *Minerals* **2018**, *8*, 520. [CrossRef]

54. Schultz, L.G. Quantitative Interpretation of Mineralogical Composition from X-ray and Chemical Data for the Pierre Shale. 1964. Available online: https://pubs.usgs.gov/pp/0391c/report.pdf (accessed on 6 August 2024).
55. Meier, L.P.; Kahr, G. Determination of the Cation Exchange Capacity (CEC) of Clay Minerals Using the Complexes of Copper (II) Ion with Triethylenetetramine and Tetraethylenepentamine. *Clays Clay Miner.* **1999**, *47*, 386–388. [CrossRef]
56. Villarroel-Rocha, J.; Barrera, D.; García Blanco, A.A.; Roca Jalil, M.E.; Sapag, K. Importance of the αs-plot Method in the Characterization of Nanoporous Materials: Enhanced Reader. *Adsorpt. Sci. Technol.* **2013**, *31*, 165–183. [CrossRef]
57. Thommes, M.; Kaneko, K.; Neimark, A.V.; Olivier, J.P.; Rodriguez-Reinoso, F.; Rouquerol, J.; Sing, K.S.W. Physisorption of gases, with special reference to the evaluation of surface area and pore size distribution (IUPAC Technical Report). *Pure Appl. Chem.* **2015**, *87*, 1051–1069. [CrossRef]
58. Hatch, C.D.; Wiese, J.S.; Crane, C.C.; Harris, K.J.; Kloss, H.G.; Baltrusaitis, J. Water adsorption on clay minerals as a function of relative humidity: Application of BET and Freundlich adsorption models. *Langmuir* **2012**, *28*, 1790–1803. [CrossRef]
59. Akin, I.D.; Likos, W.J.; Asce, M. Evaluation of Isotherm Models for Water Vapor Sorption Behavior of Expansive Clays. *J. Perform. Constr. Facil.* **2017**, *31*, D4016001. [CrossRef]
60. Roca Jalil, M.E.; Baschini, M.; Sapag, K. Influence of pH and antibiotic solubility on the removal of ciprofloxacin from aqueous media using montmorillonite. *Appl. Clay Sci.* **2015**, *114*, 69–76. [CrossRef]
61. Mohammed, A.-O. Ciprofloxacin: Analytical Profile. *Profiles Drug Subst. Excip. Relat. Methodol.* **2004**, *31*, 179–207.
62. Powell, M.F. Lidocaine and lidocaine hydrochloride. *Anal. Profiles Drug Subst.* **1986**, *15*, 761–779.
63. Roca Jalil, M.E.; Sanchez, M.; Pozo, M.; Soria, C.O.; Vela, L.; Gurnik, N.; Baschini, M. Assessment of natural and enhanced peloids from the Copahue thermal system (Argentina): Effects of the drying procedure on lidocaine adsorption. *Appl. Clay Sci.* **2020**, *196*, 105751. [CrossRef]
64. Eberl, D.D.; Jones, B.F.; Khoury, H.N. Mixed-layer kerolite/stevensite from the Amargosa Desert, Nevada. *Clays Clay Miner.* **1982**, *30*, 321–326. [CrossRef]
65. Christidis, G.E.; Koutsopoulou, E. A simple approach to the identification of trioctahedral smectites by X-ray diffraction. *Clay Miner.* **2013**, *48*, 687–696. [CrossRef]
66. Madejová, J.; Gates, W.P.; Petit, S. IR Spectra of Clay Minerals. *Dev. Clay Sci.* **2017**, *8*, 107–149.
67. Guggenheim, S.; Koster van Groos, A.F. Baseline studies of the clay minerals society source clays: Thermal analysis. *Clays Clay Miner.* **2001**, *49*, 433–443. [CrossRef]
68. Sakizci, M.; Erdoğan Alver, B.; Alver, Ö.; Yörükoğullari, E. Spectroscopic and thermal studies of bentonites from Ünye, Turkey. *J. Mol. Struct.* **2010**, *969*, 187–191. [CrossRef]
69. Grim, R.; Rowland, R. Differential thermail analysis of clay minerals and other hydrous materials. Part 1. *Am. Mineral. J. Earth Planet. Mater.* **1942**, *27*, 746–761.
70. Grabowska-Olszewska, B.; Osipov, V.; Sokolov, V. *Atlas of the Microstructure of Clay Soils*; Panstwowe Wydawnictwo Naukowe: Varsovia, Poland, 1984.
71. El Mouzdahir, Y.; Elmchaouri, A.; Mahboub, R.; ElAnssari, A.; Gil, A.; Korili, S.A.; Vicente, M.A. Interaction of stevensite with Cd^{2+} and Pb^{2+} in aqueous dispersions. *Appl. Clay Sci.* **2007**, *35*, 47–58. [CrossRef]
72. González-Santamaría, D.E.; López, E.; Ruiz, A.; Fernández, R.; Ortega, A.; Cuevas, J. Adsorption of phenanthrene by stevensite and sepiolite. *Clay Miner.* **2017**, *52*, 341–350. [CrossRef]
73. Giles, C.H.; Smith, D. A General Treatment and Classification of the Solute Adsorption Isotherm. *J. Colloid. Interface Sci.* **1974**, *47*, 755–765. [CrossRef]
74. Lambert, J.F. Organic pollutant adsorption on clay minerals. *Dev. Clay Sci.* **2018**, *9*, 195–253.
75. Li, Z.; Hong, H.; Liao, L.; Ackley, C.J.; Schulz, L.A.; MacDonald, R.A.; Mihelich, A.L.; Emard, S.M. A mechanistic study of ciprofloxacin removal by kaolinite. *Colloids Surf. B Biointerfaces* **2011**, *88*, 339–344. [CrossRef] [PubMed]
76. Genç, N.; Dogan, E.C.; Yurtsever, M. Bentonite for ciprofloxacin removal from aqueous solution. *Water Sci. Technol.* **2013**, *68*, 848–855.
77. Gutiérrez-Sánchez, P.; Hrichi, A.; Garrido-Zoido, J.M.; Álvarez-Torrellas, S.; Larriba, M.; Victoria Gil, M.; Ben Amor, H.; García, J. Natural clays as adsorbents for the efficient removal of antibiotic ciprofloxacin from wastewaters: Experimental and theoretical studies using DFT method. *J. Ind. Eng. Chem.* **2024**, *134*, 137–151. [CrossRef]

minerals

MDPI

Article

Enrichment Mechanism of Polymetallic Elements at the Base of the Niutitang Formation in Southeast Chongqing

Guozhi Wang [1,2,3], Can Zhang [3,4], Dayong Liu [1,2,3,*], Linfei Qiu [5], Ziying Li [5] and Ping'an Peng [1,2,3]

[1] State Key Laboratory of Organic Geochemistry, Guangzhou Institute of Geochemistry, Chinese Academy of Sciences, Guangzhou 510640, China; wangguozhi23@mails.ucas.ac.cn (G.W.); pinganp@gig.ac.cn (P.P.)
[2] CAS Center for Excellence in Deep Earth Science, Guangzhou 510640, China
[3] College of Earth and Planetary Sciences, University of Chinese Academy of Sciences, Beijing 100049, China; 17665052978@163.com
[4] Real Estate Registration Center of Zhaoqing City, Zhaoqing 526060, China
[5] Beijing Research Institute of Uranium Geology, China National Nuclear Corporation, Beijing 100029, China; qlf0602@163.com (L.Q.); liziying@briug.cn (Z.L.)
* Correspondence: liudayong@gig.ac.cn

Abstract: Polymetallic enrichment layers are commonly found at the base of the Lower Cambrian and extensively distributed across the Upper Yangtze Platform, yet their genetic models remain controversial. This study systematically collected samples from a typical section in the southeastern Chongqing region for mineral, organic, and inorganic analyses. It investigates the relationship between the abundance of various trace metal elements and organic matter at the base of the Niutitang Formation, as well as the vertical distribution characteristics of organic carbon isotopes and organic matter features. The results indicate that the Niutitang Formation shale exhibits a distinct three-part structure from bottom to top. Various metal elements are enriched in the lower interval, showing a close correlation between the abundance of polymetallic elements and the carbon isotopes of shale organic matter. The middle interval contains the highest TOC value and the lowest Ti/Al ratio, while the upper interval shows a significant decrease in organic matter abundance, with a clear positive correlation between the excess silicon content and Ti/Al ratio. Additionally, the mixing effect of deep-sea upwelling is the primary control on the formation of polymetallic enrichment layers in the lower interval, followed by the adsorption of organic matter under anoxic conditions. The sedimentary environment of the upper interval of the Niutitang Formation trends toward oxidation, with paleoclimate shifting toward colder and drier conditions, exhibiting aeolian sedimentary features that are unfavorable for the enrichment of trace metal elements. Consequently, upwelling is a key factor in the enrichment and mineralization of trace metal elements at the base of the Lower Cambrian in the Upper Yangtze region.

Keywords: Upper Yangtze region; Niutitang Formation; shale; polymetallic; upwelling; $\delta^{13}C_{org}$ value; excess silicon

1. Introduction

The Niutitang Formation, a Lower Cambrian unit in southern China, is a widely distributed, thick, and highly mature marine shale formation [1,2] in the Upper-Middle Yangtze Platform. The organic matter in the shale primarily originates from lower aquatic organisms and algae, classified as Type I organic matter [1,3–5]. The Niutitang Formation's abundant organic content makes it a key target for shale gas exploration and development. In terms of lateral distribution, it can be compared to the Qiongzhusi Formation in other regions and the Hetang Formation in the lower Yangtze region [6].

The Lower Cambrian Niutitang Formation is known for its abundance of polymetallic sulfides and rare metal elements, such as nickel, molybdenum, vanadium, and uranium [7,8]. This Formation has the potential to be a low-grade uranium deposit [9], as well as hosting other valuable metal deposits. Previous research has suggested that the enrichment of rare metal elements can be attributed to hydrothermal fluids brought by upwelling ocean currents [7,10], which also contribute to the accumulation of organic matter in the basin [9,11]. As a result, there is a positive correlation between the abundance of organic matter and rare metal elements in the shale of the Niutitang Formation [9].

One distinctive feature of the Niutitang Formation shale is the presence of "excess silica". However, the source of this excess silica is still a subject of intense debate. Some researchers argue that it is derived from hydrothermal processes [3,9,12,13], while others propose that it originates from biogenic sources [14–16]. The variability in the source of excess silica may also be influenced by co-deposition processes and subsequent tectonic activities. Previous studies have shown that Ti/Al may be closely related to aeolian deposits: Ti is often found in heavy minerals and serves as a parameter for arid climatic conditions. The positive correlation between the two in the Qiongzhusi Formation has been explained as a result of aeolian deposits bringing in more Ti and excess silica under relatively cold and dry climatic conditions [17,18]. Additionally, it can also serve as a parameter for sediment input from offshore distances under relatively humid climatic conditions.

$\delta^{13}C_{org}$ depletion has been associated with the expansion of microbially mediated carbon cycling processes in modern eutrophic lakes, high-productivity marine environments, and ancient sediments with high TOC deposited under euxinic conditions [19]. Similarly, negative $\delta^{13}C_{org}$ excursions have been reported in sediments from the Yangtze Platform and other global locations during the same period. This evidence points to the presence of bottom anoxic conditions on the Yangtze Platform during the late Cambrian Stage 2 to early Cambrian Stage 3 [20,21].

This study aims to analyze the variations and correlations of different parameters with depth along the profile of the Lower Cambrian Niutitang shale samples through relevant tests. By considering the paleogeographic location of the Lower Cambrian within the profile, the characteristics of the shale will be discussed, with a focus on the impact of upwelling events on sedimentary environments, trace metal elements, and organic matter characteristics.

2. Sample and Experiments

2.1. Geological Setting and Sample Details at the Sampling Site

The Danquan Profile is located in the Youxiu Fold Belt on the southeastern margin of the Sichuan Basin. It is characterized by open anticlines and tight synclines with compartmentalized folds. The profile is situated between two large-scale faults in the basin, creating a stable tectonic environment with rock layers that have gentle dip angles (Figure 1B).

During the Early Cambrian period in the Yangtze region, there was a notable gradation of sedimentary environments extending from northwest to southeast. This gradation began with shallow-water platform facies, transitioned through intermediary facies, and culminated in deep-water continental shelf and basin facies [22–24]. Notably, the shallow water platform facies emerged as the dominant environment. The chosen sampling site is strategically positioned in proximity to the transitional zone of the continental shelf slope (as depicted in Figure 1), a location that is particularly sensitive to sea-level fluctuations.

The underlying formation of the Niutitang Formation is the Maidiping Formation, which also belongs to the Lower Cambrian. This Formation consists of limestone with phosphatic nodules and is associated with a significant phosphogenesis event known as the Meishucun Period.

Mining activities in the area have led to the excavation of phosphate conglomerates from the Maidiping Formation, which are integrated with the base of the Niutitang Formation. This indicates that the lower part of the Niutitang Formation is fully exposed, and the collected samples are relatively fresh. A total of 31 samples were collected from the base of the Niutitang Formation, with a sampled thickness of 44 m. However, the uppermost part of the Niutitang Formation was not observed to be exposed in the area.

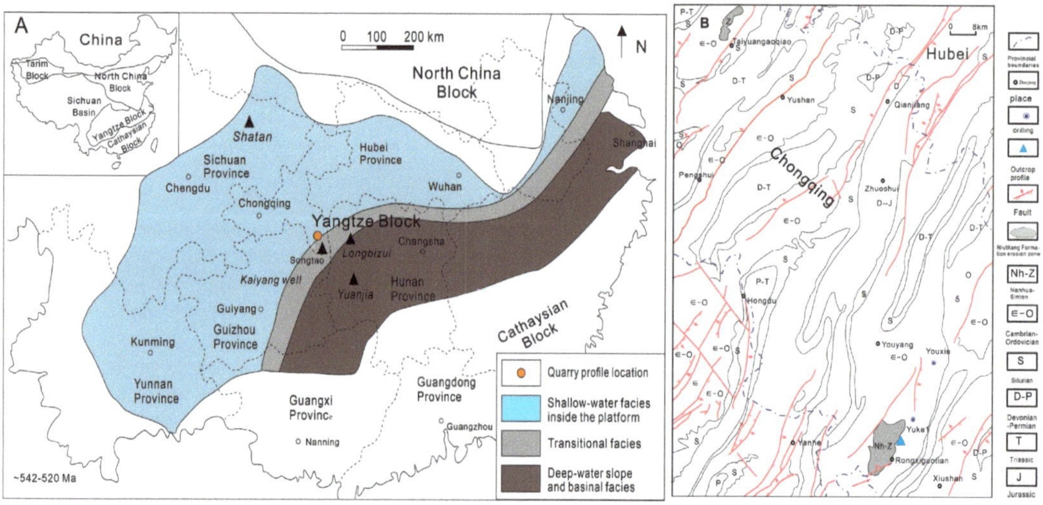

Figure 1. Simplified paleogeographic map [25] (**A**) and simplified geological map [26] (**B**) of the Yangtze Block around the Ediacaran-Cambrian boundary interval. The three black triangles in (**A**) represent typical profile locations in this area.

2.2. *Organic Geochemistry and Mineralogy Experiments*

TOC

The particle size of the samples is crushed to smaller than 100 mesh and then dried for 24 h at 60 °C. The accurately weighed samples are placed into a permeable crucible, and the total weight is recorded. Slowly add approximately 5% dilute hydrochloric acid, allowing the reaction to proceed for over 12 h. Subsequently, place the crucible with the samples onto a tray and immerse it in an 80 °C water bath for 1 h to ensure the complete removal of inorganic carbon from the shale samples. After heating, thoroughly rinse the samples with deionized water until the pH paper indicates neutrality. Additionally, rinse the tray, sample holder, and crucible to neutralize them. Once the samples are dried, add a tungsten-tin alloy combustion aid and iron powder to the surface of the treated samples in the crucible. The total organic carbon (TOC) content of the samples will be measured using the ELTRA CS 800 carbon-sulfur analyzer, ensuring a relative error of less than 2%.

Total sulfur (TS)

Before testing, the shale samples are cleaned, crushed, and ground to a particle size of 200 mesh (less than 74 μm). After drying, the powdered samples are wrapped in tin boats and placed into a VARIO EL cube elemental analyzer to obtain data with a precision of ±0.5%.

Organic carbon isotope ($\delta^{13}C_{org}$)

The sample amount is calculated to ensure a net carbon content of approximately 40 μg according to the TOC value. Then, the samples are wrapped in tin boats and measured using a Thermo Quest Finnigan DELTA plus XL isotope ratio mass spectrometer (IRMS) from Thermo Scientific, Waltham, MA, USA. The results are referenced to the VPDB standard, with an analytical error of ±0.5‰.

X-ray diffraction (XRD)

X-ray diffraction (XRD) analysis of shale can provide semi-quantitative information about the mineral content of the sample. The specific analysis method involves weighing a sample with a particle size of 200 mesh and vacuum drying it at 70 °C for 24 h. The dried sample is then analyzed for mineral content using a Bruker D8 Advance X-ray diffractometer (Bruker Corporation, Billerica, MA, USA).

The analysis is performed under the following conditions: a working voltage of 40 kV and a current of 30 mA. The scanning range is set from $2\theta = 3°$ to $85°$ with a slit width of 1 mm, and the scanning rate is 4°/min. The mineral content is calculated using the Lorenz polarization method, which corrects for peak areas of different minerals. This calculation provides semi-quantitative information about the minerals present in the sample.

2.3. Analysis of Major and Trace Elements

Analysis of Major Elements using Fusion Method: Approximately one gram of sample (with a particle size above 200 mesh) is weighed and incinerated at 1000 °C to eliminate organic matter. The remaining residue is then mixed with a flux containing lithium nitrate and other components, thoroughly blended, and subjected to high-temperature fusion. The molten material is poured into a platinum mold to form a flat glass slide, which is subsequently analyzed using a PANalytical PW2424X X-ray fluorescence spectrometer (Malvern Panalytical, Worcestershire, UK). The precision relative deviation (RD) is <5% and the accuracy relative error (RE) is <2%.

Trace element testing is conducted using two different methods: the dissolution method with HNO3-HClO4-HF-HCl and the fusion method with LiBO2/Li2B4O7.

Two dried samples of shale, each weighing 0.5 g and with a particle size above 200 mesh, are prepared.

In the dissolution method, one sample is digested using concentrated nitric acid, hydrochloric acid, and hydrofluoric acid. The mixture is then evaporated to near dryness and dissolved in diluted hydrochloric acid to a specific volume. The analysis is performed using an Agilent VISTA inductively coupled plasma emission spectrometer (Agilent Technologies, Santa Clara, CA, USA) and a Perkin Elmer Elan 9000 inductively coupled plasma mass spectrometer (PerkinElmer Corporation, Waltham, MA, USA).

In the fusion method, the other sample is mixed with a flux consisting of lithium metaborate and lithium tetraborate. The mixture is melted in a furnace at a temperature above 1025 °C, cooled, and then dissolved in nitric acid, hydrochloric acid, and hydrofluoric acid to a specific volume. The analysis is carried out using an inductively coupled plasma mass spectrometer.

Both instruments used in the analysis have a precision relative deviation (RD) of less than 10% and an accuracy relative error (RE) of less than 10%. This ensures reliable and accurate results in the trace element testing of the shale samples.

3. Results

After comparing various parameters, the Niutitang shale samples were divided into three intervals based on the Ti/Al and U/Th ratios. The lower and middle intervals exhibit relatively high U/Th values, with the lower interval displaying high Ti/Al values, while the middle interval shows noticeably lower Ti/Al values. In contrast, the upper interval has a significantly low U/Th value (Figure 2). The three intervals are roughly distributed from bottom to top at 0–8 m, 8–26 m, and 26–44 m.

Minerals: The mineral composition of the Lower Cambrian Niutitang Formation shale in the Danquan section is mainly composed of quartz, feldspar, and clay minerals. In the lower and middle intervals of the Niutitang Formation, which spans from 0 to 26 m, there are small amounts of pyrite and dolomite present. However, above 27 m from the bottom, pyrite and carbonate minerals are essentially absent (Figure 3A,B).

$\delta^{13}C_{org}$, TOC, and TS values: The lower interval of the Niutitang Formation is distinguished by negative $\delta^{13}C_{org}$ values, which show a gradual increase from −32.6‰ at the base to −31.0‰ towards the top. In the middle interval of the formation, $\delta^{13}C_{org}$ values demonstrate relative consistency, fluctuating within a narrow span from −31.4‰ to −31.1‰. Progressing to the upper interval, the $\delta^{13}C_{org}$ values vary from −31.7‰ to −31.0‰, overall exhibiting a positive trend with increasing depth (Figure 3C).

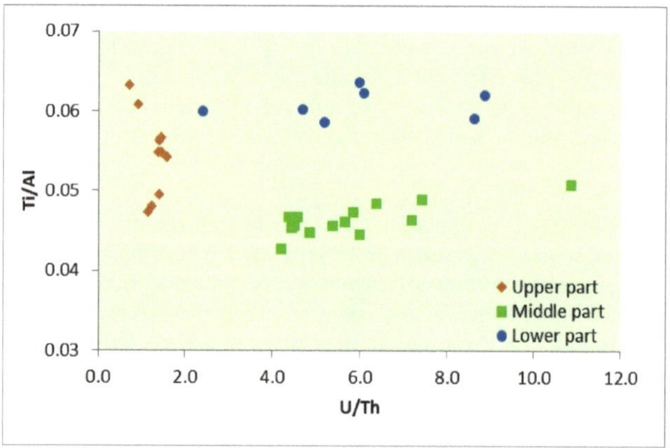

Figure 2. Correlation between Ti/Al and U/Th of Niutitang shale samples in the study area.

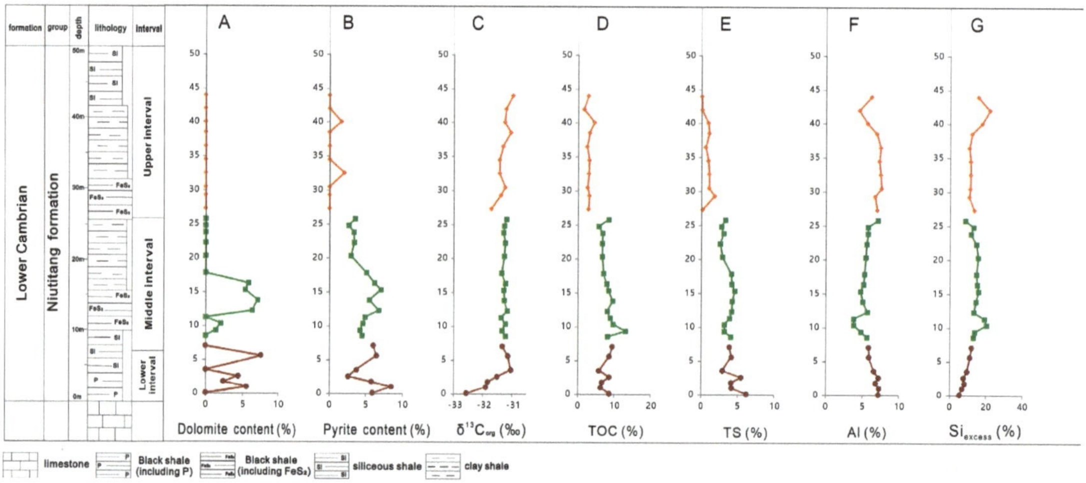

Figure 3. Mineral, $\delta^{13}C_{org}$,TOC and elemental content variations with depth in the Niutitang Formation profile of Youyang. And the minerals are dolomite (**A**); pyrite (**B**); indices related to organic matter are $\delta^{13}C_{org}$ (**C**) and TOC (**D**); indices related to elements are total sulfur (**E**); aluminum (**F**) and excess silicon (**G**).

In the Danquan section of the Niutitang Formation, the distribution of organic matter within the black shale displays marked stratigraphic variations. The middle to lower interval is characterized by a high concentration of organic matter, in contrast to the upper interval, which presents a comparatively lower abundance, as depicted in Figure 3D. These observations are consistent with the findings of Pi [7] and corroborate the stratigraphic division into the organically rich Niu2 interval and the organically lean Niu3 interval, as defined by Tan [3]. Additionally, the basal Niu1 interval, which is notable for its low total organic carbon (TOC) content, correlates with the underlying phosphatic dolomite-rich Maidiping Formation, providing a foundational context for the Niutitang Formation in this stratigraphic profile.

The variation trends of total sulfur content are similar to that of the TOC and pyrite content (Figure 3E), which indicates that sulfur is tightly associated with organic matter. The differences in TOC and total sulfur abundance are mainly attributed to the anoxic environment rich in sulfur in the lower-middle interval of the Niutitang Formation and the oxygen-deficient environment with low sulfur in the upper interval [7].

Aluminum element: The elevated aluminum concentrations, varying from 6.57% to 7.28% in the lower interval and from 4.83% to 7.88% in the upper interval, indicate a substantial terrestrial clastic contribution (Figure 3F). Conversely, the aluminum levels are primarily confined to a narrower range of 3.8% to 5.91% throughout the middle interval, with a modest peak of 7.08% at its uppermost portion, pointing to a reduced impact of terrestrial clastic influx in this section.

Excess silicon: The origins of silicon in shale can be categorized into three types: terrigenous silicon, hydrothermal silicon, and biogenic silicon. To calculate the excess silicon content ($Si_{\_excess}$), one must subtract the content of terrigenous detrital silicon using the formula:

$$Si_{\_excess} = Sis - [(Si/Al)bg \times Als] \quad (1)$$

Here, Sis represents the silicon element content in the sample, Als represents the aluminum element content in the sample, and (Si/Al)bg is 3.11 (the average content in shale) [27].

The silicon dioxide (SiO_2) content in the Niutitang Formation shale in the lower Cambrian of the DanQuan section in Youyang ranges from 51.4% to 71.1%. The $Si_{\_excess}$ varies from 5% to 22%. The lowest $Si_{\_excess}$ content is observed in the lower section of the Niutitang Formation, gradually increasing from 5.54% to 9.56% over a 5-m interval. The middle and upper sections show relatively higher $Si_{\_excess}$ content, ranging from 8.63% to 20.32% and from 10.69% to 21.99%, respectively (Figure 3G).

Trace elements: In the lower section, the concentrations of most trace metal elements, including arsenic (As), molybdenum (Mo), nickel (Ni), rhenium (Re), antimony (Sb), selenium (Se), and uranium (U), progressively diminish toward the upper part (Figure 4A–G). This trend is succeeded by a pronounced enrichment of metallic elements in the base of the middle interval. Conversely, the upper section shows a marked depletion of these trace elements.

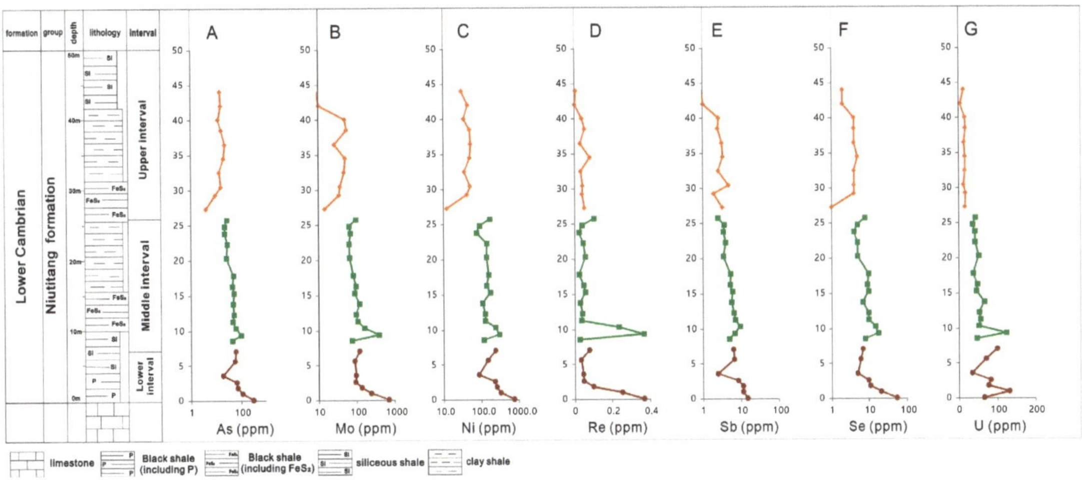

Figure 4. Trace elemental content variations with depth in the Niutitang Formation profile of Youyang. The trace elements are arsenic (**A**), molybdenum (**B**), nickel (**C**), rhenium (**D**), antimony (**E**), selenium (**F**) and uranium (**G**).

4. Discussion

4.1. Characteristics of the Sedimentary Environment of the Niutitang Formation Shale

4.1.1. Redox Conditions of Water Bodies

The U/Th ratio is a parameter used to assess the reducing conditions of water bodies, with higher U/Th values indicating stronger reducing capabilities of the water. This relationship has been established in previous studies [28–30].

In the lower interval of the Niutitang Formation, the samples generally exhibit anaerobic conditions, and most of the U/Th values are higher than 4 (except for the sample with a U/Th value of 2.4 at 3.6 m). Additionally, as the reducing conditions of the water increase, the abundance of U also increases. There is a positive correlation between total organic carbon (TOC) and U/Th (Figure 5A), indicating that stronger anaerobic conditions favor the enrichment of organic matter at middle intervals. However, in the lower interval, there is no correlation, which may be attributed to the influence of U from upwelling mixing sources on the U/Th ratio.

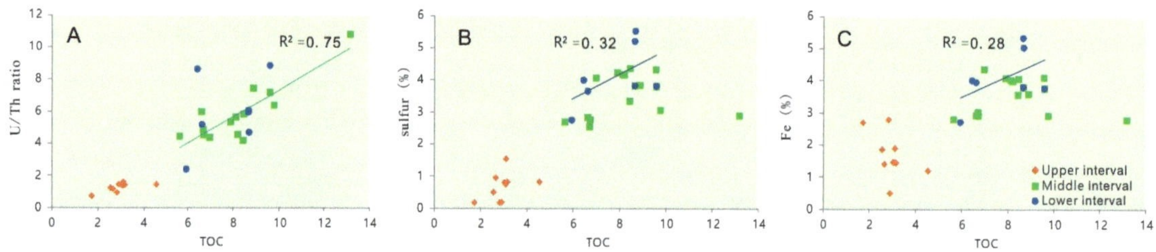

Figure 5. Correlation between TOC and U/Th ratio (**A**), sulfure content (**B**), and Fe content (**C**).

Previous studies have suggested that U/Th ratios greater than 1 indicate the influence of hydrothermal deposition during shale sedimentation [31,32]. This suggests that the lower and middle intervals of the profile may commonly be affected by hydrothermal fluids through the upwelling event, while the upper section of the Niutitang Formation shows a significant decrease in hydrothermal fluid influence.

In the lower interval of the Niutitang Formation shale, there is a weak positive correlation between TOC and total sulfur (R^2 = 0.32) and iron (R^2 = 0.28) (Figure 5B,C), indicating a rough genetic relationship between organic matter enrichment and pyrite. This relationship is attributed to the processes of anoxic sulfidation and deep-sea ironization resulting from hydrothermal mixing. Additionally, the negative correlation observed between $\delta^{13}C_{org}$ values and TS (Figure 6A), Fe (Figure 6B) in the lower section suggests that a decrease in hydrothermal $\delta^{13}C_{org}$ mixing leads to a gradual reduction in the abundance of S and Fe, supporting the theory that pyrite in this section is primarily of hydrothermal origin. On the other hand, the correlation between TOC and sulfur and iron in the middle section of organic-rich shales and the upper section of organic-poor shales is weak, suggesting that sources of S and Fe in these sections are not genetically linked to organic matter enrichment.

4.1.2. Terrestrial Input

The ratio of Ti/Al elements can be used to indicate the distance of the deposition zone from the coastline, with a higher Ti/Al ratio suggesting closer proximity to the coastline and vice versa [28,33]. Previous studies indicate that the Ti/Al ratio may be closely related to eolian sedimentation, and Ti is often found in heavy minerals and serves as a parameter for arid climate conditions [34]. The highest aluminum (Figure 7A) and titanium (Figure 7B) contents of the lower interval show that the palaeoenvironment of the lower interval is closest to the coastline. The middle interval exhibits a sharp decrease in the Ti/Al ratio at higher levels (Figure 7C), indicating a large-scale transgression event. The Ti/Al ratio is generally higher in the upper interval but still lower compared to the lower interval, with only the highest two samples showing a distance from the coastline and sea level similar to the bottom of the lower section.

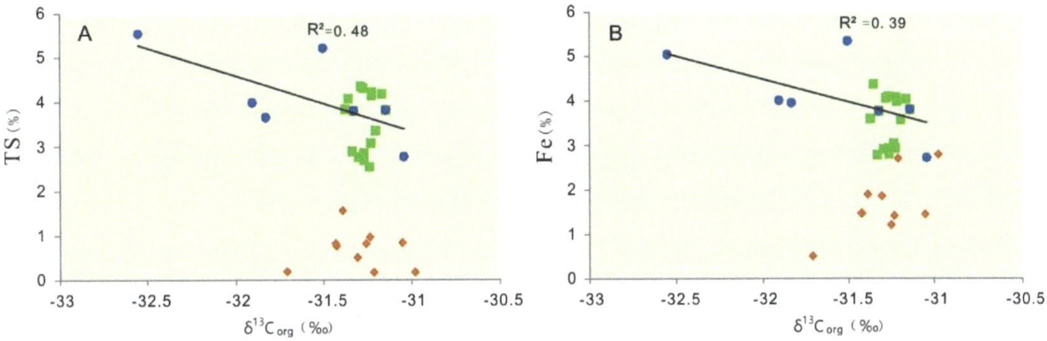

Figure 6. Correlation between $\delta^{13}C_{org}$ and sulfur (**A**) and Fe (**B**) concentrations.

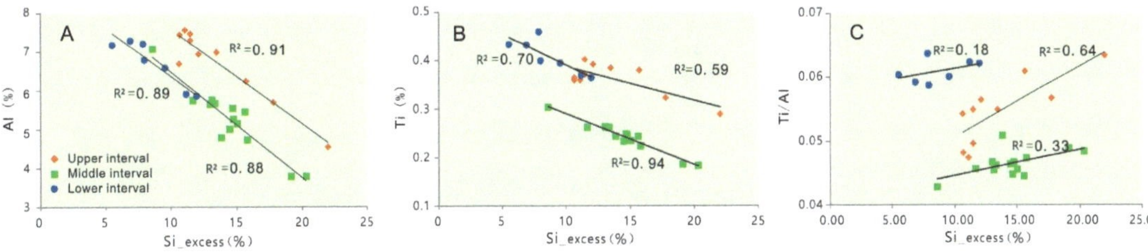

Figure 7. Correlation between $Si_{\_excess}$ and Ti concentration (**A**), Al concentration (**B**), and Ti/Al ratio (**C**).

The phosphatic dolomite of the Lower Maidiping Formation at the base of the Niutitang Formation was predominantly formed in a tidal flat environment. At the base of the Niutitang Formation, the mixing of polymetallic-rich fluids and ancient seawater, triggered by crustal extension, resulted in the deposition of abundant trace metal elements. The anaerobic conditions arising from this mixing, along with volcanic activity, led to the enrichment of polymetallic elements and the formation of shale with high TOC values. As a large-scale marine transgression followed, creating anaerobic conditions in the bottom water, the enrichment of organic matter played a crucial role in the adsorption of metal elements from seawater, serving as the primary factor in the enrichment of trace metal elements. Subsequently, during a marine regression, the shale in the upper section of the Niutitang Formation formed with relatively lower TOC and trace metal element content.

4.2. The Relationship between Negative Excursion of $\delta^{13}C_{org}$ Values and Enriched Polymetallic Elements in Lower Interval

Similar to previous studies [21], this study indicates no direct correlation between $\delta^{13}C_{org}$ and TOC (Figure 8A). However, strong correlations are observed between$\delta^{13}C_{org}$ and Mo/TOC, Ni/TOC, V/TOC, As/TOC (Figure 8B–E) as well as Sb content (Figure 8F), suggesting a close relationship between $\delta^{13}C_{org}$ and syngenetic polymetallic elements. In normal euxinic sediments, the Ni/TOC values are generally below 50 [7,35,36]. The extremely high Ni/TOC ratios in the samples of this study (ranging from 1114 to 8840) indicate the presence of enrichment mechanisms beyond normal biological accumulation. This mechanism could involve the release of nickel from hydrothermal sources into sulfide-rich bottom waters, leading to the direct formation of nickel sulfide precipitation [7]. Nickel is an important trace nutrient in water bodies and is transferred to sediments through microbial processes [7]. The correlation between Ni/TOC and $\delta^{13}C_{org}$ (R^2 = 0.88) suggests a direct causal relationship between nickel element enrichment and the $\delta^{13}C_{org}$.

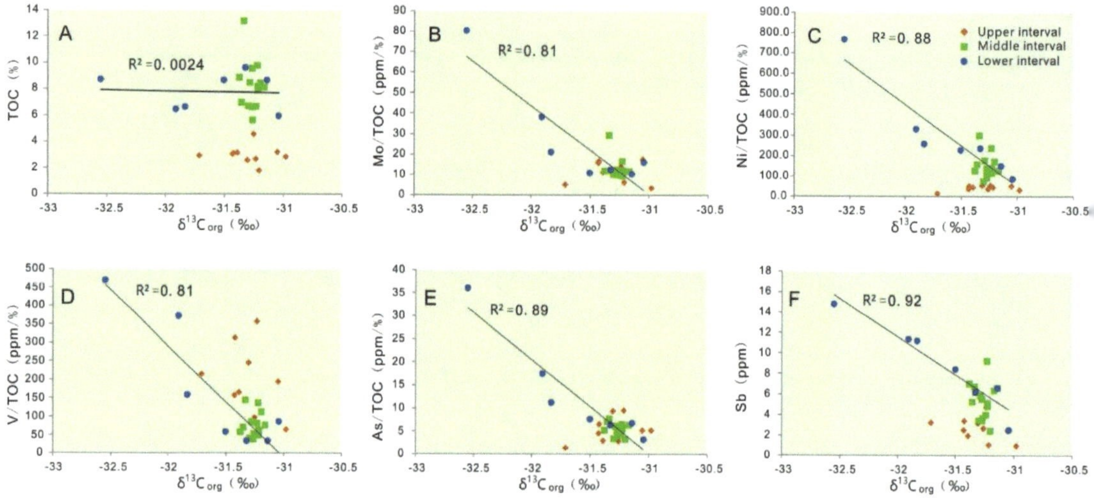

Figure 8. Correlation between $\delta^{13}C_{org}$ value with TOC and metallic parameters. The correlation between the $\delta^{13}C_{org}$ value and TOC is shown in (**A**), along with the polymetallic/TOC ratios: Mo/TOC (**B**), Ni/TOC (**C**), V/TOC (**D**), and As/TOC (**E**). (**F**) The correlation between the $\delta^{13}C_{org}$ value and antimony concentration.

The previous studies suggest a connection between the enrichment of As and Sb elements and hydrothermal fluids in the lower section of the Niutitang Formation in Guizhou and many other areas [24,37]. Similarly, in the profile of the Niutitang Formation, there is a strong correlation between the increasing $\delta^{13}C_{org}$ values from the bottom and the decreasing ratios of As/TOC (R^2 = 0.89) and Sb content (R^2 = 0.92). This correlation indicates that the negative $\delta^{13}C_{org}$ values are attributed to the influence of deep-seated hydrothermal fluids.

The $\delta^{13}C_{org}$ value of the lower section of the Niutitang Formation in the study area is −32.6‰, which is significantly lighter than the $\delta^{13}C_{org}$ value (−34.4‰) of the same stratigraphic level in the Songlin area of Guizhou [38]. This difference may be attributed to the proximity to the hydrothermal vent. The Songlin profile, located near the hydrothermal vent, exhibits a strong influence on hydrothermal activity, as evidenced by the presence of hydrothermal vent communities [39]. In contrast, the southeastern Chongqing profile shows a relatively weaker influence of hydrothermal mixing. Furthermore, the impact of hydrothermal activity gradually decreases from the bottom to the top of the southeastern Chongqing profile, ultimately becoming relatively stable in the middle section.

The occurrence of hydrothermal events and the release of methane and CO_2 gases have had a significant impact on Earth's climate and biogeochemical cycles. These events have led to changes in nutrient availability, with elements like phosphorus, nickel, iron, and vanadium being provided [40]. Metallic elements such as molybdenum, zinc, vanadium, and uranium are also released. Additionally, these events create a highly reducing environment in the bottom waters.

These hydrothermal activities have caused a notable negative shift in the $\delta^{13}C_{PDB}$ values of carbonate cement, with values as low as −48‰ being recorded. This shift in carbon isotopes has contributed to global warming [41,42]. The biogeochemical processes, particularly bacteria-mediated carbon cycling, play a crucial role in transforming inorganic carbon from hydrothermal sources into biogenic organic carbon. As a result, shale organic matter found in hydrothermally influenced shales exhibits significantly lighter $\delta^{13}C_{org}$ values.

The extensive outflow of deep-reducing hydrothermal fluids during the Ediacaran to Cambrian transition zone, which is rich in silicon and depleted in ^{13}C, can cause a negative anomaly in the carbon isotopes of organic matter. This process leads to the preservation of large quantities of organic matter in anoxic environments, giving rise to the formation of organic-rich shales.

Previous studies have revealed that there is a widespread occurrence of $\delta^{13}C_{org}$ value shifts in the lower part of the Lower Cambrian shale in the Yangtze Platform [21,43,44]. Furthermore, this negative shift in carbon isotopes observed in the Lower Cambrian of the Yangtze Platform can be globally correlated [45], as evidenced by similar observations in the Birmania Basin of northwestern India. In the Birmania Basin, the development of phosphorite is accompanied by a negative shift in organic carbon isotopes, followed by a subsequent positive shift near the Pc/C boundary, resulting in relatively heavier $\delta^{13}C_{org}$ values [46]. The findings of this study align with the observed evolutionary trend of $\delta^{13}C_{org}$ values.

From a tectonic perspective, during the late Cambrian's second stage, the Yangtze Block and the South China Block underwent rift activity as a result of the Rodinia supercontinent's breakup. This extensional tectonic activity in the Lower Cambrian allowed deep-seated hydrothermal fluids to enter the ocean basin through extensional fractures and subsequently ascend to the continental shelf [40]. Previous research has also indicated that the late Neoproterozoic to early Cambrian period was characterized by the breakup of Rodinia, leading to significant polar wander events across major continents on Earth [47,48], which led to intense extensional tectonic activity within the continents. As a result, the mixing of hydrothermal fluids exhibits a global consistency.

4.3. Relationship between Polymetallic Elements and $Si_{\_excess}$

The most vigorous polymetallic enrichment occurs at the bottom of the Niutitang shale, coinciding with the lowest levels of $Si_{\_excess}$ (Figure 9A), indicating that the polymetallic-enrich fluid originating from the depths is silicon-poor. Approximately 5 m above the base, there is a noticeable increase in $Si_{\_excess}$ content as the carbon isotopes in the organic matter become heavier, and the two exhibit a closely positive relationship (R^2 = 0.71) (Figure 5A) in the lower intervals. This trend suggests that moving upwards from the base, the mixing with polymetallic-rich fluids progressively weakens while the contribution of biogenic $Si_{\_excess}$ gradually increases. Accompanying this process is a reduction in the concentrations of hydrothermal-derived elements such as Ni, P, S, Se, and Re (Figure 9B–F).

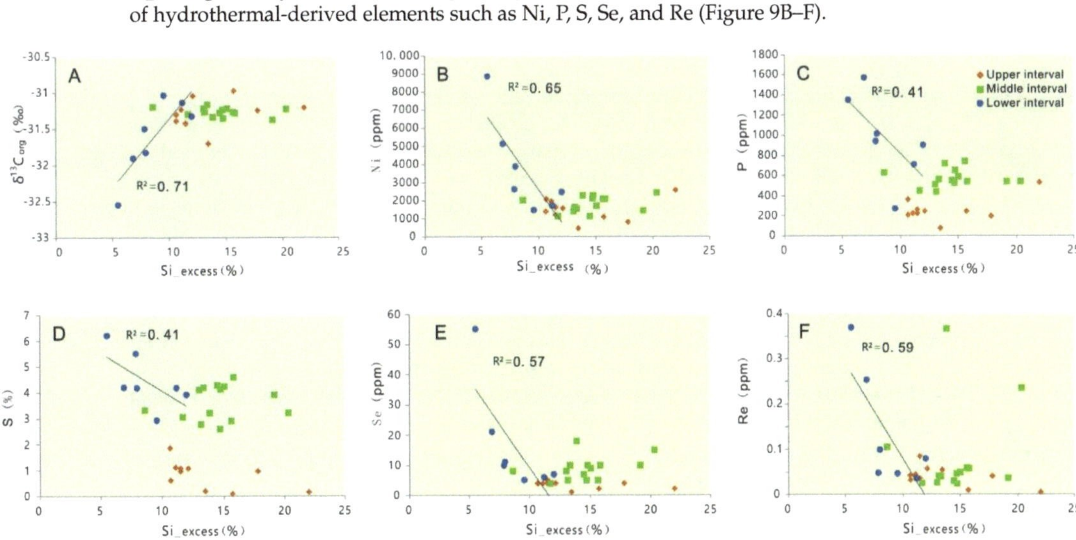

Figure 9. Correlation between $Si_{\_excess}$ with $\delta^{13}C_{org}$ value and concentration of elements. The correlation between the Si_excess and $\delta^{13}C_{org}$ value is shown in (**A**), along with the polymetallic concentrations: nickel (**B**), phosphorus (**C**), sulfur (**D**), selenium (**E**) and rhynium (**F**).

Previous studies have suggested that the $Si_{\_excess}$ content in Early Cambrian South China ranges from 20% to 30%, reaching up to 50%, with the $Si_{\_excess}$ primarily derived from hydrothermal fluids [9]. Contrary to the Lower Cambrian strata in South China as a whole, the $Si_{\_excess}$ content in this section is notably lower, and there is no clear correlation between TOC and excess silica. This suggests that the enrichment of organic matter is not directly linked to the $Si_{\_excess}$ content. In the western Hubei region, there is a significant positive correlation between $Si_{\_excess}$ and TOC in the Lower Cambrian shale, indicating that the $Si_{\_excess}$ is predominantly of biogenic origin [14].

4.4. Sources of Polymetals and the Migration Mechanisms Controlled by Upwelling

During the transition from the Neoproterozoic to the Cambrian, oceanic hypoxia led to the production of large amounts of organic matter and hydrogen sulfide (H_2S) in the ocean. When high-salinity brines reached the seafloor, rapid changes in Eh-pH conditions caused the accumulation of metals at the sediment-seawater interface, a process further enhanced during diagenesis [49,50]. Some studies suggest that multiple sources, including seafloor hydrothermal and biological contributions, may have contributed to the formation of polymetallic nickel-molybdenum-precious metal (PGE) mineralization in the black shales of the Lower Cambrian Niutou Formation in southern China [51,52]. Other research indicates that the deposition of vanadium-rich black shales in the Yangtze platform slope basin results from upwelling in deep-water basins, which first passes through mildly reducing, organic-rich North Sea Anoxic Seawater (NSASW), where vanadium (V) is reduced to vanadium (IV) and preferentially removed from seawater through the formation of organic metal complexes [53].

During the early Cambrian, hydrothermal and upwelling processes in the Upper Yangtze region affected the abundance of organic matter and the migration of trace metal elements [33,51,54,55]. In contrast, the weak upwelling in the western Sichuan region is associated with not only low organic matter abundance but also relatively low trace metal element abundance [56]. Upwelling under transgressive conditions is closely related to organic productivity. Furthermore, hydrothermal processes result in silica-rich fluids that concentrate trace metal elements. The negative correlation between the abundance of trace metal elements and excess silica in the lower section of the Niutitang Formation in the study area suggests that trace metal elements do not originate directly from hydrothermal sources but are more likely derived from upwelling migration.

The Co_{EF}*Mn_{EF} parameter is considered an effective indicator for assessing upwelling; values greater than 2 suggest a restricted marine sedimentary environment, while values below 0.5 indicate a clear influence of upwelling.

$X_{EF}(X) = (X/Al)_{sample}/(X/Al)_{ucc}$, where X_{EF} is the enrichment factor of element X, $(X/Al)_{sample}$ is the ratio of element X to Al in the sample, and $(X/Al)_{ucc}$ is the ratio of element X to Al in the average upper continental crust (UCC) [57].

Previous studies on the same profile in the research area also indicate that the lower section of the Niutitang Formation is significantly influenced by upwelling [58]. In this study, the Co_{EF}*Mn_{EF} values for the lower section of the Niutitang Formation range from 0.4 to 1.3 (Figure 10), indicating that it was affected by upwelling during sedimentation and was situated in a restricted basin environment. The influence of upwelling is most pronounced in the upper interval of Niutitang shale.

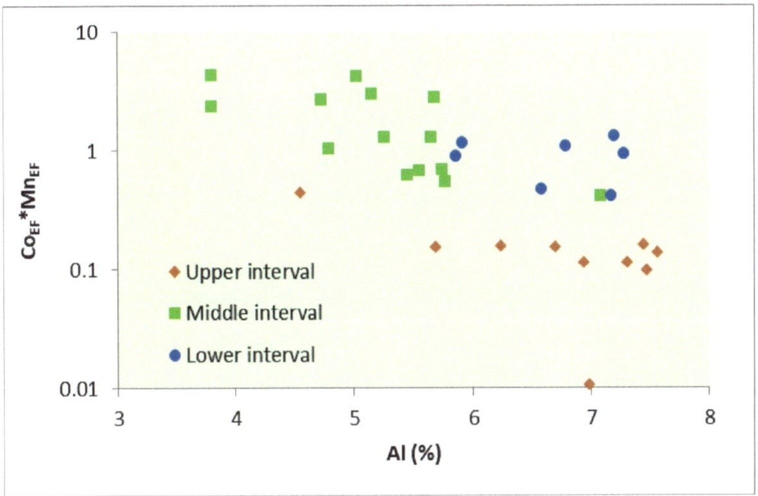

Figure 10. Correlation between aluminum content with Co_{EF}*Mn_{EF} of Niutitang shale.

5. Conclusions

1. Based on the Ti/Al ratio, accompanied by $\delta^{13}C_{org}$ values and TOC characteristics of the shale, the Danquan Niutitang shale section in the Youyang area of Southeast Chongqing in the Upper Yangtze region can be divided into three intervals: a lower interval with highest Ti/Al ratio and lightest $\delta^{13}C_{org}$ values over approximately 8 m from the base; a middle interval rich in organic matter with lowest Ti/Al ratio and relatively stable $\delta^{13}C_{org}$ values from 8 to 27 m; and an upper interval above 27 m with moderate Ti/Al ratio and low TOC and TS values;
2. The negative offset of the $\delta^{13}C_{org}$ values in the lower interval of the Niutitang shale in the study area, along with the strong negative correlation with Ni/TOC, Mo/TOC, V/TOC, U/TOC, as well as P and S, indicates that the mixing of upwelling fluids with anoxic sedimentary environment is the controlling factor for the negative drift of $\delta^{13}C_{org}$ values and the enrichment of trace metal elements in the lower interval of the Niutitang Formation;
3. From the bottom to the top of the lower interval of $Si_{_excess}$Niutitang Formation, as the influence of hydrothermal mixing weakens, the $Si_{\_excess}$ content gradually increases, suggesting that $Si_{\_excess}$ is primarily biogenic in origin rather than sourced from the hydrothermal fluids which are not silicon-rich;
4. Previous studies and the present research both indicate that the mixing of upwelling and local marine sediments is the primary factor for the enrichment of polymetals in the lower section of the Niutitang Formation shale in the study area.

Author Contributions: Conceptualization, D.L. and P.P.; methodology, C.Z.; software, G.W.; validation, G.W., D.L. and Z.L.; formal analysis, C.Z.; investigation, C.Z.; resources, P.P.; data curation, D.L.; writing—original draft preparation, D.L.; writing—review and editing, G.W. and D.L.; visualization, G.W.; supervision, L.Q.; project administration, L.Q.; funding acquisition, Z.L. All authors have read and agreed to the published version of the manuscript.

Funding: This research was supported by the National Natural Science Foundation of China-Enterprise Joint Fund (Grant No. U2167210), Special project of CAS (THEMSIE04010104) and Self-Directed Research of the State Key Laboratory of Organic Geochemistry (SKLOG2024-03). This is contribution No. IS 3560 from GIGCAS.

Data Availability Statement: An Excel document was uploaded in the attachment, containing all the data used in the article.

Conflicts of Interest: The authors declare no conflict of interest.

References

1. Li, D.; Li, R.; Tan, C.; Zhao, D.; Xue, T.; Zhao, B.; Khaled, A.; Liu, F.; Xu, F. Origin of silica, paleoenvironment, and organic matter enrichment in the Lower Paleozoic Niutitang and Longmaxi formations of the northwestern Upper Yangtze Plate: Significance for hydrocarbon exploration. *Mar. Pet. Geol.* **2019**, *103*, 404–421. [CrossRef]
2. Liu, Z.; Zhuang, X.; Teng, G.; Xie, X.; Yin, L.; Bian, L.; Feng, Q.; Algeo, T. The lower Cambrian Niutitang Formation at Yangtiao (Guizhou, SW China): Organic matter enrichment, source rock potential, and hydrothermal influences. *J. Pet. Geol.* **2015**, *38*, 411–432. [CrossRef]
3. Tan, J.; Wang, Z.; Wang, W.; Hilton, J.; Guo, J.; Wang, X. Depositional environment and hydrothermal controls on organic matter enrichment in the lower Cambrian Niutitang shale, southern China. *AAPG Bull.* **2021**, *105*, 1329–1356. [CrossRef]
4. Wu, C.; Tuo, J.; Zhang, M.; Sun, L.; Qian, Y.; Liu, Y. Sedimentary and residual gas geochemical characteristics of the Lower Cambrian organic-rich shales in Southeastern Chongqing, China. *Mar. Pet. Geol.* **2016**, *75*, 140–150. [CrossRef]
5. Kříbek, B.; Sýkorová, I.; Pašava, J.; Machovič, V. Organic geochemistry and petrology of barren and Mo–Ni–PGE mineralized marine black shales of the Lower Cambrian Niutitang Formation (South China). *Int. J. Coal Geol.* **2007**, *72*, 240–256. [CrossRef]
6. Xie, G.; Liu, S.; Xia, G.; Hao, W. Mineralogy and geochemical investigation of Cambrian and Ordovician–Silurian shales in South China: Implication for potential environment pollutions. *Geol. J.* **2018**, *55*, 477–500. [CrossRef]
7. Pi, D.; Liu, C.; Shields-Zhou, G.A.; Jiang, S.Y. Trace and rare earth element geochemistry of black shale and kerogen in the early Cambrian Niutitang Formation in Guizhou province, South China: Constraints for redox environments and origin of metal enrichments. *Precambrian Res.* **2013**, *225*, 218–229. [CrossRef]
8. Jiang, S.; Chen, Y.; Ling, H.; Yang, J.; Feng, H.; Ni, P. Trace-and rare-earth element geochemistry and Pb–Pb dating of black shales and intercalated Ni–Mo–PGE–Au sulfide ores in Lower Cambrian strata, Yangtze Platform, South China. *Miner. Depos.* **2006**, *41*, 453–467. [CrossRef]
9. Jiang, T.; Jin, Z.; Liu, G.; Liu, Q.; Gao, B.; Liu, Z.; Nie, H.; Zhao, J.; Wang, R.; Zhu, T.; et al. Source analysis of siliceous minerals and uranium in Early Cambrian shales, South China: Significance for shale gas exploration. *Mar. Pet. Geol.* **2019**, *102*, 101–108. [CrossRef]
10. Xia, W.; Yu, B.; Sun, M. Depositional setting and enrichment mechanism of organic matter of the black shales of Niutitang Formation at the bottom of Lower Cambrian, in well YuKe1, Southeast Chongqing. *Mineral. Petrol.* **2015**, *35*, 70–80. (In Chinese with English Abstract) [CrossRef]
11. Zhao, K.; Jiang, S.; Chen, W.; Chen, P.; Ling, H. Mineralogy, geochemistry and ore genesis of the Dawan uranium deposit in southern Hunan Province, South China. *J. Geochem. Explor.* **2014**, *138*, 59–71. [CrossRef]
12. Zhao, W.; Xie, Z.; Wang, X.; Zhen, A.; Wei, G.; Wang, Z.; Wang, K. Sinian gas sources and effectiveness of primary gas-bearing system in Sichuan Basin, SW China. *Pet. Explor. Dev.* **2021**, *48*, 1089–1099. (In Chinese with English Abstract) [CrossRef]
13. Liu, T.; Jiang, Z.; Liu, W.; Zhang, K.; Xie, X.; Yin, L.; Huang, Y. Effect of hydrothermal activity on the enrichment of sedimentary organic matter at Early Cambrian in the Xiuwu Basin. *Pet. Geol. Recovery Effic.* **2018**, *25*, 68–76. (In Chinese with English Abstract) [CrossRef]
14. Ma, Y.; Lu, Y.; Liu, X.; Zhai, G.; Wang, Y.; Zhang, C. Depositional environment and organic matter enrichment of the lower Cambrian Niutitang shale in western Hubei Province, South China. *Mar. Pet. Geol.* **2019**, *109*, 381–393. [CrossRef]
15. Li, Y.; Zhang, C.; Chi, G.; Duo, J.; Li, Z.; Song, H. Black and red alterations associated with the Baimadong uranium deposit (Guizhou, China): Geological and geochemical characteristics and genetic relationship with uranium mineralization. *Ore Geol. Rev.* **2019**, *111*, 102981. [CrossRef]
16. Niu, X.; Yan, D.; Zhuang, X.; Liu, Z.; Li, B.; Wei, X.; Xu, H.; Li, D. Origin of quartz in the lower Cambrian Niutitang Formation in south Hubei Province, upper Yangtze platform. *Mar. Pet. Geol.* **2018**, *96*, 271–287. [CrossRef]
17. Gao, P.; He, Z.; Lash, G.G.; Zhou, Q.; Xiao, X. Controls on silica enrichment of lower cambrian organic-rich shale deposits. *Mar. Pet. Geol.* **2021**, *130*, 105126. [CrossRef]
18. Yeasmin, R.; Chen, D.; Fu, Y.; Wang, J.; Guo, Z.; Guo, C. Climatic-oceanic forcing on the organic accumulation across the shelf during the Early Cambrian (Age 2 through 3) in the mid-upper Yangtze Block, NE Guizhou, South China. *J. Asian Earth Sci.* **2017**, *134*, 365–386. [CrossRef]
19. Guo, Q.; Strauss, H.; Zhu, M.; Zhang, J.; Yang, X.; Lu, M. High resolution organic carbon isotope stratigraphy from a slope to basinal setting on the Yangtze platform, south China: Implications for the Ediacaran–Cambrian transition. *Precambrian Res.* **2013**, *225*, 209–217. [CrossRef]
20. Goldberg, T.; Strauss, H.; Guo, Q.; Liu, C. Reconstructing marine redox conditions for the Early Cambrian Yangtze Platform: Evidence from biogenic sulphur and organic carbon isotopes. *Palaeogeogr. Palaeoclimatol. Palaeoecol.* **2007**, *254*, 175–193. [CrossRef]
21. Wang, K.; Deng, J.; Hao, X. The geochemical behavior of uranium and mineralization: South China uranium province as an example. *Acta Petrol. Sin.* **2020**, *36*, 35–43. [CrossRef]
22. Steiner, M.; Li, G.; Qian, Y.; Zhu, M.; Erdtmann, B.D. Neoproterozoic to early Cambrian small shelly fossil assemblages and a revised biostratigraphic correlation of the Yangtze Platform (China). *Palaeogeogr. Palaeoclimatol. Palaeoecol.* **2007**, *254*, 67–99. [CrossRef]

23. Chen, D.; Wang, H.; Qing, H.; Yan, D.; Li, R. Hydrothermal venting activities in the Early Cambrian, South China: Petrological, geochronological and stable isotopic constraints. *Chem. Geol.* **2009**, *258*, 168–181. [CrossRef]
24. Gao, P.; He, Z.; Li, S.; Lash, G.G.; Li, B.; Huang, B.; Yan, D. Volcanic and hydrothermal activities recorded in phosphate nodules from the Lower Cambrian Niutitang Formation black shales in South China. *Palaeogeogr. Palaeoclimatol. Palaeoecol.* **2018**, *505*, 381–397. [CrossRef]
25. Zhu, M.; Zhang, J.; Yang, A.; Li, G.; Steiner, M.; Erdtmann, B.D. Sinian-Cambrian stratigraphic framework for shallow-to deep-water environments of the Yangtze Platform: An integrated approach. *Progress. Nat. Sci.* **2003**, *13*, 951–960. [CrossRef]
26. Zhao, Z.; Yu, Q.; Zhou, X.; Liu, W.; He, J.; Xie, Y. Shale gas accumulation conditions of the Niutitang Formation of Lower Cambrian in Qianjiang Area, Chongqing. *Geol. Sci. Technol. Inf.* **2017**, *36*, 122–129. (In Chinese with English Abstract)
27. Holdaway, H.K.; Clayton, C.J. Preservation of shell microstructure in silicified brachiopods from the Upper Cretaceous Wilmington Sands of Devon. *Geol. Mag.* **1982**, *119*, 371–382. [CrossRef]
28. Zhang, K.; Jiang, Z.; Yin, L.; Gao, Z.; Wang, P.; Song, Y.; Jia, C.; Liu, W.; Liu, T.; Xie, X.; et al. Controlling functions of hydrothermal activity to shale gas content-taking lower Cambrian in Xiuwu Basin as an example. *Mar. Pet. Geol.* **2017**, *85*, 177–193. [CrossRef]
29. Chen, X.; Ling, H.F.; Vance, D.; Shields-Zhou, G.A.; Zhu, M.; Poulton, S.W.; Och, M.L.; Jiang, S.; Cremonese, L.; Archer, C. Rise to modern levels of ocean oxygenation coincided with the Cambrian radiation of animals. *Nat. Commun.* **2015**, *6*, 7142. [CrossRef]
30. Leventhal, J. Comparative geochemistry of metals and rare earth elements from the Cambrian alum shale and kolm of Sweden. In Proceedings of the Sediment-Hosted Mineral Deposits: Proceedings of a Symposium, Beijing, China, 30 July–4 August 1988; Wiley: Hoboken, NJ, USA, 1990; pp. 203–215. [CrossRef]
31. Zhang, Y.; Yan, D.; Zhang, F.; Li, X.; Qiu, L.; Zhang, Y. Stratigraphic sequences, abundance anomalies and occurrences of as, Sb, Au, Ag in the lower cambrian Niutitang formation in kaiyang phosphate mine area. *Acta Petrol. Sin.* **2016**, *32*, 3252–3268. (In Chinese with English Abstract)
32. Yang, J.; Yi, F.C.; Hou, L.J. Genesis and petrogeochemistry characteristics of Lower Cambrian black shale series in northern Guizhou. *Acta Mineral. Sin.* **2004**, *24*, 285–289. (In Chinese with English Abstract) [CrossRef]
33. Cao, H.; Wang, Z.; Dong, L.; Xiao, Y.; Hu, L.; Chen, F.; Wei, K.; Chen, C.; Song, Z.; Wu, L. Influence of hydrothermal and upwelling events on organic matter accumulation in the gas-bearing lower Cambrian shales of the middle Yangtze Block, South China. *Mar. Pet. Geol.* **2023**, *155*, 106373. [CrossRef]
34. LaGrange, M.T.; Konhauser, K.O.; Catuneanu, O.; Harris, B.S.; Playter TL Gingras, M.K. Sequence stratigraphy in organic-rich marine mudstone successions using chemostratigraphic datasets. *Earth-Sci. Rev.* **2020**, *203*, 103137. [CrossRef]
35. Dong, L.; Huang, Y.; Li, W.; Duan, C.; Luo, M.; Ren, S. Understanding Hydrothermal Activity and Organic Matter Enrichment with the Geochemical Characteristics of Black Shales in Lower Cambrian, Northwestern Hunan, South China. *Lithosphere* **2022**, *2022*, 2241381. [CrossRef]
36. Scott, C.; Lyons, T.W.; Bekker, A.; Shen, Y.; Poulton, S.W.; Chu XAnbar, A.D. Tracing the stepwise oxygenation of the Proterozoic ocean. *Nature* **2008**, *452*, 456–459. [CrossRef]
37. Wu, D.; Purnomo, B.J.; Sun, S. As and Sb speciation in relation with physico-chemical characteristics of hydrothermal waters in Java and Bali. *J. Geochem. Explor.* **2017**, *173*, 85–91. [CrossRef]
38. Li, R.; Lu, J.; Zhang, S.; Lei, J. Organic carbon isotopic composition of Sinian and Early Cambrian black shales. *Sci. China Ser. D* **1999**, *29*, 351–357. (In Chinese with English Abstract) [CrossRef]
39. Yang, R.; Zhu, L.; Gao, H.; Zhang, W.; Jiang, L.; Wang, Q.; Bao, M. A study on characteristics of the hydrothermal vent and relating biota at the Cambrian bottom in Songlin, Zunyi County, Guizhou province. *Geol. Rev.* **2005**, *51*, 481–492. (In Chinese with English Abstract)
40. Awan, R.S.; Liu, C.; Gong, H.; Dun, C.; Tong, C.; Chamssidini, L.G. Paleo-sedimentary environment in relation to enrichment of organic matter of Early Cambrian black rocks of Niutitang Formation from Xiangxi area China. *Mar. Pet. Geol.* **2020**, *112*, 104057. [CrossRef]
41. Bristow, T.F.; Bonifacie, M.; Derkowski, A.; Eiler, J.M.; Grotzinger, J.P. A hydrothermal origin for isotopically anomalous cap dolostone cements from south China. *Nature* **2011**, *474*, 68–71. [CrossRef]
42. Sahoo, S.K.; Planavsky, N.J.; Kendall, B.; Wang, X.; Shi, X.; Scott, C.; Anbar, A.D.; Lyons, T.W.; Jiang, G. Ocean oxygenation in the wake of the Marinoan glaciation. *Nature* **2012**, *489*, 546–549. [CrossRef] [PubMed]
43. Jiang, G.; Wang, X.; Shi, X.; Xiao, S.; Zhang, S.; Dong, J. The origin of decoupled carbonate and organic carbon isotope signatures in the early Cambrian (ca. 542–520 Ma) Yangtze platform. *Earth Planet. Sci. Lett.* **2012**, *317*, 96–110. [CrossRef]
44. Cremonese, L.; Shields-Zhou, G.; Struck, U.; Ling, H.F.; Och, L.; Chen, X.; Li, D. Marine biogeochemical cycling during the early Cambrian constrained by a nitrogen and organic carbon isotope study of the Xiaotan section, South China. *Precambrian Res.* **2013**, *225*, 148–165. [CrossRef]
45. Chang, C.; Hu, W.; Fu, Q.; Cao, J.; Wang, X.; Yao, S. Characterization of trace elements and carbon isotopes across the Ediacaran-Cambrian boundary in Anhui Province, South China: Implications for stratigraphy and paleoenvironment reconstruction. *J. Asian Earth Sci.* **2016**, *125*, 58–70. [CrossRef]
46. Maheshwari, A.; Sial, A.N.; Mathur, S.C. Carbon isotope fluctuations through the neoproterozoic-lower Cambrian Birmania basin, Rajasthan, India. *Carbonates Evaporites* **2002**, *17*, 53–59. [CrossRef]
47. Kirschvink, J.L.; Ripperdan, R.L.; Evans, D.A. Evidence for a large-scale reorganization of Early Cambrian continental masses by inertial interchange true polar wander. *Science* **1997**, *277*, 541–545. [CrossRef]

48. Kirschvink, J.L.; Raub, T.D. A methane fuse for the Cambrian explosion: Carbon cycles and true polar wander. *Comptes Rendus Geosci.* **2003**, *335*, 65–78. [CrossRef]
49. Fu, Y.; Yang, Z.; Li, C.; Xia, P. Enrichment of platinum group elements in Lower Cambrian polymetallic black shale, SE Yangtze Block, China. Front. *Earth Sci.* **2021**, *9*, 651948. [CrossRef]
50. Mao, J.; Lehmann, B.; Du, A.; Zhang, G.; Ma, D.; Wang, Y. Re-Os dating of polymetallic Ni-Mo-PGE-Au mineralization in lower cambrian black shales of South China and its geologic significance. *Econ. Geol.* **2002**, *97*, 1051–1061. [CrossRef]
51. Jin, C.; Li, C.; Algeo, T.J.; Wu, S.; Cheng, M.; Zhang, Z.; Shi, W. Controls on organic matter accumulation on the early-Cambrian western Yangtze Platform, South China. *Mar. Pet. Geol.* **2020**, *111*, 75–87. [CrossRef]
52. Jiang, S.Y.; Yang, J.H.; Ling, H.F.; Chen, Y.Q.; Feng, H.Z.; Zhao, K.D.; Ni, P. Extreme enrichment of polymetallic Ni–Mo–PGE–Au in Lower Cambrian black shales of South China: An Os isotope and PGE geochemical investigation. *Palaeogeogr. Palaeoclimatol. Palaeoecol.* **2007**, *254*, 217–228. [CrossRef]
53. Fu, Y.; Dong, L.; Li, C.; Qu, W.; Pei, H.; Qiao, W.; Shen, B. New Re-Os isotopic constrains on the formation of the metalliferous deposits of the Lower Cambrian Niutitang formation. *J. Earth Sci.* **2016**, *27*, 271–281. [CrossRef]
54. Zhang, G.; Chen, D.; Ding, Y.; Huang, T. Controls on Organic Matter Accumulation from an Upper Slope Section on the Early Cambrian Yangtze Platform, South China. *Minerals* **2023**, *13*, 260. [CrossRef]
55. Liu, Z.; Yan, D.; Yuan, D.; Niu, X.; Fu, H. Multiple controls on the organic matter accumulation in early Cambrian marine black shales, middle Yangtze Block, South China. *J. Nat. Gas. Sci. Eng.* **2022**, *100*, 104454. [CrossRef]
56. Zhang, Q.; Liu, E.; Pan, S.; Wang, H.; Jing, Z.; Zhao, Z.; Zhu, R. Multiple Controls on Organic Matter Accumulation in the Intraplatform Basin of the Early Cambrian Yangtze Platform, South China. *J. Mar. Sci. Eng.* **2023**, *11*, 1907. [CrossRef]
57. Tribovillard, N.; Algeo, T.J.; Lyons, T.; Riboulleau, A. Trace metals as paleoredox and paleoproductivity proxies: An update. *Chem Geol.* **2006**, *232*, 12–32. [CrossRef]
58. Zhu, G.; Zhao, K.; Li, T.; Zhang, Z.; Tang, S.; Wang, P. Anomalously high enrichment of mercury in early Cambrian black shales in South China. *J. Asian Earth Sci.* **2021**, *216*, 104794. [CrossRef]

minerals

MDPI

Article

Enhancing Mineral Exploration Programs Through Quantitative XRD: A Case Study from the Gumsberg Polymetallic Sulphide Deposits, Sweden

Sheida Makvandi [1,*], Evelien Rost [1], Thomas Witzke [1], Matteo Pernechele [1] and Hein Raat [2]

1 Application Competence Center, Malvern Panalytical B.V., 7602 EA Almelo, The Netherlands; evelien.rost@malvernpanalytical.com (E.R.); thomas.witzke@malvernpanalytical.com (T.W.); matteo.pernechele@malvernpanalytical.com (M.P.)

2 EMX Royalty Corporation, 10001 W. Titan Road, Littleton, CO 80125, USA; hraat@emxroyalty.com

* Correspondence: sh.makvandi@gmail.com or sheida.makvandi@malvernpanalytical.com

Abstract: As challenges in precious and base metal exploration intensify due to the diminishing availability of high-grade ore deposits, rising demand, energy costs, and stricter regulations towards net-zero carbon activities, advanced techniques to enhance exploration efficiency are becoming increasingly critical. This study demonstrates the effectiveness of quantitative X-ray diffraction (QXRD) with Rietveld refinement, coupled with multivariate statistical analysis (including agglomerative hierarchical clustering, principal component analysis, and fuzzy analysis), in characterizing the complex mineralogy of strata-bound volcanic-associated limestone-skarn Zn-Pb-Ag-(Cu-Au)-type sulphide deposits (SVALS). Focusing on 113 coarse rejects from the Gumsberg project located in the Bergslagen mining district in central Sweden, the research identified five distinct mineralogical clusters corresponding to polymetallic base metal sulphide mineralization, its proximal alteration zones, and variably metamorphosed host rocks. The results reveal significant sulphide mineralization, ranging from disseminated to massive occurrences of sphalerite, pyrrhotite, pyrite, and galena, with trace amounts of secondary minerals like anglesite in certain samples indicating weathering processes. The study also identifies rare minerals such as armenite, often overlooked in traditional geological logging. These findings underscore the potential of QXRD to enhance resource estimation, optimize exploration strategies, and contribute to more efficient and sustainable mineral exploration programs. The accuracy of QXRD was cross-validated with geological logs and geochemical data, confirming its reliability as a mineralogical discrimination tool.

Keywords: quantitative XRD; Rietveld; cluster analysis; principal component analysis; HighScore Plus; polymetallic ore deposits; massive sulphides; precious metals; armenite; Gumsberg

1. Introduction

Accurate mineralogical characterization is the key to understanding the geological history and processes that led to the formation of mineral deposits, thereby guiding exploration strategies and optimizing resource extraction. This detailed analysis can reveal valuable information about the concentration of economically important minerals and the quality and quantity of ore. Furthermore, precise mineral quantification supports metallurgical processes by determining the best methods for mineral extraction and processing, ultimately enhancing the efficiency and profitability of mining operations.

Quantitative X-ray diffraction (QXRD) using the Rietveld method has emerged as a mineralogical characterization tool offering several advantages over traditional methods like optical microscopy and floor-standing SEM-EDS (Scanning Electron Microscopy–Energy Dispersive X-ray Spectroscopy) or MLA (Mineral Liberation Analyzer) [1–6]. Unlike optical microscopy, which depends heavily on the operator's expertise and visual judgment, QXRD provides objective and reproducible identification and quantification of mineral phases by analyzing diffraction patterns against extensive reference libraries. Compared to floor-standing SEM-EDS or MLA, QXRD allows the rapid processing of multiple samples. Additionally, the sample preparation for XRD is simpler, and the system operation requires minimal training [4–6]. Additionally, compact XRD systems are affordable and can be easily accommodated by most mining sites or exploration companies. It is important to note that, XRD, despite being a powerful mineralogical tool, has limitations when it comes to identifying specific mineral sub-members. While the closest matching structure is selected to explain certain peaks in the XRD pattern, the identified mineral names from supergroups (e.g., amphiboles, chlorite, biotite, zeolite) generally represent the broader group rather than specific sub-members. This limitation arises due to gaps in XRD databases, which often lack full structural coverage of the entire compositional range of these minerals. Additionally, factors like exsolution, twinning, cleavage, and parting can further complicate the interpretation of XRD data. In such cases, information from other techniques, such as XRF data to constrain the chemistry, become crucial for the cross-validation and confirmation of specific subgroup minerals.

Among various types of ore deposits, polymetallic sulphide deposits are paramount in supplying essential metals like zinc, lead, and copper [7], which are integral to modern society and play a significant role in various industrial applications, including construction, electronics, and transportation. With the demand for copper projected to double to 40 million tons in the next 20 years, the mining industry must invest heavily in discovering new mineral deposits to bridge this supply gap.

Despite their importance, the exploration of base metals presents numerous challenges. These deposits are often located in complex geological settings, necessitating advanced geophysical and geochemical techniques for accurate detection and delineation. The typically deep and remote locations of mineral deposits further escalate the costs and logistical difficulties of exploration. Additionally, managing environmental and regulatory concerns is crucial to ensure sustainable and responsible mining practices. Overcoming these obstacles requires adopting advanced, cost-effective, and time-efficient technologies to ensure the success of exploration programs and/or mining activities.

To exemplify the potential of QXRD and its practical application in exploration and mining programs, this study focuses on the mineralogical characterization of 113 coarse reject samples collected in the framework of the Gumsberg project (2016), located in the Bergslagen mining district in Central Sweden. The Gumsberg project is notable for its historic production, the apparent lack of modern exploration and drilling, and its high-grade styles of volcanogenic massive sulphide mineralization that have not been fully appreciated.

1.1. Geological Context

The Gumsberg property is located in the northern part of the Bergslagen district, a classical mining region mined for centuries. This district hosts thousands of mineral occurrences, prospects, and historic mine workings, with small to large iron oxide and polymetallic base metal sulphide mineralization being the most abundant and historically significant (Figure 1).

The Bergslagen district is an intensely mineralized part of a 280 $\times$ 300 km, Early Proterozoic felsic magmatic province characterized by polyphase deformation and medium- to high-grade metamorphism during the Svecokarelian orogeny (1.9–1.8 Ga). The region is interpreted as an intra-continental arc or back-arc extensional basin that developed above an active subduction zone [8–10]. The stratigraphy is marked by felsic meta-volcanic succession enveloped by siliciclastic metasedimentary rocks [10]. Marble interbeds in the upper meta-volcanic succession, which are former limestone, host many of the district's mineral deposits.

Various generations of syn-orogenic to post-orogenic intrusive rocks were emplaced between 1.91 and 1.75 Ga [9]. Significant hydrothermal alteration affected the supracrustal rocks [9,11] before extensive ductile deformation and regional metamorphism under upper greenschist to amphibolite facies [12,13]. This led to significant changes in the deposits, including metal recrystallization and syn-tectonic remobilization [13–17].

The polymetallic sulphide deposits at Bergslagen are difficult to classify in terms of the major internationally recognized genetic classes of mineral deposits. Allen et al. [13] proposed that strata-bound volcanic-associated limestone-skarn Zn-Pb-Ag-(Cu-Au) sulphide deposits (SVALS) at Bergslagen were formed by the syn-volcanic subseafloor replacement of these shallow marine stromatolitic limestones, whereas the stratiform ash-siltstone (SAS) Zn-Pb-Ag polymetallic sulphide ores were formed by the deposition of metalliferous hydrothermal sediments between laminated, distal, deeper-water carbonate and felsic siltstone facies. Notable examples of SVALS-type deposits include the active Garpenberg Mine, Falun, Sala, and Stollberg deposits, which are commonly developed as systems of stratabound lenses, ranging in style from massive to vein networks and sulphide-cemented breccias with clasts of altered wall rock [18]. These deposits are typically rich in Zn, Pb, Ag, and sometimes Cu-Au. Sphalerite is the dominant mineral, with zones containing argentiferous galena, and in some areas, the mineralization shifts to chalcopyrite-rich zones with high pyrite and/or pyrrhotite contents [13].

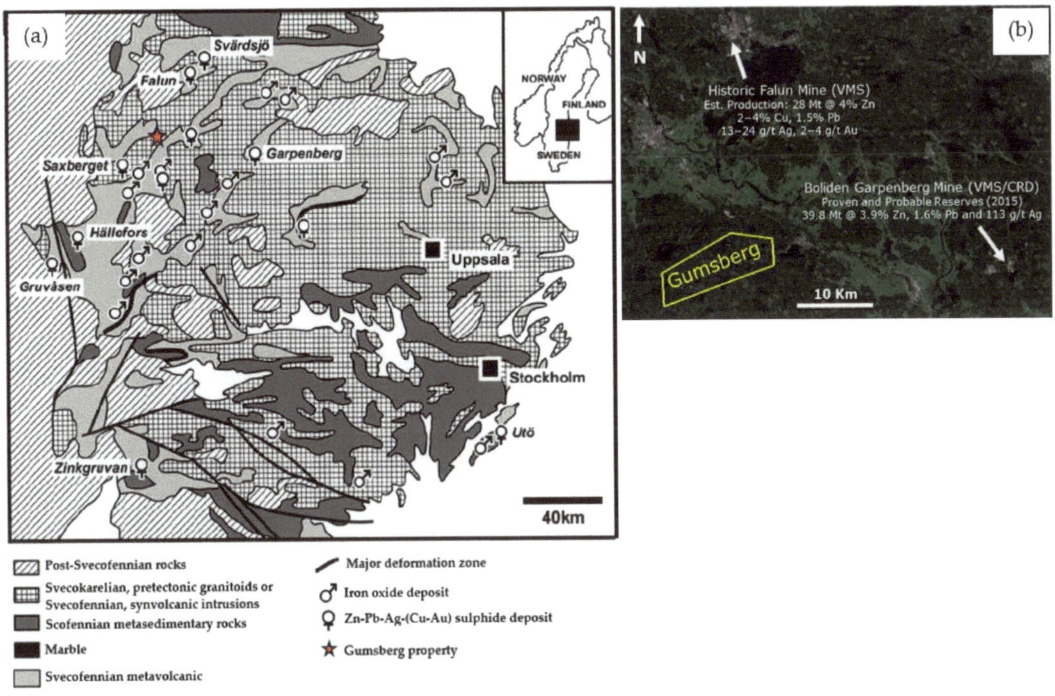

Figure 1. (**a**) Simplified bedrock geology of the Bergslagen ore province, showing principal mineral deposits (modified after Allen et al. [13]). (**b**) Location of the Gumsberg property in yellow in relation to the Falun Mine and Boliden's Garpenberg Mine, within the Bergslagen mining district in central Sweden [19].

1.2. Local Geology

The Gumsberg project is located in the historic Bergslagen district, 30 km south of the world class Falun Mine (Zn-Pb-Cu-Au) and 30 km west of Boliden's expanding Garpenberg Mine (Zn-Pb-Cu-Au-Ag; Figure 2). The property contains various polymetallic base metal sulphide prospects including over 30 historic mines that were actively mined from the 13th century until the early 1900s. This includes Östra Silverberg, Europe's largest silver mine from 1300 to 1590. Massive sulphide mineralization within the Gumsberg property is locally rich in silver and gold and dominated by sphalerite and galena. The prospects vary from steeply dipping remobilized massive sulphide veins, sub-seafloor replacement style hydrothermal mineralization, and exhalative mineralization to more disseminated feeder zone/stockwork style mineralization along a northeast trend [13,19]. Historical mining efforts suggest the mineralization is, partly structurally controlled, laterally continuous and forms pipe-like, moderate to steeply southeast plunging mineralized intervals [13].

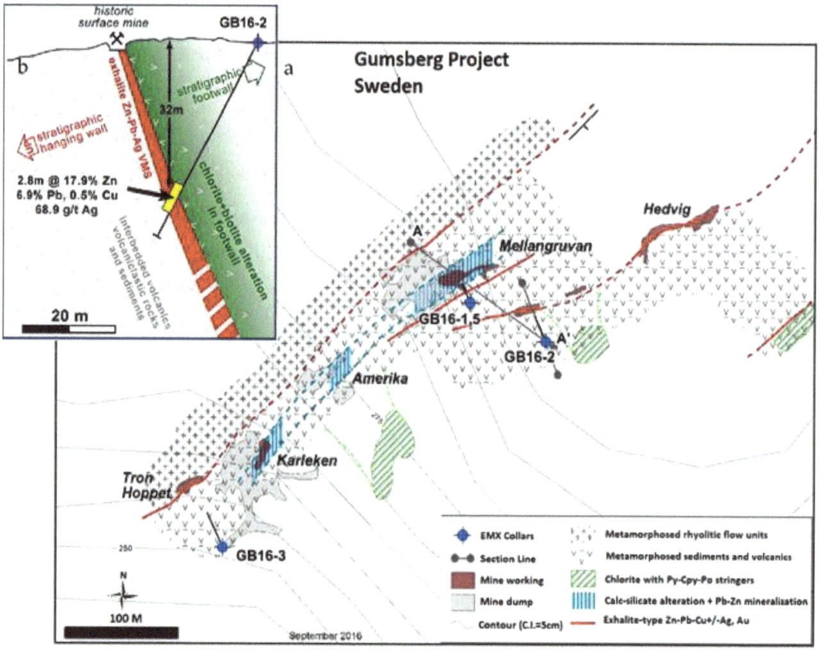

Figure 2. (**a**) Location of the four 2016 drill holes with high-grade intercepts (GB16-1, GB16-2, GB16-3, and GB16-5) within the Gumsberg property, part of the EMX Royalty Corp. exploration program in the Bergslagen mining district, southern-central Sweden. (**b**) Cross-section of drill hole GB16-2, looking towards the northeast [19].

Exploration drilling conducted by EMX Royalty Corp., a Canadian mineral prospect generator, in 2016, confirmed significant intervals of polymetallic Ag-Zn-Pb mineralization, with notable results from the Vallberget–Loberget trend (Figure 2). The Gumsberg property is now recognized for its 27 km of stratigraphy prospective for Ag-Zn-Pb mineralization, underscoring its potential for significant base and precious metal resources [19]. Figure 2 depicts the location of high-priority drill targets in the Vallberget–Loberget trend (GB16-1 to GB16-5 except GB16-4). The target drill hole GB-16-4 was a target in Gumsgruvan, another trend within the Gumsberg property, ~1–1.5 km NW of the Vallberget area.

2. Materials and Methods

2.1. Samples Description

This study analyzes 113 coarse reject samples from five reconnaissance diamond drill holes conducted by EMX Royalty Corp. in 2016. The high-priority drill targets were identified through comprehensive geological mapping and sampling. The drill holes were meticulously logged, and the selected samples provide a basis for this analysis, supported by detailed geological logs and chemical data, allowing for reliable cross-validation of QXRD results [19]. Figure 2 shows the location of drill holes GB16-1, GB16-2, GB16-3, and GB16-5 in the Vallberget–Loberget trend within the Gumsberg property. Figure S1a in the Supplementary Materials provides a plan view outcrop map indicating the locations of the drill holes. The drill hole GB-16-4 is located in the Gumsgruvan area, ~1–1.5 km NW of Vallberget–Loberget. The sample descriptions, including the type of host rocks, ore mineralogy, and alteration mineral assemblages, are presented in Table 1. Sub-samples from all collected specimens were prepared and analyzed for trace element compositions in 2016 at ALS Global in Ireland. The analysis utilized Inductively Coupled Plasma Atomic Emission Spectroscopy (ICP-AES) with the ME-MS41 and ME-OG analytical packages, following an aqua regia digestion method. A summary of the analytical results is presented in Table 2. Table 2.

Table 1. Overview of geology and mineralogy of the 5 studied drill holes.

Drill Hole ID	Depth (m)	Host Rock Type	Ore Mineralogy	Alteration Mineralogy
GB16-1	80–129	Metasedimentary rocks (mostly metatuff and leptite) characterized by metavolcanic dykes	Replacement style Zn-Pb-Ag mineralization; Disseminated to massive sphalerite; Disseminated pyrite (traces)	Meta-breccia of clinozoisite and chlorite schist fragments; Quartz ± albite ± K-feldspar
including	86.5–90.2	Metavolcanics	Massive sphalerite and disseminated pyrite (<1%)	Amphibole and chlorite
GB16-1B	120–129	Fine grained tuffaceous laminated rocks underlain by massive, hard siliceous rock	Wispy zones of base metals. Disseminated pyrite; Galena and sphalerite in skarn rock, in breccia and carbonate veins with pyrrhotite on fractures	Quartz, K-feldspar, biotite, sericite, chlorite, and traces of garnet
GB16-2	31.3–31.8 36.9–39.7	Meta-sedimentary rock (arkose) Massive sulphide (sphalerite) zone + 20% wall rocks	Hematite 5%–10% Massive sphalerite ± pyrite, galena	Actinolite, biotite, albite K-feldspar, chlorite, biotite, sericite, actinolite
GB16-3	52.5–61.7	Relatively massive volcanic sandstone and siltstone metamorphosed to chlorite grade	Sulphide (1%–2%); Disseminations of pyrite (traces)	Quartz, albite, K-feldspar, chlorite, actinolite
including	57.1–57.6 59.1–59.7	Massive sulphide	Massive sulphide (fault banded)	K-feldspar, chlorite, actinolite
GB16-4	48–92.5	Metavolcanics porphyritic, lithic tuffs	Disseminations of pyrite (1.00%–5.00%) ± pyrrhotite, magnetite	Quartz, amphibole, chlorite, sericite, cordierite ± epidote
GB16-5	26.1–29.1	Tuffaceous fine-grained rock (Rhyolite?); Fine laminated siliceous rock grades into silica zone; Quartz breccia	Some pyrite within deformed bands of quartz-clinozoisite	Quartz, biotite, chlorite, clinozoisite
Including:	27.1–28.1	Massive sulphide	Massive sulphide (2/3 sphalerite, 1/3 galena, some chalcopyrite) to sulphide disseminations	Quartz, chlorite, iron oxides
Including:	43.6–52.2	Multiple stages of discordant breccias, some zones heterolithic with quartz vein fragments; Fine sediment/tuffs	Zn replacement zones, richer in clinozoisite rich lithologies	Clinozoisite ± chlorite ± quartz ± sericite

A total of 29 coarse reject samples were selected from the drill holes GB16-1 and GB16-1B. The drill hole GB16-1 encounters metasedimentary rocks, metavolcanics, and fine-grained sediments. The presence of metavolcanic dykes and fractured textures indicates a complex geological history with multiple stages of deformation and mineralization (Tables 1 and 2). The dissemination of pyrite is observed, along with higher-grade intervals of pyrite, indicating potential zones of economic interest. The alteration minerals, such as clinozoisite, chlorite, and quartz, suggest a metamorphic overprint on the original mineralization, which is characteristic of the SVALS-type replacement-style sulphide mineralization in the region [13].

Table 2. High-grade mineralized zones in the sampled drill hole sections [19]. [x]: the true width is unknown. [y]: the true width is estimated to be 80%–90% of reported interval.

Hole ID	Total Length (m)	Intercept Depth		Intercept Length (m)	Pb (wt.%)	Zn (wt.%)	Ag (g/t)
		From	To				
GB16-1 [x]	152.5	84.5	92.7	8.2	0.1	4.6	2.3
including		86.5	90.2	3.7	0.1	8.9	4.4
GB16-1B		120	129	9	0.2	0.6	3.6
GB16-2 [y]	46.2	36.9	39.7	2.8	6.9	17.9	68.8
GB16-3 [y]	58.8	52.5	61.7	9.2	0.5	3.7	7.3
including		57.1	57.6	0.5	4.1	28.5	51.4
including		59.1	59.7	0.6	2.3	16.4	39.3
GB16-4 [y]	110	No Significant Intercepts					
GB16-5 [y]	65	26.1	29.1	3	3	9.2	12.8
including		27.1	28.1	1	8.8	26.7	34.9
including		43.6	52.2	8.6	0.04	1.6	1.7

Fourteen samples were selected from the drill hole GB16-2 for this study. This hole intersects massive sulphide zones, particularly sphalerite-rich, which are fault-bounded. High grades of Zn (17.9%) and Pb (6.9%) with significant Ag (68.8 g/t) indicate substantial sulphide mineralization (Tables 1 and 2). The alteration assemblage of K-feldspar, chlorite, and biotite is consistent with intense hydrothermal activity typically associated with volcanogenic massive sulphide (VMS) deposits [19].

Fifteen samples from the study set were selected from the drill hole GB16-3. This drill hole shows relatively massive volcanic sandstone and siltstone, with brecciation indicating structural complexity (Table 1). Notable sulphide mineralization, with a high zinc content (up to 28.5% in specific intervals), suggests a robust sulphide system (Table 2). The presence of quartz, albite, and chlorite indicates hydrothermal alteration consistent with VMS mineralization processes.

From the drill hole GB16-4, 39 samples were studied for their mineralogical characteristics. The metavolcanic porphyritic and lithic tuffs characterized by disseminations of pyrite and occasional pyrrhotite and magnetite suggest dispersed sulphide mineralization in the host rocks (Table 1). Alteration minerals such as quartz and amphibole indicate moderate hydrothermal alteration.

Sixteen of the studied samples were from the drill hole GB16-5. The presence of tuffaceous fine-grained rock and quartz breccia points to a volcanic origin with significant fracturing. The high zinc (up to 26.7%) and lead (up to 8.8%) contents in the massive sulphide zones, along with significant silver (34.9 g/t), highlight the economic potential (Table 2).

The alteration minerals, including quartz and biotite, reflect intense hydrothermal activity. EMX Royalty Corp. [19] proposed that GB16-1 hosts replacement-style Zn-Pb-Ag mineralization, while GB16-2 and GB16-3 exhibit exhalative-style VMS mineralization. At GB16-5, the mineralization transitions from an exhalative style between 26.1 and 29.1 m to a replacement style between 43.6 and 52.2 m.

2.2. Analytical Methods

2.2.1. X-ray Powder Diffraction

X-ray powder diffraction is a versatile analytical method used for identifying and quantitatively determining the crystalline phases present in powdered and bulk samples. In this study, XRD scans of the 113 samples were collected at the Application Competence Centre, Malvern Panalytical, in the Netherlands, using the Minerals Edition of the Aeris compact X-ray diffractometer. This 600 W system was equipped with a cobalt-anode X-ray tube, a goniometer radius of 145 mm, 0.04 rad Soller slits, a 1/4° divergence slit, a 23 mm mask, a low beam-knife position, a step size of 0.02°, and an acquisition time of 79 s/step. The Bragg–Brentano measurement covered a range of 6–80° 2θ, allowing the detection of clay minerals and heavy minerals such as Fe oxides/hydroxides and sulphides. Cobalt radiation was used to prevent the emission of fluorescence, which occurs when using Cu radiation to analyze Fe, Cu, and/or Zn-containing materials. This choice improves the penetration depth of X-rays in the sample and enhances counting statistics. The use of a linear PIXcel^{1D} Medipix3 detector with an active length of 5.54° 2θ enables a scan acquisition time of just a few minutes.

2.2.2. XRD Data Analysis

Phase identification of all samples was performed using the all-in-one Malvern Panalytical's software suite, HighScore Plus, version 5.2 [20], with the ICDD PDF-5+ database [21]. The abundance of identified minerals in the analyzed samples was then determined using the Rietveld method [22]. Modern XRD quantification techniques, such as Rietveld analysis, provide a compelling alternative to classical peak intensity or area-based methods, as they do not require any standards or monitors. This method offers impressive accuracy and speed of analysis. The Rietveld method calculates structure factors based on atom positions in a crystal structure model, which can be adjusted during refinement [20,22]. However, the structure is often left unchanged and used to compute theoretical peaks and a profile. In this study, the Rietveld refinement included parameters such as scale factors, sample displacement, background coefficients, unit cell parameters, and Cagliotti W. For major minerals (over 10 wt.%) with preferential orientation, orientation parameters were also refined. To improve the fitting for abundant minerals, additional peak width functions (Cagliotti U and V) were refined when necessary. The final refinement model of an XRD scan, with proper parameter constraints, can serve as a template for analyzing similar samples, enabling automated and efficient quantitative XRD analysis of numerous samples. It is worth mentioning that understanding the exact crystal structure of all minerals present in the samples from the polymetallic base metal sulphide mineralization and associated alteration zones is essential for accurate Rietveld refinements.

2.2.3. Extended Cluster Analysis of XRD Data in HighScore Plus

Cluster analysis (CA) is designed to simplify the analysis of large datasets. By automatically sorting all closely related scans from an experiment into distinct groups, it identifies the most representative scan and the most outlying scans within each group [23,24]. Samples clustered together exhibit similar mineralogy, making it necessary to identify and quantify phases only in the most representative scan of each cluster. This allows for the creation of an efficient analytical routine, where other scans in the group can be automatically analyzed according to the established routine. This method is particularly effective for analyzing multiple samples from the same type of rocks or lithology and in quality control environments.

The CA tool in HighScore Plus exemplifies its comprehensive integration, allowing users to perform statistical analysis of XRD scans within the same software [1–4]. This eliminates the need for additional statistical software, streamlining workflows and enhancing efficiency in handling and interpreting large datasets. The module employs various algorithms, including correlation matrix creation, agglomerative hierarchical CA, grouping, and scan visualization. The process involves the following steps:

Comparison of Scans: All scans in a document are compared, resulting in a correlation matrix that represents the dissimilarities between data points of any given pair of scans.

Agglomerative Hierarchical Cluster Analysis: Scans are classified into different groups based on their similarity, producing a dendrogram where each scan starts as an individual cluster on the left side. Clusters merge stepwise until they form a single group.

Optimal Grouping Estimation: The best possible number of separate clusters is estimated using the KGS test or the largest relative step on the dissimilarity scale. This number can also be manually adjusted. The most representative scan and the two most outlying scans within each cluster are then identified and marked.

In addition to hierarchical clustering, three independent tools—Principal Components Analysis (PCA), Metric Multi-Dimensional Scaling (MMDS), and t-Stochastic Neighbor Embedding (t-SNE)—can be used to define clusters. These tools provide pseudo-three-dimensional plots for visualizing the clusters.

In this study, raw XRD patterns were analyzed using cluster analysis (CA) and principal component analysis (PCA) in HighScore Plus, without any correction or normalization, as all data were acquired on the same diffractometer. The CA algorithm classified the XRD patterns based solely on peak positions, ignoring peak intensities. Euclidean distances and the average linkage method were utilized for clustering [20]. The integration of CA and PCA could significantly enhance data analysis efficiency, and when combined with Rietveld refinement, this approach revealed hidden structural details and enabled comprehensive mineralogical characterization across multiple samples.

3. Results

3.1. XRD Pattern Analysis and PCA Results

The XRD patterns of 113 samples were analyzed using an extended approach in HighScore Plus. This method incorporated agglomerative hierarchical clustering, visualized in a dendrogram, along with three-dimensional principal component analysis (3D-PCA) and fuzzy analysis to classify and interpret the data. The 3D-PCA results are shown in Figure 3, with the corresponding scree plot displayed in Figure 3a. The scree plot illustrates the eigenvalues of the principal components, where a sharp decline after the first few components indicates that these capture the majority of the dataset's variance. This aids in determining the optimal number of components to retain (PC1 to PC3), as components beyond the "elbow" (PC4 to PC18) contribute minimally to the total variance (Figure 3a).

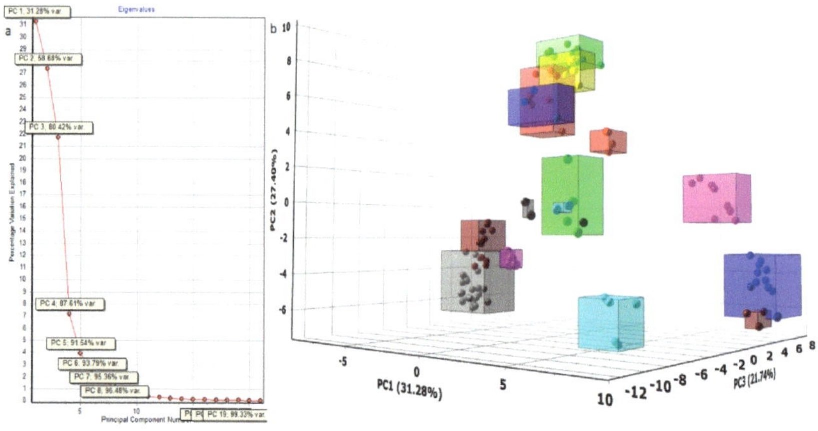

Figure 3. (**a**) Scree plot illustrating the eigenvalues of the principal components from the PCA, with the first three components accounting for 80.42% of the total variance (PC1: 31.28%, PC2: 27.40%, PC3: 21.74%). The x-axis denotes the principal components (PC1, PC2, ..., PC21), while the y-axis represents the eigenvalues, indicating the percentage of variance explained. (**b**) 3D-PCA plot identifying fifteen clusters, and one non-clustered sample based on the analysis of 113 XRD scans. The x-axis displays the PCA scores, representing the transformed coordinates of the original data points in the reduced-dimensional space defined by the principal components. The colored clusters show data points as spheres, with each sphere representing a scan within that cluster.

As a result, the first three principal components collectively captured 80.4% of the total variance in the dataset, leading to the identification of 15 distinct mineralogical clusters, with 1 sample standing alone as a non-clustered outlier (Figure 3b). The most representative sample from each cluster was easily identified and it is marked with *** next to the scan name in the dendrogram (Figure S1b in the Supplementary Materials; this figure is not included here due to the large number of samples, which would make the legend and sample names difficult to read). Figure 4 illustrates both an overlay and individual plots of the 15 representative scans alongside the non-clustered outlier. These 16 XRD patterns were imported into a new document in HighScore Plus to facilitate further analysis and the development of Rietveld models. These models were then used for automated data analysis of other samples within the same cluster.

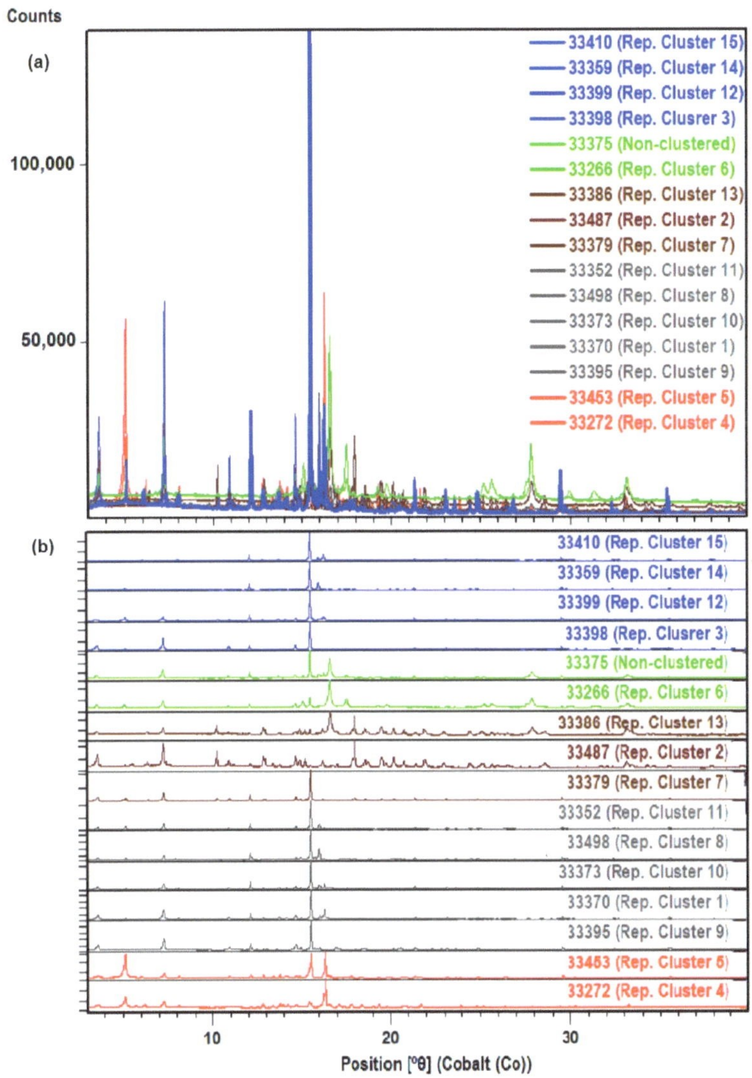

Figure 4. (**a**) An overlay view of the 15 representative and 1 outlier XRD patterns identified by the PCA in Figure 3b. (**b**) All 16 XRD patterns displayed separately, ordered and color-coded according to the 2nd cluster analysis (CA2) results.

3.2. *Cluster Composition and Rietveld Refinement*

As shown in Figures 3b and S2, certain clusters exhibit greater compositional similarity to one another, as indicated by the short distances between clusters in the latent variable space. This is also supported by fuzzy analysis (Table S1 in Supplementary Materials) revealing that only five samples have probabilities (class membership) coefficients above 0.85 for Cluster 4, indicating they exclusively belong to this group, while four other samples are uniquely associated with Cluster 6. Additionally, the fuzzy analysis shows a probabilities coefficient above 0.85 for the non-clustered sample, confirming its status as an outlier. The remaining 103 samples exhibit membership coefficients between 0.15 and 0.85, suggesting that they could belong to more than one cluster.

Based on the extended CA results, it was possible to reduce the number of clusters by merging those with similar compositions. To further refine and optimize the selection of representative patterns for developing Rietveld models, a second cluster analysis (CA2) was conducted on the representative scans identified in the initial analysis (Figure 5). As a result, the 16 identified XRD patterns representing the 113 studied samples, including the outlier, were consolidated into five new groups. However, the dendrogram visualizing the agglomerative hierarchical clustering results reveals that samples within each group are connected by relatively longer tie bars compared to the first cluster analysis (CA1; Figure S2), indicating a greater degree of mineralogical variation within these groups.

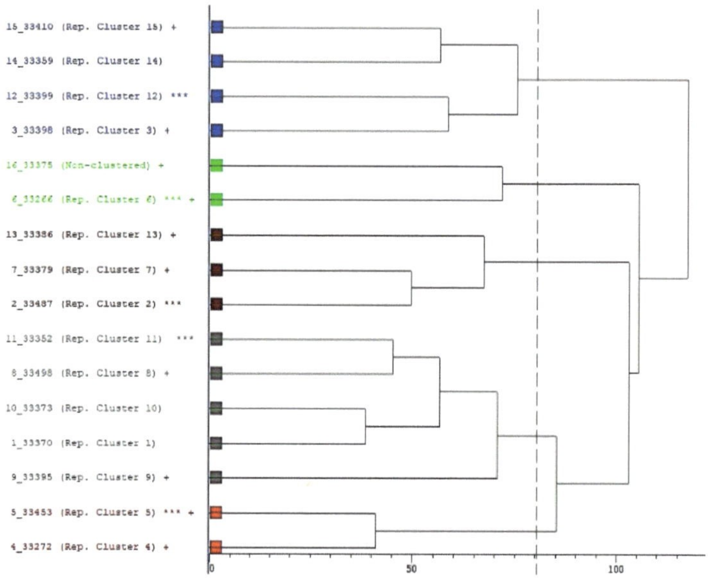

Figure 5. Visualization of agglomerative hierarchical clustering of XRD patterns representative of 15 clusters and the outlier resulted from CA1 in a tree-like fashion (dendrogram). In the dendrogram, the x-axis displays the dissimilarity of the tie bars, with each color representing a distinct cluster (Group 1: blue; Group 2: green; Group 3: brown; Group 4: gray; Group 5: red). The most representative scan and the two most outlying scans within each cluster are respectively marked by *** and +.

3.3. Fuzzy Clustering Prior to Rietveld Quantification

To explore the potential transitions between different mineralogical groups, fuzzy clustering was employed in HighScore Plus. Table 3 presents the membership matrix (M), detailing the calculated probabilities for each measurement. Notably, none of the analyzed samples exhibit a membership coefficient exceeding 0.85 for any of the five clusters, indicating that no sample exclusively belongs to a single cluster. However, all samples show coefficients above 0.48 for a particular cluster, suggesting a strong likelihood of association with that group [3,20]. Samples marked with an "x" in the mixture column likely contain mineral assemblages that are partially similar to those in other clusters. This mineralogical variation could be seen in the Rietveld quantification models (Figure 6). Moreover, individual Rietveld models for the five identified groups are provided in the Supplementary Materials (Figures S1c–S5C). The full-pattern Rietveld quantification was performed on XRD scans of five samples, each representing one of the groups identified in CA2 (Figure 6). This method provided a comprehensive evaluation of the mineralogical composition within each group. By concentrating the Rietveld analysis on these representative samples, the mineralogical characterization of other samples within the corresponding group was both facilitated and automatically executed.

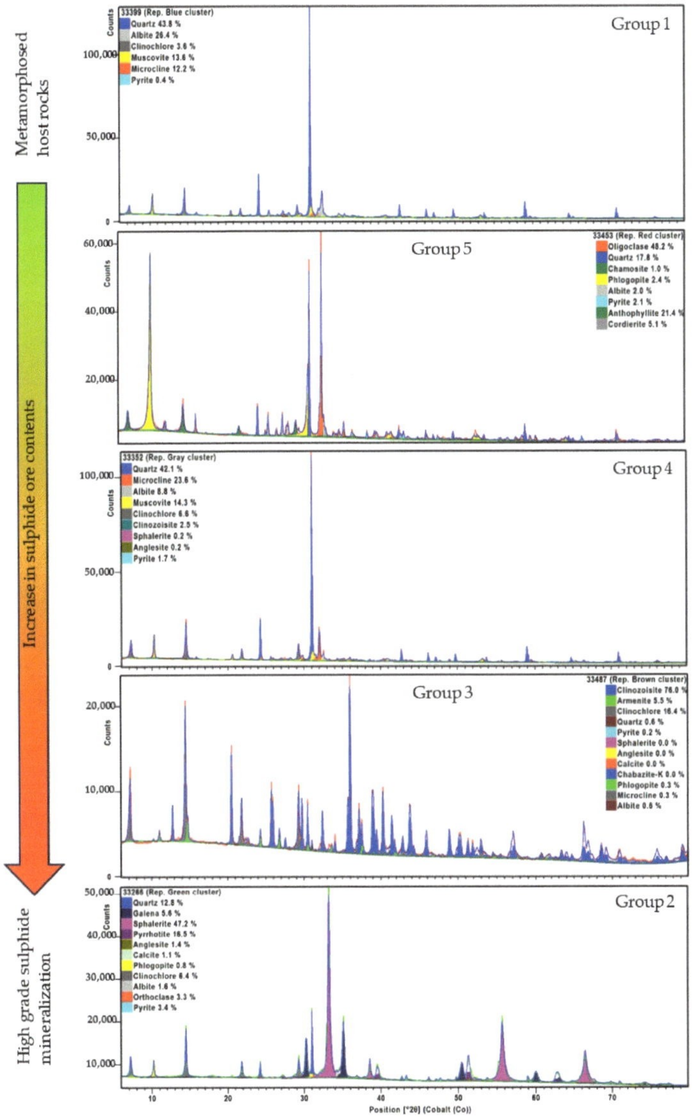

Figure 6. Full Rietveld pattern qualification of XRD scans for five samples, each representing one of the groups identified in CA2 (Groups 1 to 5 are highlighted in blue, green, brown, gray, and red, respectively). The graphs are arranged to illustrate the mineralogical changes from host rocks toward mineralization.

Table 3. Fuzzy clustering results for samples representing 15 clusters and 1 outlier following CA1. Probability coefficients are provided, with colors corresponding to the different clusters identified in Figures 4 and 5.

Sample ID	Rep. CA1	Cluster 1	Cluster 2	Cluster 3	Cluster 4	Cluster 5	Mixture
33370	Cluster 1	0.21	0.16	**0.7**	0.16	0.31	x
33487	Cluster 2	0.06	0.09	0.2	**0.87**	0.16	x
33398	Cluster 3	**0.48**	0.17	0.3	0.14	0.12	x
33272	Cluster 4	0.13	0.16	0.29	0.21	**0.68**	x
33453	Cluster 5	0.13	0.13	0.35	0.18	**0.71**	x
33266	Cluster 6	0.07	**0.58**	0.16	0.2	0.16	
33379	Cluster 7	0.13	0.01	0.44	**0.69**	0.2	x
33498	Cluster 8	0.14	0.05	**0.68**	0.21	0.16	
33395	Cluster 9	0.15	0.09	**0.57**	0.25	0.17	x
33373	Cluster 10	0.27	0.38	**0.69**	0.12	0.2	x
33352	Cluster 11	0.2	0.08	**0.83**	0.24	0.16	
33399	Cluster 12	**0.77**	0.13	0.2	0.1	0.14	
33386	Cluster 13	0.07	0.36	0.1	**0.67**	0.11	
33359	Cluster 14	**0.6**	0.03	0.24	0.06	0.02	
33410	Cluster 15	**0.65**	0.1	0.15	0.04	0.12	
33375	Non-clustered	0.26	**0.83**	0.33	0.15	0.1	

3.4. Variation in Mineralogical Characteristics Revealed by the QXRD Analysis

Phase identification and Rietveld quantification of the five XRD patterns identified as representative of all studied samples through a stepwise cluster analysis (CA1 and CA2) emphasize their distinct mineralogical compositions. As a result, Sample 33399, representing Group 1 in CA2 (Figure 5; also representative of Cluster 12 in CA1), is primarily composed of quartz, albite, microcline, clinochlore, and muscovite, with trace amounts of pyrite.

Unlike Group 1, for Group 2, the mineralogical analysis shows the highest degree of sulphide mineralization among the studied samples (Figure 6, Table 1). Rietveld quantification of Sample 33266, representative of this group (also representative of Group 6 in the initial cluster analysis), revealed a composition of 47.2% sphalerite, 16.5% pyrrhotite, 5.6% galena, and 3.4% pyrite, with a chemical composition inferred from stoichiometric formulas indicating 28.7% Zn and 5.8% Pb. Similarly, another member of Group 2, Sample 33375, exhibited high concentrations of Zn (20.8%) and Pb (2.8%).

Group 3 is defined by the variable dominance of clinozoisite, ranging from 36.0% to 76.0%, and the unique presence of minor amounts of armenite (2.1% to 5.5%). While sulphide ores are largely absent in the QXRD analysis reveals that Sample 33487, a representative of Group 3 in CA2 and Cluster 2 in CA1, lacks significant sulphide ores (also Cluster 2 in CA1), this sample is primarily composed of clinozoisite (76%), armenite (5.5%), and clinochlore (16.4%).

As illustrated in Figure 5, the three members of Group 3 (highlighted in brown) exhibit varying degrees of similarity and dissimilarity, as indicated by the differing lengths of the tie bars connecting them. Samples 33487 and 33379 (representing Cluster 7 in CA1) are linked by a short tie bar, indicating close compositional similarity, though Sample 33379 has lower clinozoisite (36.4%) and armenite (2.1%) contents and a notably higher proportion of quartz (39%) compared to the other two group members. In contrast, Sample 33386, the third member of Group 3, is connected by a longer tie bar in the dendrogram to its groupmates. Rietveld quantification showed that this sample has a distinct mineralogy with 15.1% sphalerite and dominant clinozoisite (72.4%).

Sample 33352, representative of Group 4, was retrieved from the drill hole GB16-1B as a coarse reject. The QXRD results revealed a mineral assemblage predominantly consisting of quartz, microcline, albite, clinochlore, clinozoisite, muscovite, pyrite, and trace amounts of sphalerite (Figure 6). The QXRD results for Sample 33453, a Group 5 representative, also revealed a composition of 21.4% anthophyllite, 5.1% cordierite, 1% chamosite, 2.4% phlogopite, 17.8% quartz, 48.2% oligoclase, 2.0% albite, and 2.0% pyrite. The dendrogram in Figure 5 connects Group 5 to Group 1 by the longest tie bar, reflecting significant compositional dissimilarity between these groups (Figure 5).

4. Discussion

This study's results showed that stepwise hierarchical clustering and PCA of 113 XRD patterns greatly simplified data analysis by reducing the dataset into 16 representative patterns, grouped into five meaningful clusters (Figure 5 and Figure S2 in the Supplementary Materials). Fuzzy analysis served as a valuable complementary tool, not only further validating the distinct identities of certain clusters (e.g., Clusters 4 and 6 in CA1), but also revealing that the compositional boundaries between some clusters are not sharply defined, as indicated by the overlap in membership coefficients for most samples (Table 3 and Table S1 in the Supplementary Materials). The compositional overlap between different groups can be attributed to shared host rock types and/or geological events, such as hydrothermal activities or metamorphism, which play a role in the formation and/or alteration of mineral assemblages in this area [13–19]. As illustrated in Figure 5, Group 1 consists of four samples from coarse rejects of three different drill holes, hosted by metasedimentary, metavolcanic, and tuffaceous rocks (Tables 1 and 4). Despite their overall similarity, CA2 revealed internal variations: Samples 33410 (GB16-4) and 33359 (GB16-1B) are closely linked, and both are connected to Samples 33398 and 33399 (GB16-2) by a longer tie bar. This subgrouping reflects their geological origins, with the GB16-2 samples sharing metasedimentary host rocks, while the other two samples are primarily associated with metavolcanic rocks [19].

Table 4. Summary of ore mineral composition and chemistry as determined by Rietveld analysis of XRD patterns (QXRD), compared with the base metal content analyzed by ICP-AES. An asterisk (*) indicates the representative sample of each group. Accounting for the Pb content in the absence or scarcity of galena. Acronyms: G: group; Sph—sphalerite; Gn—galena; Py—pyrite; Armn—armenite; Phy—pyrrhotite. The colors in the 2nd CA clusters' column match those in Figure 5, while the columns presenting the ICP-AES, chemistry, and mineralogy by XRD–Rietveld results use a color scale, with red indicating the highest values and green the lowest.

2nd CA	DH	Depth (m)	Sample ID	ICP-AES (%)		Chemistry_XRD-Rietveld (%)					Mineralog_XRD-Rietveld (%)				
				Zn	Pb	Zn	Pb	Fe	S	Ba	Sph	Gn	Py	Armn	Pyh
G1	GB16-1B	139.0–140.0	33359 (Rep. Cluster 14)	0.8	0.3	0.0	0.0	1.0	0.4	0.0	0.0	0.0	0.8	0.0	0.0
	GB16-2	29.5–31.3	33398 (Rep. Cluster 3)	0.1	0.0	0.0	0.0	2.1	0.2	0.0	0.0	0.0	0.4	0.0	0.0
		31.3–31.8	33399 (Rep. Cluster 12)	0.1	0.0	0.0	0.0	1.2	0.2	0.0	0.0	0.0	0.4	0.0	0.0
	GB16-4	48.0–49.5	33410 (Rep. Cluster 15)	0.0	0.0	0.0	0.0	0.2	0.0	0.0	0.0	0.0	0.4	0.0	0.0
G2	GB16-2	36.9–39.7	33266 (Rep. Cluster 6)	9.2	0.3	28.7	5.8	15.8	24.6	0.0	47.2	5.6	3.4	0.0	16.5
	GB16-3	59.1–59.7	33375 (Non-clustered)	16.3	2.3	20.8	2.9	4.9	13.3	0.0	34.2	1.6	3.1	0.0	0.0
G3	GB16-1	79.5–80.5	33379 (Rep. Cluster 7)	0.0	0.0	0.0	0.1	4.9	0.0	0.0	0.0	*	0.0	0.1	0.0
		86.5–87.9	33386 (Rep. Cluster 13)	10.7	0.1	10.1	0.0	4.2	5.0	0.6	15.1	0.0	0.0	4.7	0.0
	GB16-5	46.7–47.2	33487 (Rep. Cluster 2)	0.4	0.0	0.0	0.0	0.0	6.5	0.1	0.7	0.0	0.2	5.5	0.0
G4	GB16-1	128.0–129.0	33352 (Rep. Cluster 11)	0.4	0.2	0.1	0.1	2.4	1.0	0.0	0.2	*	1.7	0.0	0.0
	GB16-1	120.0–122.0	33498 (Rep. Cluster 8)	0.1	0.4	0.0	0.0	2.0	1.0	0.0	0.0	0.0	1.9	0.0	0.0
	GB16-2	27.4–28.0	33395 (Rep. Cluster 9)	0.1	0.0	0.0	0.0	3.7	1.0	0.0	0.0	0.0	1.9	0.0	0.0
	GB16-3	55.2–57.1	33370 (Rep. Cluster 1)	0.2	0.0	0.0	0.5	2.3	0.9	0.0	0.0	*	1.5	0.0	0.0
	GB16-3	58.1–59.1	33373 (Rep. Cluster 10)	2.3	0.3	0.9	0.3	2.1	1.3	0.0	1.3	*	1.6	0.0	0.0
G5	GB16-4	66.0–67.0	33272 (Rep. Cluster 4)	0.0	0.0	0.0	0.0	5.1	1.3	0.0	0.0	0.0	2.4	0.0	0.0
		56.5–58.0	33453 (Rep. Cluster 5)	0.0	0.0	0.0	0.0	3.4	1.1	0.0	0.0	0.0	2.0	0.0	0.0

Similarly, Group 4 consists of five samples from three different drill holes, predominantly hosted by metasedimentary rocks. CA2 analysis also revealed internal variations, with Samples 33352 and 33498 (both from GB16-1 but collected at different depths and hosted by poorly mineralized metasedimentary rocks containing mostly disseminated pyrite) forming a distinct subgroup. This subgroup is connected by a longer tie bar to Samples 33370 and 33373 (GB16-3), which originates from volcanic sandstone and siltstone with disseminated pyrite and sphalerite. Sample 33395 (GB16-2) links these two subgroups with another long tie bar. Additionally, the QXRD results in Table 4 strongly correlate with the CA findings, demonstrating that distinct mineralogical and compositional characteristics of the samples were key factors in their grouping. Additionally, variations within the groups reflect the individual mineralogy of each sample.

4.1. Mineralogical Signature of Host Rocks

The mineralogical composition of certain clusters demonstrates a strong correlation with their associated host rocks. For example, the mineralogy of Sample 33399, representative of Group 1, closely mirrors that of its fine-grained, tuffaceous laminated host rocks [19]. The presence of quartz, K- and Na-feldspars, chlorite, and muscovite (Figure 5) is consistent with host rocks that have undergone greenschist facies metamorphism. Additionally, the trace amounts of pyrite detected in this sample by QXRD align with the localized occurrence of base metal zones, where pyrite dissemination is a common feature in the host rocks [19]. This mineralogical alignment reinforces the relationship between cluster composition and geological processes. As shown in Table 4, the chemical composition of this sample, calculated from the stoichiometric formula of the identified minerals, strongly correlates with the ICP-AES results.

Although Group 5 exhibits significant compositional dissimilarity compared to Group 1, as illustrated by the longest tie bar in Figure 5, it is nonetheless characterized by a mineral assemblage (Figure 6) that reflects the typical mineralogy of Gumsberg metamorphosed host rocks (Table 1). Sample 33453, representative of this group, is a coarse reject from the drill hole GB16-4 at a depth of 48.0–92.0 m within metavolcanic porphyritic lithic tuffs (Table 1). These rocks have undergone intense hydrothermal alteration, which was later overprinted by strong polyphase deformation and metamorphism during the Svecokarelian orogeny [9,11–13].

Although the given sequence of events complicates the interpretation of the local geology and the understanding of paragenetic sequence of minerals [15], the significant presence of oligoclase and quartz in Sample 33453 (Figure 6) suggests that the original volcanic material was likely felsic in composition, consistent with the felsic metavolcanic rocks that characterize the region's lithotectonic unit [15–19]. The occurrence of anthophyllite and cordierite indicates that these rocks have undergone high-grade metamorphism, likely reaching amphibolite facies conditions [25]. Anthophyllite, a product of magnesium-rich environments, points to the mafic components within the original volcanic tuffs, while cordierite's presence suggests that the rocks were subjected to high temperatures, possibly due to proximity to syn-orogenic or post-orogenic intrusions. The minor amounts of phlogopite and chamosite further indicate a history of both prograde and retrograde metamorphism, with chamosite hinting at a retrograde phase as the region cooled and the higher-grade minerals began to break down [25]. Additionally, the presence of pyrite, although minor, suggests that the rocks were subjected to hydrothermal alteration, which aligns with the known history of significant hydrothermal activity in the Bergslagen district, which played a crucial role in metal recrystallization and remobilization during regional metamorphism [13,18,19].

4.2. Mineralogical Signature of Proximal Alteration Zones Revealed by the QXRD Data

Unlike Groups 1 and 5, which reflect the mineralogical signature of host rocks, Groups 3 and 4 represent a transition from metamorphic alteration zones to proximal sulphide mineralization (Figure 6). The QXRD results reveal a mineral assemblage predominantly consisting of quartz, microcline, and albite, alongside clinochlore, clinozoisite, muscovite, pyrite, and trace amounts of sphalerite (Figure 6). This mineralogy aligns with the host environment (Table 1), reflecting a history of intense hydrothermal alteration, sulphide deposition, followed by deformation and metamorphism during the Svecokarelian orogeny [15]. The abundance of silicate minerals suggests derivation from siliceous to tuffaceous rocks, while the presence of clinozoisite, clinochlore, and muscovite indicates a metamorphic overprint [10–13]. The occurrence of pyrite and sphalerite suggests proximity to underlying massive sulphide zones. Figure 7 compares the XRD patterns of Group 3 samples with mineral variations corresponding to massive clinozoisite and replacement-style sphalerite mineralization [13–17]. As illustrated, armenite variably occurs in these samples.

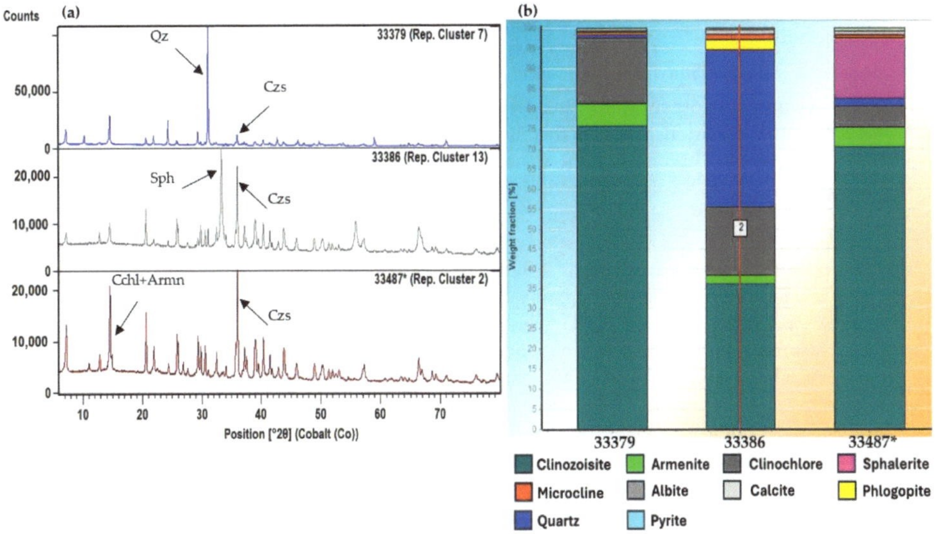

Figure 7. (**a**) 2D view of the distinct XRD patterns for samples clustered into Group 3 (highlighted in brown) in CA2. The most intense peak of each key mineral is marked. Acronyms—Qz: quartz; Czs: clinozoisite; Sph: sphalerite; Cchl: clinochlore; Armn: armenite. Armenite and clinozoisite have overlapping peaks. (**b**) Histograms illustrate the variable abundance of identified minerals in these samples. * Indicates the Group's representative scan.

In contrast to QXRD, the geological log data do not report the presence of armenite in the Group 3 samples. The ICP-AES analysis also measured average Ba concentrations of 0.2% and 0.5%, respectively, in some samples from the drill holes GB16-1 (80.5 to 92.7 m depth) and GB16-5 (28.1 to 47.2 m depth). Additionally, elevated Ba concentrations, typically occurring as barite and/or Ba-bearing feldspars, have been observed in other studied areas from the Bergslagen district [26,27].

Armenite is a rare double-ring silicate hydrate from the milarite-osumilite group containing barium, aluminum, and calcium with the formula of $BaCa_2Al_6Si_9O_{30}{\cdot}2H_2O$. It forms under low- to medium-grade metamorphic conditions, typically involving significant fluid–rock interaction, where barium-rich fluids interact with calcium-rich host rocks [28–31]. Skarn zones are common environments for its formation [28–31]. Dario et al. [28] reported the occurrence of armenite in the Rosas mine area (Mitza Sermentus mineworks, southwest Sardinia). They identified armenite-bearing samples collected along the contact zone between a sulphide-mineralized skarn vein and the black phyllite host rock. The black phyllite matrix primarily consists of muscovite, chamosite, and quartz, with feldspars, clinozoisite, titanite, and calcite as accessory phases. Gaspar and Inverno [30] also documented the occurrence of armenite alongside clinozoisite in the distal strata-bound scheelite skarns at the Riba de Alva Mine in Northeastern Portugal. They observed that armenite was present within the mineralized skarn zones, where it was sometimes intergrown with or replacing clinozoisite. Armenite also occurs in outcrops of leucocratic gneiss composed of celsian, margarite, quartz, white mica, and zoisite in the Berisal Complex, Swiss Central Alps. The estimated conditions for its formation were temperatures between 280 and 320 °C and pressures ranging from 2.0 to 2.6 kbar [30]. Dario et al. [29] emphasized three key reasons why armenite merits further investigation beyond its apparent rarity and distinctive crystal structure: (1) Genesis Insight: Understanding armenite's formation could provide valuable information about the pressure–temperature–fluid conditions in the rocks where it occurs, which are often mineralized. (2) Potential Underestimation: Armenite is likely more common than previously believed, as it is difficult to identify using standard techniques and may frequently be overlooked in geological studies. (3) Indicator of Ba-rich Deposits: Given that armenite formation requires a primary barium source, its presence could indicate the proximity of Ba-rich mineral deposits.

4.3. QXRD Insights into the Mineralogical Signature of Massive Sulphides

In contrast to other groups, the mineralogy of Group 2 (in green), as determined by Rietveld quantification, indicates the highest grade of sulphide mineralization among the studied samples. This observation is consistent with both the geological log data [19] and the grade analysis results obtained by ICP-AES (Tables 1 and 2). The Rietveld analysis showed that Sample 33266, which represented Group 6 in the initial cluster analysis, predominantly consists of sphalerite, followed by pyrrhotite galena, and pyrite (Figure 6). The chemical composition inferred from the stoichiometric formulas of the identified minerals indicates this sample contains 28.7% Zn, and 5.8% Pb (Table 4). Similarly, another member of Group 2, Sample 33375, which was not clustered with any other samples in the CA1 due to its unique mineral composition, also shows high concentrations of Zn (20.8%) and Pb (2.8%). However, the ICP-AES analysis reported lower Zn contents for both samples in Group 2. Overall, XRD effectively captures the general trends in element concentrations, such as Zn, Pb, and Fe, as compared to ICP-AES. The discrepancy between the results from these two methods can be attributed to mineralogical heterogeneity within the mineralized zone, along with sample heterogeneity and the absence of homogenization procedures. These factors lead to variations in the compositions of the sub-samples analyzed by ICP-AES and XRD.

The stoichiometric calculation also reveals the presence of two Pb-bearing minerals in Sample 33266: galena and anglesite. Of the total 5.8% Pb measured in this sample by QXRD (Table 4), 4.8% is associated with the 5.6% abundance of galena, while approximately 1.0% is attributed to anglesite, which has an abundance of 1.4% in this sample (Figure 6). The occurrence of anglesite in sulphide mineralization zones raises questions about its origin. Anglesite ($PbSO_4$) typically occurs in two distinct forms, as identified by Zeng et al. [32]: (1) primary hydrothermal anglesite, which forms under low Pb/high S conditions, and (2) secondary supergene anglesite, which develops in high Pb/low S environments. Primary hydrothermal anglesite is found in the pores of precipitated hydrothermal fluids, often intergrown with pyrite and sphalerite. In contrast, secondary supergene anglesite forms through the low-temperature alteration of galena (PbS), frequently surrounding relict galena.

While further investigation using optical microscopy or SEM is needed to confirm the presence or absence of galena relics within anglesite, the occurrence of anglesite alone in some of the studied samples, particularly in association with sphalerite and pyrite (Group 2 and Group 3; Figure 6) likely suggests a primary hydrothermal origin. However, as shown in Figure 7, the presence of clinozoisite, albite, clinochlore, phlogopite, quartz, and trace amounts of calcite alongside anglesite and the sulphide ores reflects the region's complex geological history, characterized by hydrothermal mineralization, metamorphic overprinting, and subsequent weathering.

This finding serves as a reminder that while common minerals are consistently documented in geological log data, minor minerals like anglesite, which can have significant implications for mineral exploration, might not always be identified or noted. This oversight occurs because these minor minerals often appear in smaller quantities or are less visually distinctive, making them easier to overlook during routine logging processes. The experience of the logging geologist or petrographer plays a crucial role in being alert to the presence of certain minerals in specific geological settings.

Moreover, Table 4 indicates an elevated Fe content in Group 2 samples and other samples may correspond to various Fe-bearing phases, it is crucial to identify the specific minerals contributing to the overall Fe content (or other elements of interest). This level of insight allows for a more accurate interpretation of geochemical data, which is critical for precise economic resource estimation. The stochiometric calculation (enabled by the chemistry calculator tool included in HighScore Plus [3,20]) provides detailed insights into the sample composition, enabling a thorough analysis of mineral phases and their elemental contributions. For instance, in Sample 33266, the stochiometric calculation reveals that Fe is distributed across several mineral phases: 3% of Fe is present within the structure of sphalerite, approximately 10.2% in pyrrhotite, around 1.0% in clinochlore, and 1.6% in pyrite.

5. Conclusions

This study demonstrates the effectiveness of QXRD coupled with Rietveld refinement and extended cluster analysis in characterizing the complex mineralogy of the Gumsberg polymetallic sulphide deposit. The analysis identified five distinct mineralogical clusters, offering valuable insights into geological processes and alteration zones. Two clusters revealed mineralogical signatures of metamorphic host rocks, while two others indicated different proximal hydrothermal-metamorphic alteration zones—one with a variable base metal mineralization signature and another featuring massive clinozoisite and sub-sea floor replacement-style sphalerite mineralization. The fifth cluster also demonstrated the mineralogical characteristics of polymetallic sulphides. The detection of significant sulphide mineralization, particularly high levels of sphalerite, pyrrhotite, and galena, highlights the economic potential of the region. The presence of rare minerals like armenite, often missed in traditional methods, underscores the benefits of QXRD.

QXRD offers significant advantages as a complementary tool to traditional techniques such as optical microscopy, which rely on the petrography of a limited number of thin sections and the experience of the petrographer. QXRD allows for the rapid, high-throughput analysis of every sample, making it a time- and cost-efficient tool for exploration and mining. The straightforward sample preparation for QXRD further enhances its practicality. QXRD results can quickly (within minutes) identify samples needing further investigation, streamlining decision-making and improving the overall efficiency of exploration and mining programs. In this study, the accuracy of QXRD was cross-validated with geological logs and geochemical data obtained using the ICP-AES technique, confirming its reliability as a mineralogical discrimination tool.

Supplementary Materials: The following supporting information can be downloaded at: https://www.mdpi.com/article/10.3390/min14111100/s1, Figure S1: (a) Plan view outcrop map with location of drill holes including two cross-sections: (b) AA′ through Vallberget-Loberget, and (b) BB′ through Mellangruvan [17]. The historic holes, before 2016, are shown in gray. The red dots are the intersections for the 2016 drill holes. Note: the location of drill hole GB-16-4 is not seen in the images as it was a target in Gumsgruvan, another trend within the Gumsberg property, ~1–1.5 km NW of the Vallberget area. Figure S2: Dendrogram illustrating the agglomerative hierarchical clustering of 114 analyzed XRD patterns (1st cluster analysis). The XRD patterns are grouped into 15 clusters, with one outlier (sample 33375) remaining unclustered. Table S1: Results of fuzzy clustering of 113 XRD scans of studied samples. Probability/membership coefficients are provided. A coefficient $\geq$ 0.85 for any of the 15 clusters indicates that the sample exclusively belongs to that cluster; Figure S3c: Full Rietveld pattern quantification of XRD scan of sample 33399 representative of Group 1 in the 2nd cluster analysis (CA2); Figure S2c. Full Rietveld pattern quantification of XRD scan of sample 33453 representative of Group 5 in the 2nd cluster analysis (CA2). Figure S3c. Full Rietveld pattern quantification of XRD scan of sample 33352 representative of Group 4 in the 2nd cluster analysis (CA2): Figure S4C. Full Rietveld pattern quantification of XRD scan of sample 33379 representative of Group 3 in the 2nd cluster analysis (CA2); Figure S5C: Full Rietveld pattern quantification of XRD scan of sample 33266 representative of Group 2 in the 2nd cluster analysis (CA2).

Author Contributions: Conceptualization—S.M. and E.R.; Formal analysis—S.M., E.R. and T.W.; Investigation—S.M. and E.R.; Methodology—S.M., E.R., T.W. and M.P.; Validation—S.M.; Visualization—S.M.; Resources—H.R.; Writing—original draft preparation, S.M.; Writing—review and editing, S.M., E.R., T.W., M.P. and H.R. All authors have read and agreed to the published version of the manuscript.

Funding: This research received no external funding.

Data Availability Statement: Data are contained within the article.

Acknowledgments: We would like to extend our gratitude to EMX Scandinavia AB/EMX Royalty Corp. for their collaboration and for providing samples along with geological and geochemical databases. T. Degen and M. Sommariva are also thanked for their technical advice and support.

Conflicts of Interest: Sheida Makvandi, Evelien Rost, Thomas Witzke, and Matteo Pernecheleare are employees of the Application Competence Center. Also, Hein is an employee of Raat EMX Royalty Corporation. The paper reflects the views of the scientists and not the company.

References

1. Pernechele, M.; López, Á.; Davoise, D.; Maestre, M.; König, U.; Norberg, N. Value of Rapid Mineralogical Monitoring of Copper Ores. *Minerals* **2021**, *11*, 1142. [CrossRef]
2. König, U.; Degen, T.; Norberg, N. PLSR as a New XRD Method for Downstream Processing of Ores: Case Study: Fe_2 + Determination in Iron Ore Sinter. In *Powder Diffraction*; Cambridge University Press: Cambridge, UK, 2014; Volume 29, pp. S78–S83.
3. König, U. Nickel Laterites—Mineralogical Monitoring for Grade Definition and Process Optimization. *Minerals* **2021**, *11*, 1178. [CrossRef]
4. Makvandi, S.; Cauzid, J.; Tarantola, A.; Melfos, V.; Voudouris, P. XRD Clustering and Quantitative Analysis as a Fingerprinting Technique to Study Ore Deposits. In Proceedings of the 17th SGA Biennial Meeting 2023, Zürich, Switzerland, 28 August–1 September 2023.
5. Makvandi, S.; Pagé, P.; Tremblay, J.; Girard, R. Exploration for Platinum-Group Minerals in till: A New Approach to the Recovery, Counting, Mineral Identification and Chemical Characterization. *Minerals* **2021**, *11*, 264. [CrossRef]
6. Makvandi, S.; Beaudoin, G.; McClenghan, M.B.; Quirt, D.; Ledru, P. PCA of Fe-oxides MLA data as an advanced tool in provenance discrimination and indicator mineral exploration: Case study from bedrock and till from the Kiggavik U deposits area (Nunavut, Canada). *J. Geochem. Explor.* **2019**, *197*, 199–211. [CrossRef]
7. Franklin, J.M.; Gibson, H.L.; Jonasson, I.R.; Galley, A.G. Volcanogenic Massive Sulphide Deposits. In Proceedings of the 12th Biennial SGA Meeting, Uppsala, Sweden, 12–15 August 2013; pp. 523–560.
8. Barbosa, L.; Tiu, G.; Jasson, N.; Wanhainen, C.; Lilja, L.; Ghorbani, Y. Gold occurrence in the footwall of the Lappberget Deposit, Garpenberg Mine, Sweden. *Ore. Geol. Rev.* **2024**, *171*, 106174. [CrossRef]
9. Stephens, M.B. *Synthesis of the Bedrock Geology in the Bergslagen Region, Fennoscandian Shield, South-Central Sweden*; Sveriges Geologiska Undersokning (SGU): Uppsala, Sweden, 2009.
10. Stephens, M.B.; Jansson, N.F. Paleoproterozoic (1.9–1.8 Ga) syn-orogenic magmatism, sedimentation and mineralization in the Bergslagen lithotectonic unit, Svecokarelian Orogen. *Geol. Soc. Lond. Mem.* **2020**, *50*, 155–206. [CrossRef]
11. Allen, R.L.; Jansson, N.; Ripa, M. Bergslagen: Geology of the Volcanic-and Limestone-Hosted Base Metal and Iron Oxide Deposits. Excursion Guidebook Swe 4. In Proceedings of the 12th Biennial SGA Meeting, Uppsala, Sweden, 12–15 August 2013.
12. Vivallo, W. The geology and genesis of the Proterozoic massive sulfide deposit at Garpenberg, central Sweden. *Econ. Geol.* **1985**, *80*, 17–32. [CrossRef]
13. Allen, R.L.; Lundstrom, I.; Ripa, M.; Christofferson, H. Facies analysis of a 1.9 Ga, continental margin, back-arc, felsic caldera province with diverse Zn-Pb-Ag-(Cu-Au) sulfide and Fe oxide deposits, Bergslagen region, Sweden. *Econ. Geol.* **1996**, *91*, 979–1008. [CrossRef]
14. Jansson, N. The Origin of Iron Ores in Bergslagen, Sweden, and Their Relationships with Polymetallic Sulphide Ores. Ph.D. Thesis, Luleå Tekniska Universitet, Luleå, Sweden, 2011.
15. Jansson, N.F.; Allen, R.L. Multistage ore formation at the Ryllshyttan marble and skarn-hosted Zn-Pb-Ag-(Cu)+magnetite deposit, Bergslagen, Sweden. *Ore Geol. Rev.* **2015**, *69*, 217–242. [CrossRef]
16. Andersson, S.S.; Jonsson, E.; Hogdahl, K. Metamorphism and deformation of a Palaeoproterozoic polymetallic sulphide–oxide mineralisation: Hornkullen, Bergslagen, Sweden. *GFF* **2016**, *138*, 410–423. [CrossRef]
17. Kampmann, T.C.; Stephens, M.B.; Ripa, M.; Hellstrom, F.A.; Majka, J. Time constraints on magmatism, mineralisation and metamorphism at the Falun base metal sulphide deposit, Sweden, using U-Pb geochronology on zircon and monazite. *Precambrian Res.* **2016**, *278*, 52–68. [CrossRef]
18. Fahlvik, A.; Kampmann, T.C.; Jansson, N.F. Hydrothermal alteration, lithogeochemical marker units and vectors towards mineralisation at the Svärdsjö Zn-Pb-Cu deposit, Bergslagen, Sweden. *GFF* **2022**, *144*, 177–195. [CrossRef]

19. EMX Royalty Corp. *Gumsberg-Au-Cu (Zn-Pb) VMS Projects, Bergslagen District, Southern Sweden. Technical Marketing Report*; EMX Royalty Corp.: Vancouver, BC, Canada, 2016.
20. Degen, T.; Sadki, M.; Bron, E.; König, U.; Nénert, G. The High Score Suite. In *Powder Diffraction*; Cambridge University Press: Cambridge, UK, 2014; Volume 29, pp. S13–S18.
21. Gates-Rector, S.; Blanton, T. The Powder Diffraction File: A Quality Materials Characterization Database. *Powder Diffr.* **2019**, *34*, 352–360. [CrossRef]
22. Rietveld, H.M. A Profile Refinement Method for Nuclear and Magnetic Structures. *J. Appl. Crystallogr.* **1969**, *2*, 65–71. [CrossRef]
23. Kelley, L.A.; Gardner, S.P.; Sutcliffe, M.J. An automated approach for clustering an ensemble of NMR-derived protein structures into conformationally related subfamilies. *Protein Eng. Des. Sel.* **1996**, *9*, 1063–1065. [CrossRef]
24. Liao, B.; Chen, J. The application of cluster analysis in X-ray diffraction phase analysis. *J. Appl. Crystallogr.* **1992**, *25*, 336–339. [CrossRef]
25. Morrison, S.M.; Prabhu, A.; Hazen, R.M. An evolutionary system of mineralogy, Part VIII: The evolution of metamorphic minerals. *Am. Mineral.* **2024**, *109*, 1760–1784. [CrossRef]
26. Jansson, N.F.; Allen, R.L.; Skogsmo, G.; Turner, T. Origin of Palaeoproterozoic, sub-seafloor Zn-Pb-Ag skarn deposits, Sala area, Bergslagen, Sweden. *Miner. Depos.* **2021**, *57*, 455–480. [CrossRef]
27. Camitz, J.; Rauséus, G.; Jönberger, J.; Persson, L.; Sopher, D.; Bastani, M. *Characterisation of Mining Waste in Central and Southern Bergslagen, Sweden*; Geological Survey of Sweden: Uppsala, Sweden, 2024.
28. Dario, F.; Luca, D.M.; Antonio, A.; Fabrizio, C.; Antonio, F.; Stefano, N. Armenite: A Really Rare Mineral? In Proceedings of the 3rd European Mineralogical Conference EMC 2020, Cracow, Poland, 6–10 September 2020.
29. Gaspar, L.M.; Inverno, C.M.C. Mineralogy and metasomatic evolution of distal strata-bound scheelite skarns in the Riba de Alva Mine, Northeastern Portugal. *Econ. Geol.* **2000**, *95*, 1259–1275. [CrossRef]
30. Hetherington, C.J.; Mullis, J.; Graeser, S.; Gieré, R. Formation of Armenite in the Berisal Complex, Simplon Region, Switzerland. *Schweiz. Mineral. Petrogr. Mitteilungen* **2003**, *83*, 243–259.
31. Ray, G.L.E.; Webster, I.C.L. *Skarns in British Columbia. Geological Survey of British Columbia*; Geological Survey Branch: Reston, VA, USA, 1997.
32. Zeng, Z.; Chen, Z.; Qi, H. Two Processes of Anglesite Formation and a Model of Secondary Supergene Enrichment of Bi and Ag in Seafloor Hydrothermal Sulfide Deposits. *J. Mar. Sci. Eng.* **2022**, *10*, 35. [CrossRef]

minerals

MDPI

Article

Metasomatic to Hydrothermal Genesis of Natural Calcium Silicate Hydrates (C-S-H): Evidence from Lessini Mountains, Veneto Volcanic Province, Italy

Michele Mattioli [1,*], Matteo Giordani [1] and Franco Filippi [2]

1 Department of Pure and Applied Sciences, University of Urbino Carlo Bo, 61029 Urbino, Italy; matteo.giordani@uniurb.it
2 Independent Researcher, Strada Vicinale Monte Crocetta 33, 36100 Vicenza, Italy; filippifranco43@gmail.com
* Correspondence: michele.mattioli@uniurb.it

Abstract: We report the occurrence of natural calcium silicate hydrates (C-S-H) from the Grolla quarry in the Lessini Mountains of Northern Italy. These minerals are formed by basic and ultrabasic magma interacting with carbonate rocks. The mineral assemblage includes thaumasite, xonotlite, tobermorite, and plombierite, often intergrown with other silicates, as well as minor amounts of carbonates and sulfates. Common zeolites in this area include chabazite, phillipsite/harmotome, natrolite, and thomsonite. Although less abundant, these zeolites are typically associated with calcite, fluoroapophyllite, and barite. The Grolla quarry outcrop allows for the study of the in situ complex crystalline overgrowths and specific crystal chemistry of rare natural mineral phases, such as C-S-H minerals, formed under metasomatic to hydrothermal conditions.

Keywords: C-S-H; tobermorite; plombierite; xonotlite; thaumasite; zeolites; Grolla quarry

1. Introduction

Calcium silicate hydrate minerals (C-S-H, the dashes denoting their versatile stoichiometry) [1] are natural and synthetic phases that are of great scientific and technological interest. In particular, C-S-H (according to cement shorthand, C = CaO, S = SiO_2 and H = H_2O) are the main binding phases in many artificial materials, such as aerated concrete (with or without additives), mortars and cement paste, making up 60%–70% of a fully hydrated paste. Due to their critical role in the safety of building structures and construction, C-S-H minerals have been extensively studied, as reflected in the abundant literature [2–6]. C-S-H minerals have a variable composition, are often X-ray amorphous, and can include aluminum and other guest ions. Aluminum incorporation likely affects the chemical and mechanical properties of C-S-H [7]. Improving the strength and durability of hydrated cement—by addressing factors like porosity, creep, shrinkage, and aging—requires a deep understanding of the C-S-H structure at the crystalline level and the atomic scale up to 100 nm [8,9].

Although C-S-H crystalline phases are the main components of artificial materials such as cement, they are rare as minerals of natural origin, especially due to their particular environment and condition of formation. Only a small number of restricted geological environments are known for their natural C-S-H phases, and the main occurrences come from ultrabasic rocks associated with ophiolite complexes involved in low-temperature serpentinization process, or hornfels (contact metamorphic zones with high temperature and low pressure) around igneous intrusions in limestones and calcareous shales [10–12].

During the metasomatic processes that take place in these conditions, significant quantities of volatiles (especially water) can be mobilized, inducing the formation of C-S-H phases such as xonotlite, tobermorite, plombierite, afwillite, hillebrandite, and scawtite [13–15]. This family of minerals is particularly interesting for their various structural arrangements and the peculiar transformation processes in which they are involved (e.g., dehydration). For example, the natural mineral tobermorite is often used as a model to study the molecular structure of C-S-H [16]. Investigating the long-term alteration and carbonation (10s or 100s of thousands of years) of these minerals also makes it very important to simulate man-made Portland cement alteration and, for example, to the sealing performance of well cement systems [4,12]. Moreover, C-S-H minerals are able to exchange cations and have potential applications in waste disposal [17]. Therefore, obtaining data on the crystallization of C-S-H under natural conditions and its transformations is particularly important for understanding cement mixtures' hardening and durability mechanisms.

In this study, different samples of C-S-H phases naturally crystallized in the Lessini Mountains, Northern Italy, were mineralogically and chemically characterized to describe the mineralization within a rare and unusual geological context. Various methodologies were employed, including binocular microscope observations, environmental scanning electron microscope (ESEM) observations, chemical composition analysis (SEM-EDS), and X-ray powder diffraction (XRPD) analysis. The origins of the natural C-S-H phases and potential toxicological aspects are also discussed.

2. Geological Background

The Veneto Volcanic Province covers about 2000 square kilometers in Northern Italy (Figure 1). Due to tectonic tension in the South Alpine foreland, volcanic activity began in the Tertiary period, mostly underwater [18]. Volcanic eruptions happened in bursts from the Late Paleocene to the Miocene, each lasting briefly and separated by inactive phases with shallow-water carbonate deposits [19–21]. This volcanic area has four main districts: the Lessini Mountains, Marostica Hills, Berici Hills, and Euganei Hills. The main rock types include volcaniclastic rocks, hyaloclastites, pillow lavas, and lava flows of mafic to ultramafic composition. In the Euganei Hills, acidic rocks like rhyolite are also found.

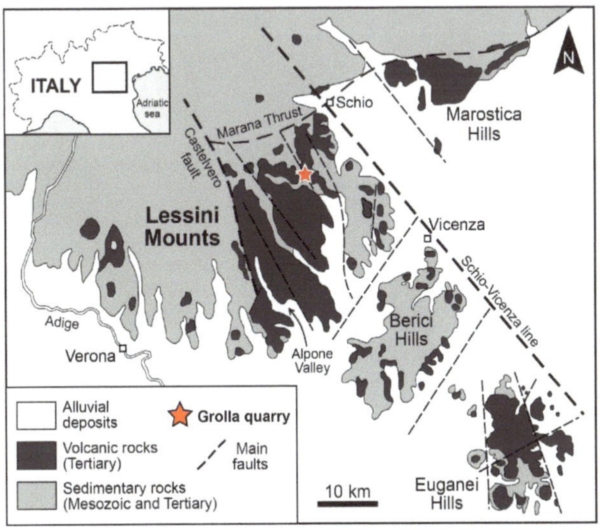

Figure 1. Simplified geological map of the Veneto Volcanic Province highlighting the Lessini Mountains and the location of the Grolla quarry.

The Lessini Mountains stretch from Mt. Baldo in the west to the Schio-Vicenza tectonic line in the east, within a north-northwest-oriented extensional structure called the Alpone-Agno graben. The western fault defines the Castelvero boundary, which was key in shaping the region's environment from the Late Paleocene to the Middle Eocene. This fault separated a sediment-dominated western area with thin volcanic layers from an eastern area rich in volcanic rocks. With layers up to 400 m thick, the Lessini Mountains form the largest volcanic sequence in the Veneto Volcanic Province [20]. The area's volcanism includes mainly tuffs and lava flows, with column-like eruptive necks, hyaloclastites, and pillow lavas. Common rock types are basanite and alkali olivine basalt, with smaller amounts of transitional basalts, nephelinite, hawaiites, trachy-basalt, and basaltic andesites. Studies indicate that the magmatism here has within-plate characteristics, likely forming in a tensional setting as a foreland response to the Alpine orogeny [22–24].

The volcanic rocks of the Lessini Mountains are often extensively altered by intense hydrothermal activity. This process results in significant secondary mineralization, filling numerous cavities and vesicles within the volcanic rocks and forming new mineral phases [25]. Among these, zeolites are the most abundant group, including minerals such as analcime, chabazite, phillipsite, harmotome, and gmelinite. Rare zeolites with distinctive features, such as willhendersonite and yugawaralite, have also been identified in the area [25–27]. Additionally, in recent years, other fibrous zeolites (including carcinogenic erionite, and potentially toxic offretite, mordenite, and ferrierite) have been discovered in several localities of the Lessini Mountains, highlighting their particular importance in toxicological research [28–33].

A rare but significant type of mineralization in the Lessini area is associated with thermometamorphism, a process occurring at very high temperatures and low pressures. This phenomenon affects sedimentary rocks near volcanic bodies, creating a contact zone (or aureole) around the magma. New minerals often form within this aureole, influenced by the type of rocks present, the chemical composition of circulating fluids, and the prevailing P-T conditions. A distinctive form of thermometamorphism occurs when basic magmas, hydrothermally altered volcanic rocks, and carbonate rocks interact. During metasomatism, chemical exchange between rocks releases volatile substances, mainly water. This likely leads to the formation of hydrated calcium silicates like xonotlite, tobermorite, and plombierite, making the Grolla quarry a globally important site for studying these minerals.

3. Materials and Methods

3.1. Materials

All the samples analyzed in this study were collected from the Grolla quarry in Cornedo Vicentino (VI), Lessini Mountains, Italy. Grolla quarry is an active site that produces calcareous materials for construction, prized for their excellent technical and mechanical properties. Large veins and volcanic conduits were uncovered during quarrying operations, cutting through a sequence of carbonate sedimentary rocks. Tertiary basic magmatism led to the formation of abundant crystals in the contact zones between hydrothermally altered basic rocks and calcareous rocks. These crystals range in size from macroscopic (>1 cm) to microscopic (<<1 mm) and are found growing in veins, cavities, and vesicles within deeply metasomatized rocks. The cavities and vesicles vary from millimeters to centimeters and may be fully or partially filled with secondary minerals. In some cases, a single mineral species occupies the space, while in others, multiple species coexist in paragenetic associations. Over 150 samples were collected from various quarry sectors, with 20 selected for detailed investigation to represent the diversity observed in the field.

3.2. Optical and Scanning Electron Microscopy

Observations of the morphology and physical properties were conducted directly on the cavities and vesicles using a Leica Wild Mod. M10 binocular optical microscope at the Department of Pure and Applied Sciences, University od Urbino, Italy, with IntraLux 5000-1 fiber optic illumination, and a DCM-510 Digital image acquisition system.

Small-scale morphological observations were conducted using an environmental scanning electron microscope (ESEM) FEI Quanta 200 FEG, at the Department of Pure and Applied Sciences, University od Urbino, Italy, equipped with an energy-dispersive X-ray spectrometer (EDS) for qualitative microchemical analysis. The settings included a 25 kV accelerating voltage, adjustable beam diameter, a working distance of 10–12 mm, and a tilt angle of 0°. The ESEM operated in low vacuum mode, with the chamber pressure between 0.80 and 0.90 mbar. Images were captured using a single-shot detector (SSD) or an Everhart–Thornley secondary electron detector (ETD). Microchemical analysis was conducted using a JEOL 6400 SEM at the University of Parma, Italy, following the methods outlined in the references [34,35]. The operating conditions were voltage 15 kV, beam current 1.2 nA, and counting time 100 s. The errors are 2%–5% for the major elements and 5%–10% for the minor ones. Compositions of elements, oxides, and silicates were used as standards.

3.3. X-Ray Powder Diffraction

Pure crystals were carefully selected from each sample under a binocular microscope, then disaggregated, finely ground in an agate mortar, and placed in a 0.7 mm side-opened aluminum sample holder. X-ray powder diffraction (XRPD) data were collected using a Philips X'Change PW1830 powder diffractometer at the Department of Pure and Applied Sciences, University od Urbino, Italy, set to 35 kV accelerating voltage and 30 mA beam current, with CuKα radiation (λ = 1.54506 Å). Measurements were made in Bragg–Brentano geometry from 2° to 65° at 2θ, with a 0.01° step size and 2.5 s per step to obtain high-intensity patterns. The diffractometer used a 1° maximum divergence compensating slit, a 0.2 mm receiving slit, and a graphite crystal monochromator. Semi-quantitative XRPD analysis was conducted using X'Pert Quantify and X'Pert HighScore Plus software, version 5.2 release, with quartz as the internal calibration standard. Peak width variations were minimal (within ±2% of the average value), and each sample was measured three times. Detailed X-ray diffraction analyses with extended exposure times (up to 24 h) were performed for each sample to determine the mineralogical composition, assess the separated crystals' quality, and rule out the presence of impurities.

4. Results and Discussion

A summary of the investigated samples and their related mineralogical composition is reported in Table 1. The minerals identified at the Grolla quarry are predominantly silicates, with smaller amounts of carbonates and sulfates. The most common silicates belong to the zeolite group, including chabazite, phillipsite/harmotome, natrolite, and thomsonite. Though generally minor, these zeolites are typically associated with varying amounts of calcite, fluoroapophyllite, and barite. This mineral assemblage is consistent with similar occurrences reported from other parts of the Lessini Mountains [25]. However, unlike other mineralized areas in the Lessini Mountains, the Grolla quarry uniquely yields calcium silicate hydrate minerals (C-S-H) alongside the previously mentioned minerals. Specifically, we identified thomsonite, xonotlite, tobermorite, and plombierite (Table 1). The powder XRD patterns of these C-S-H natural phases are shown in Figure 2. The narrow peaks indicate high crystallinity. Table 2 lists their chemical compositions, which match the ideal stoichiometry. Backscattered electron imaging during ESEM analysis showed no unusual grains or secondary phases in the investigated samples. The C-S-H minerals described below were positively identified using X-ray powder diffraction, optical data, and chemical microanalyses. While stoichiometries and optical data resembling other phases (e.g., afwillite, scawtite, jennite, clinotobermorite) were obtained, definitive identification was not possible due to the absence of clear X-ray data.

Table 1. Summary of the investigated samples from Grolla quarry and their related mineralogical composition, identified through optical microscope, environmental scanning electron microscope (ESEM), and X-ray powder diffraction (XRPD) analyses. Calcium silicate hydrates are in **bold**.

Sample	Mineralogical Composition
CG1	Phillipsite, Thomsonite, Chabazite, Calcite
CG2	Barite, Chabazite
CG3	Chabazite
CG4	**Xonotlite**, Apofillite, Chabazite, Calcite
CG5	Apofillite, Natrolite, Calcite
CG6	Thomsonite
CG7	**Xonotlite**, Chabazite, Thomsonite
CG8	Chabazite, Phillipsite/Armotomo
CG9	Chabazite, Apophyllite, Calcite
CG11	**Thaumasite**, **Xonotlite**
CG12	Thomsonite, Calcite
CG13	**Xonotlite**, Calcite
CG14	**Tobermorite**, **Plombierite**, **Xonotlite**, Calcite
CG16	Thomsonite, Chabazite, Calcite
CG17	**Xonotlite**, **Tobermorite**, Fluorapophyllite
CG18	Thomsonite, Calcite
CG19	**Tobermorite**, **Plombierite**, Thomsonite, Calcite
CG20	**Thaumasite**, Fluorapophyllite, Chabazite, Calcite
CG21	Natrolite, Calcite
CG30	Chabazite
CG39	**Tobermorite**, **Plombierite**, Calcite

4.1. *Xonotlite* $Ca_6Si_6O_{17}(OH)_2$

Xonotlite was first identified by Rammelsberg in 1866 in carbonate rocks that had undergone contact metamorphism with magma in Tetela de Xonotla, Mexico (whence the mineral is given its name) [36]. Since then, although it is a rare mineral, xonotlite has been discovered to fill veins and cavities in heat-altered rocks, and less commonly due to metasomatic reactions in rodingites [37]. Xonotlite has a monoclinic crystal structure, typically forming thin, needle-like (acicular) or fibrous crystals a few millimeters in size. Its structure consists of chains of tetrahedra alternating with calcium–oxygen layers [38]. These layers contain calcium ions in two coordination types: seven-fold (CaO_7) and octahedral (CaO_6) [39]. This arrangement allows xonotlite to exist in at least six structural variations (four ordered and two disordered forms) or polytypes [40,41].

Xonotlite is common in the Grolla quarry and appears in various forms (Figure 3a,b). It often forms very thin, needle-like crystals grouped into silky, fibrous clusters. These are generally white inside but may appear white to pink on the surface. Another frequent type is small, transparent needle-like crystals sometimes found in small bundles. Xonotlite also forms isolated, silky-looking spheroidal clusters with glassy crystals. The crystals sometimes grow in parallel, forming fibrous white layers that follow the rock's surface. In other cases, the crystals are so fine that they appear as white, earthy coatings visible only under a microscope. Under an electron microscope, xonotlite crystals are always thin and fibrous, ranging from a few microns to hundreds of microns in length, with diameters smaller than one micron (Figure 4). Larger crystals sometimes show their monoclinic shape, dominated by prism-like features. In samples from the Grolla quarry, xonotlite is often found in aggregates, alongside large fluoroapophyllite crystals and small calcite rhombohedrons. In all investigated samples, xonotlite always forms on the surface of other minerals and seems to develop in the final stages of crystallization.

Table 2. Representative chemical composition of the natural C-S-H minerals from the Grolla quarry. H_2O has been calculated by difference.

	Thaumasite								Xonotlite					
	CG11-01	CG11-02	CG11-03	CG20-01	CG20-02	CG20-06	CG20-07	CG20-08	CG17-01	CG17-02	CG17-03	CG17-04	CG17-08	CG17-09
SiO_2	9.85	9.81	10.02	9.98	9.75	9.91	9.88	10.05	50.15	49.61	49.24	50.44	50.25	49.98
Al_2O_3	bdl	bdl	bdl	bdl	bdl	bdl	bdl	bdl	0.05	bdl	0.04	0.02	0.05	0.02
FeO_{tot}	0.1	0.11	0.23	0.1	0.25	0.2	0.12	0.2	0.07	0.21	0.04	0.11	0.15	0.05
MnO	0.04	0.02	bdl	0.04	bdl	bdl	bdl	0.05	bdl	bdl	bdl	bdl	bdl	bdl
MgO	bdl	bdl	bdl	bdl	bdl	bdl	bdl		0.02	0.12	bdl	0.22	0.21	0.15
CaO	26.95	26.94	26.55	26.78	26.81	26.88	26.64	26.88	47.45	46.66	47.17	48.13	47.74	47.61
P_2O_5	0.42	0.41	0.55	0.38	0.42	0.46	0.35	0.44	bdl	bdl	bdl	bdl	bdl	bdl
SO_3	12.27	12.29	12.21	12.43	12.77	12.35	12.21	12.85	bdl	bdl	bdl	bdl	bdl	bdl
CO_2	7.1	7.51	6.85	7.88	7.41	7.15	6.55	6.97	bdl	bdl	bdl	bdl	bdl	bdl
(F, Cl)	0.25	0.28	0.3	0.18	0.22	0.24	0.35	0.21	bdl	bdl	bdl	bdl	bdl	bdl
H_2O	43.02	42.63	43.29	42.23	42.37	42.81	43.9	42.35	2.31	3.4	3.55	1.1	1.65	2.21
Total	100	100	100	100	100	100	100	100	100	100	100	100	100	100
	Tobermorite										**Plombierite**			
	CG19-01	CG19-02	CG19-03	CG19-06	CG14-01	CG14-02	CG14-04	CG14-05	CG14-09	CG14-10	CG14-11	CG39-01	CG39-02	CG39-03
SiO_2	45.12	45.31	45.22	45.19	45.41	44.53	45.68	44.84	43.81	44.12	43.05	41.91	41.76	41.05
TiO_2	bdl	0.02	bdl	0.02	bdl	bdl	bdl	bdl	bdl	bdl	bdl	bdl	bdl	bdl
Al_2O_3	0.31	0.39	0.41	0.38	0.31	0.25	0.28	0.26	0.15	0.11	0.23	1.05	1.14	1.05
CaO	35.11	34.22	34.06	34.52	35.15	34.18	34.55	34.61	30.47	31.09	32.6	33.12	33.17	33.84
H_2O	19.46	20.06	20.31	19.89	19.13	21.04	19.49	20.29	25.57	24.68	24.12	23.92	23.93	24.06
Total	100	100	100	100	100	100	100	100	100	100	100	100	100	100

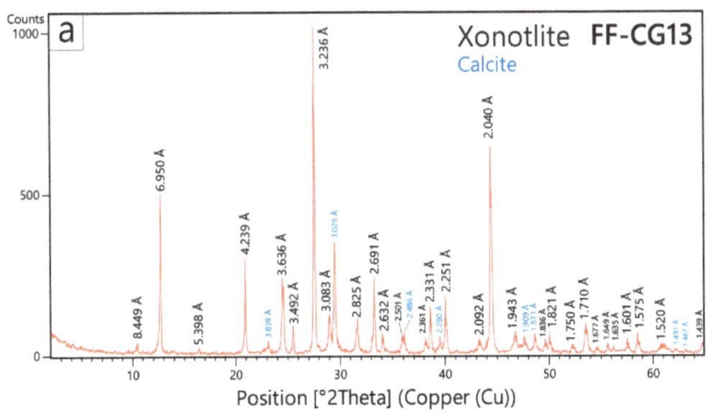

Figure 2. *Cont.*

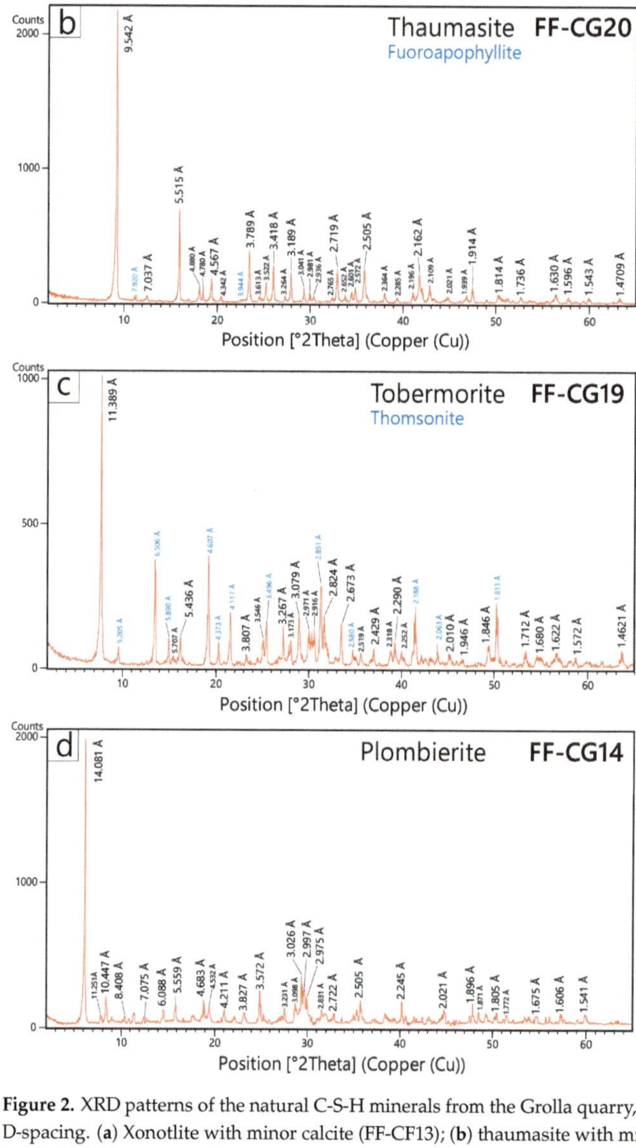

Figure 2. XRD patterns of the natural C-S-H minerals from the Grolla quarry, with the indication of D-spacing. (**a**) Xonotlite with minor calcite (FF-CF13); (**b**) thaumasite with minor fluoroapophyllite (FF-CG20); (**c**) Tobermorite with subordinate thomsonite from (FF-CG19); (**d**) plombierite (FF-CG14). The d-spacing values of minor minerals are in blue.

The chemical analysis of xonotlite from Grolla quarry shows a composition close to its ideal formula (Table 2). It is a hydrated calcium silicate, with 49.24–50.44 wt% SiO_2 and 46.66–48.13 wt% CaO. Other elements, such as Al_2O_3, FeO, and MgO, are only present in trace amounts, and no significant variations in composition were observed across the different crystal forms.

Figure 3. (**a**) Fibrous xonotlite crystals grouped in millimetric silky-looking spheroidal to botryoidal clusters (CG11); (**b**) needle-like xonotlite crystals grouped into silky, fibrous clusters (CG13); (**c**) prismatic vitreous crystals of thaumasite with a pseudo-hexagonal section (CG20); (**d**) tobermorite in sub-spherical, fibrous-radial forms creating white, silky, botryoidal aggregates (CG19); (**e**,**f**) plombierite in white botryoidal aggregates with a silky sheen, formed by small, sub-spherical fibrous-radial crystals (CG19).

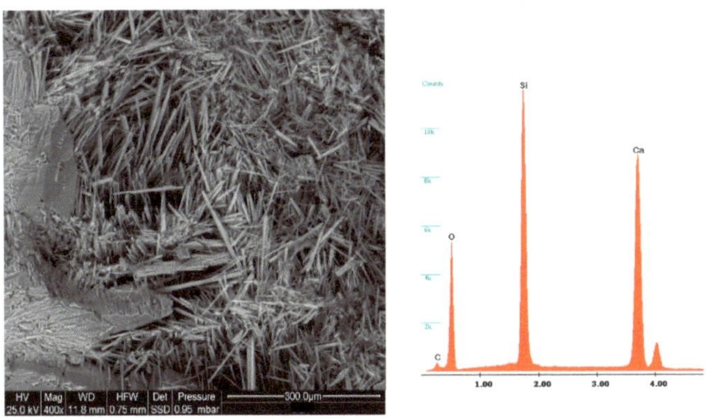

Figure 4. Typical morphologies of xonotlite crystals from the Grolla quarry. The crystals vary from flattened prismatic shapes to highly elongated acicular forms, with some developing into strongly fibrous structures. The co-growth of xonotlite with fluoroapophyllite crystals is visible, with fluoroapophyllite appearing in the left part of the image.

4.2. *Thaumasite* $Ca_3Si(OH)_6(CO)_3(SO_4)\cdot 12H_2O$

Thaumasite is a rare, naturally occurring mineral discovered in 1878 by Baron von Nordenskiöld at the Areskuta copper mine in Sweden. Its name comes from the Greek word thaumazein, meaning "to be surprised", due to its unusual composition, which includes carbonate, sulfate, and hexahydrosilicate groups. A very particular crystal chemical formula characterizes thaumasite: it contains ~42% H_2O, 7% CO_2, 13% SO_3, 28% CaO, and 10% SiO_2, giving it a very low density (~1.89 g/cm^3). Thaumasite formation requires a source of these ions. The crystal-chemical structure of thaumasite was solved only recently (late 1900s) through X-ray diffraction analysis of single crystals [42–44]. It consists of columns of $[Ca_3Si(OH)_6(H_2O)_{12}]^{4+}$ parallel to [001] and linked to $CO_3{}^{2-}$ and $SO_4{}^{2-}$ groups and water molecules via hydrogen bonds. Thaumasite is the only known mineral with silicon in six-fold coordination with hydroxyl groups (OH) under low-pressure conditions [45,46]. Thaumasite belongs to the ettringite group and forms a secondary phase at low temperatures in mafic igneous and metamorphic rocks. It is often found alongside minerals like ettringite, zeolites, calcite, gypsum, and tobermorite. In concrete, thaumasite is significant due to its role in sulfate attack, where sulfate ions react with cement compounds (C-S-H, portlandite) to form gypsum, ettringite, or thaumasite [47–50]. These reactions degrade concrete in sulfate-rich environments.

Thaumasite in the Grolla quarry is typically found in small, fibrous to acicular crystals or prismatic shapes with a hexagonal section (Figure 3c). The more common forms are fibrous-radial aggregates, whitish in color, forming sub-spherical or botryoidal shapes, and chaotic clusters of acicular crystals, transparent to translucent, forming fan-like or intertwined aggregates, especially with calcite, barite, and fluorapophyllite crystals (Figure 5). Microscopic observations show that thaumasite is one of the first minerals to crystallize on cavity and vesicle walls, along with calcite and fluoroapophyllite. Less commonly, thaumasite can form thin, whitish patinas or crusts with a silky sheen, often hard to distinguish from minerals like xonotlite or plombierite without detailed analysis.

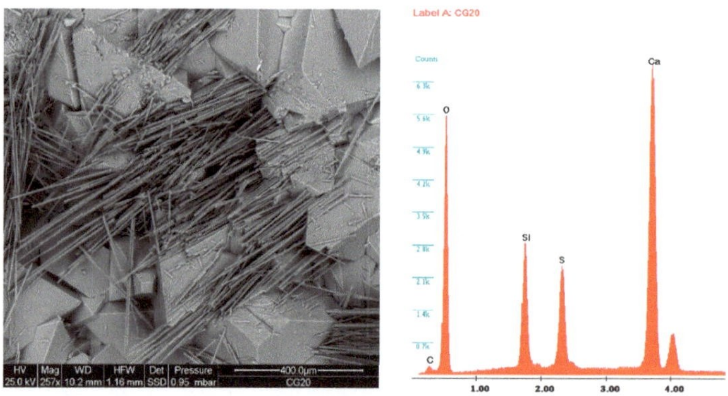

Figure 5. Thaumasite crystals associated with fluoroapophyllite from the Grolla quarry exhibit a very elongated prismatic to acicular habit, indicating their fibrous nature and growth patterns. The EDS spectrum is related to thaumasite crystals, with the typical chemical composition given by silicon, calcium, carbon, and sulfur.

Thaumasite from the Grolla quarry is chemically consistent (Table 2). Calcium is always the most abundant cation (CaO ~26–27 wt%), while SiO_2 is in the order of 10 wt%. The SO_3 group is present in significant quantities (12–13 wt%), as is CO_2 (7–8 wt%). The H_2O content is between 42 and 43 wt%, and the halogen (F, Cl) content is less than 1%, as is that of phosphorus and iron.

4.3. Tobermorite (Tobermorite-11 Å) $Ca_5Si_6O_{17}$ $5H_2O$

Tobermorite was first identified by Heddle in 1880 in samples from several locations in Scotland, especially near Tobermory on the Isle of Mull [51]. X-ray diffraction studies confirmed its status as a distinct mineral in the mid-20th century, revealing a characteristic basal reflection at 11.3 Å [52–54]. Significant progress has been made in understanding the structure of minerals in the tobermorite group over the past few decades. Studies by Merlino et al. [55,56] and Bonaccorsi et al. [57] have determined their crystal structures, providing insights into this group's chemical and structural differences.

The minerals in the tobermorite group are hydrated calcium silicates similar to the crystalline phases (C-S-H) formed during cement hydration. Structurally, tobermorites consist of layers of calcium polyhedra (with seven-fold coordination) arranged parallel to the (001) plane [11]. These layers are connected to single chains of "wollastonite-type" tetrahedra on both sides. Their classification is based on water content, which affects the spacing between layers (basal distance) and, thus, their structure. Three types of tobermorites were named based on hydration levels: plombierite (highly hydrated, 14 Å basal distance), tobermorite (moderately hydrated, 11 Å basal distance), and riversideite (low hydration, 9 Å basal distance). Recently, advances in understanding their crystal chemistry have led to a standardized naming system and the creation of the Tobermorite Supergroup, named after the most common mineral in this category [11,58]. According to IMA (International Mineralogical Association), the Tobermorite Supergroup includes tobermorite group minerals (obermorite and kenotobermorite), clinotobermorite group minerals (clinotobermorite, kenoclinotobermorite, and kalitobermorite), and two additional minerals (plombierite and riversideite). Although plombierite and riversideite are not officially recognized names, they are widely used in the scientific literature and considered reference species [58].

Recent research has clarified tobermorite's crystal chemistry and thermal behavior [59,60]. Its calcium content varies between 4 and 5 atoms per formula unit (apfu). Lower calcium levels may result from non-homogeneous material or calcium-poor phases, while higher levels may occur due to associations with calcium-rich phases like xonotlite or calcite. Aluminum can also replace silicon in the structure, reaching up to 1 apfu, though aluminum-free samples are rare and found mainly in South Africa's Kalahari Manganese Field [61]. Tobermorite mixes two end-member compositions: $Ca_4Si_6O_{15}(OH)_2{\cdot}5H_2O$ and $Ca_5Si_6O_{17}{\cdot}5H_2O$. The slight difference in composition influences its thermal behavior: a higher calcium content corresponds to "normal" thermal behavior, while low-Ca samples behave "anomalously" [60].

In the samples from the Grolla quarry, the tobermorite appears in several forms. It can form as (1) colorless needle-like crystals, transparent to translucent, often in fibrous-radial aggregates; (2) whitish fibrous crystals exhibiting pseudo-radial growth or small bundles (Figure 6a); (3) sub-spherical, fibrous-radial forms creating small, silky, botryoidal aggregates (Figure 3d); (4) intertwined needle-like crystals forming chaotic aggregates; and (5) thin coatings or vein fillings, appearing fine-grained, whitish, and silky to earthy in luster. Like thaumasite, microscopic evidence suggests that tobermorite also forms during the early stages of crystallization in mineralized cavities.

The chemical composition of Grolla quarry tobermorite is uniform and matches its theoretical formula (Table 2). It contains ~45 wt% SiO_2, 34–35 wt% CaO (4–5 apfu), and low aluminum (<1 wt% Al_2O_3), indicating minimal silicon–aluminum substitution. Its water content ranges between 19 and 21 wt%.

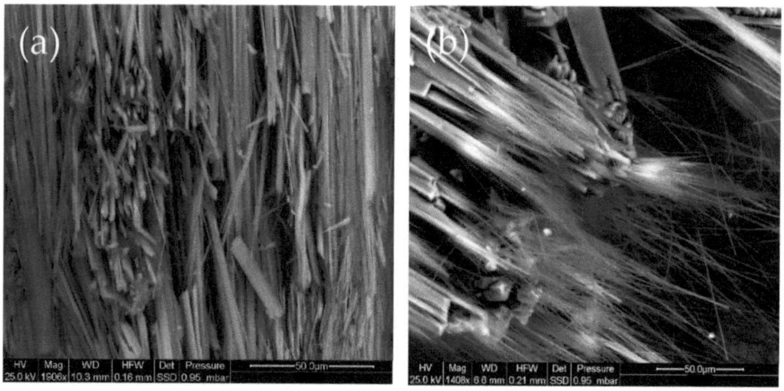

Figure 6. (**a**) Prismatic to fibrous crystals of tobermorite forming small bundles; (**b**) prismatic-acicular crystals of plombierite displaying a striking extremely fibrous termination, reflecting their characteristic morphology.

4.4. *Plombièrite (Tobermorite-14 Å)* $Ca_5Si_6O_{16}(OH)_2\ 7H_2O$

Plombierite belongs to the Tobermorite Supergroup, as noted in the nomenclature guidelines of the CNMNC (Commission on New Minerals, Nomenclature and Classification) and the IMA (International Mineralogical Association). While minerals in the tobermorite and clinotobermorite groups are officially recognized, plombierite and riversideite are not fully acknowledged. Still, they are considered reference species due to their established use in the scientific literature.

Plombierite was first used in 1858 to describe a silicate gel formed by thermal waters in Roman-era cementitious materials in Plombières, France [62]. In the mid-1900s, McConnell analyzed a similar material from Ballycraigy, Ireland, using X-ray diffraction [52]. The material was identified as a hydrated calcium silicate (C-S-H) group member, with the most hydrated phase classified as plombierite. It was later found that plombierite is a variety of tobermorite with a basal spacing of 14 Å, now known as tobermorite-14 Å. Despite this, the name plombierite remains widely used.

The structure of plombierite has been solved using a specimen from Crestmore (California, USA) [57]. Like other hydrated calcium silicates, plombierite consists of layers of calcium polyhedra (CaO_7) parallel to (001). These layers are linked by single tetrahedra (SiO_4) chains, arranged along the b-axis, and stacked along the c-axis. The layers are connected by extra calcium ions and water molecules, which fill cavities in the structure. The water content of plombierite determines its basal spacing, giving it the highest hydration level among the phases in the Tobermorite Supergroup, with a basal distance of 14 Å [60,63]. In this structure, the layers are separated by calcium ions and nearly double the water molecules found in the less hydrated 11 Å phases (five water molecules versus three). Concerning the ideal chemical formula, the composition of plombierite shows limited variability, with minor changes in the calcium content (usually 4.5–5 atoms per formula unit) and small aluminum–silicon substitutions.

In the Grolla quarry, plombierite generally appears as colorless needle-like crystals, transparent to translucent, often forming fibrous-radial aggregates (Figure 3e,f), or whitish acicular crystals, sometimes in pseudo-radial clusters or small bundles displaying striking extremely fibrous termination (Figure 6b). More frequently, plombierite forms botryoidal aggregates with a silky sheen, formed by small, sub-spherical fibrous-radial crystals, or it can be found as thin coatings or vein fillings, fine-grained with a silky to earthy luster, and whitish in color. Plombierite forms later than other minerals in all observed samples and often appears at the ends of tobermorite crystals.

Plombierite from the Grolla quarry shows slight variability across samples, and two main groups can be distinguished based on the major element composition (Table 2). The first group shows slightly higher silica (SiO_2 43–44 wt%) and lower calcium (CaO 30–33 wt%) and aluminum (Al_2O_3 0.1–0.2 wt%) contents, while in the second group, silica varies from 41 to 42 wt%, calcium is in the range 33–34 wt% and aluminum is more than 1 wt%. The water content in all the samples ranges between 19wt% and 21wt%.

4.5. Origin of C-S-H Minerals and Potential Implications

Natural C-S-H minerals typically form through heat-driven metamorphic reactions in contact zones between silicate and carbonate rocks [63,64]. Recent research highlights significant carbonation reactions in calcium silicate hydrate-bearing rocks [13,65]. These reactions generally occur in low-temperature hydrated calcium and aluminum silicate minerals surrounding high-temperature metamorphic calcium or aluminum silicate cores. Basic and ultrabasic magmas, rich in calcium, magnesium, and iron silicates like olivine, serpentine, pyroxene, and plagioclase, are especially reactive with dissolved CO_2 in water, making them well suited for forming solid carbonates [66–68]. Unfortunately, the data collected from the investigated samples do not allow for the reconstruction of a precise genetic crystallization sequence of the C-S-H minerals, as additional information is required. However, based on the data and observations of the samples and their growth conditions, an early crystallization phase of C-S-H minerals at the Grolla quarry, possibly linked to high-temperature metasomatic events, can be proposed. These events can occur in the contact zones between basic and ultrabasic melts associated with the magmatism of the Veneto Volcanic Province and the surrounding carbonate rocks. Thaumasite and tobermorite likely crystallized during this phase, along with calcite and fluorapophyllite. Later, a lower-temperature phase, possibly linked to hydrothermal processes that have already been widely described in this area [25], could have led to the formation of xonotlite and plombierite, along with sulfates (e.g., baryte) and zeolites such as chabazite, phillipsite/harmotome, natrolite, and thomsonite.

Finally, a final consideration emerges from the discovery, in the Grolla quarry samples, of C-S-H minerals with an extremely fibrous appearance. These characteristics raise worries about potential adverse health effects. Small particles, especially those with a fibrous shape, are known to cause toxic and cancerous lung effects [69–72]. Most scientific knowledge about the toxicity and cancer-causing effects of fibrous particles comes from research on asbestos, the most extensively studied mineral fiber. However, recent studies on erionite and other fibrous zeolites have grown significantly, providing new insights into the toxicology of fibrous particles [73–75]. The presence of fibrous minerals is particularly relevant for carbonate and volcanic rocks, which are widely mined and processed in the same quarries, with the finished products distributed for various uses. The discovery of C-S-H minerals with a highly fibrous habit in the Lessini Mountains suggests the need for future studies to quantify potentially airborne fibers for a preliminary risk assessment.

5. Concluding Remarks

This study provides a comprehensive analysis of natural calcium silicate hydrates (C-S-H) from the Grolla quarry in the Lessini Mountains of Northern Italy. The main conclusions are as follows:

- The minerals identified at the Grolla quarry are predominantly silicates, with subordinate carbonates and sulfates. The most common silicates are zeolites, including chabazite, phillipsite/harmotome, natrolite, and thomsonite, often associated with calcite, fluoroapophyllite, and barite.

- Unlike other mineralized areas in the Lessini Mountains, the Grolla quarry uniquely yields calcium silicate hydrate minerals (C-S-H) such as thomsonite, xonotlite, tobermorite, and plombierite. Most of the C-S-H minerals exhibit an acicular to extremely fibrous habit.
- The origin of the main phases in the Grolla quarry can be linked to metasomatic events and hydrothermal stages. This site offers a unique opportunity to study complex crystalline overgrowths and the specific crystal chemistry of rare natural mineral phases, such as C-S-H minerals.
- Finally, the presence of C-S-H phases with a highly fibrous structure raises concerns for human health, especially given the number of quarries and mining activities in the Lessini Mountains.

Author Contributions: M.M.: supervision, conceptualization, investigation, data curation, writing—original draft preparation, writing—review and editing, and funding acquisition. M.G.: formal analysis, data curation, and writing—review and editing. F.F.: sample collection, data curation, and writing—review and editing. All authors have read and agreed to the published version of the manuscript.

Funding: This research was conducted under the framework of the 2022 research programs of the Department of Pure and Applied Sciences of the University of Urbino Carlo Bo (project "New asbestiform natural fibers: mineralogical and physical-chemical characterization of fibrous zeolites", responsible M. Mattioli).

Data Availability Statement: Data are contained within the article. The data presented in this study can be seen in the content above.

Conflicts of Interest: The authors declare no conflicts of interest. The funders had no role in the study's design, in the collection, analyses, or interpretation of the data, in the writing of the manuscript, or in the decision to publish the results.

References

1. Scrivener, K.L.; Kirkpatrick, R.J. Innovation in use and research on cementitious material. *Cem. Concr. Res.* **2008**, *38*, 128–136 [CrossRef]
2. Richardson, I.G.; Groves, G.W. Models for the composition and structure of calcium silicate hydrate (C-S-H) gel in hardened tricalcium silicate pastes. *Cem. Concr. Res.* **1992**, *22*, 1001–1010. [CrossRef]
3. Richardson, I.G. The nature of C-S-H in hardened cements. *Cem. Concr. Res.* **1999**, *29*, 1131–1147. [CrossRef]
4. Richardson, I.G. The calcium silicate hydrates. *Cem. Concr. Res.* **2008**, *38*, 137–158. [CrossRef]
5. Picker, A.; Nicoleau, L.; Burghard, Z.; Bill, J.; Zlotnikov, I.; Labbez, C.; Nonat, A.; Cölfen, H. Mesocrystalline calcium silicate hydrate: A bioinspired route toward elastic concrete materials. *Sci. Adv.* **2017**, *3*, e1701216. [CrossRef]
6. Tang, S.; Wang, Y.; Geng, Z.; Xu, X.; Yu, W.; A, H.; Chen, J. Structure, Fractality, Mechanics and Durability of Calcium Silicate Hydrates. *Fractal Fract.* **2021**, *5*, 47. [CrossRef]
7. Faucon, P.; Petit, J.C.; Charpentier, T.; Jacquinot, J.F.; Adenot, F. Silicon substitution for aluminum in calcium silicate hydrates. *J. Am. Ceram. Soc.* **1999**, *82*, 1307–1312. [CrossRef]
8. Hou, D.; Zhao, T.; Ma, H.; Li, Z. Reactive molecular simulation on water confined in the nanopores of the valcium silicate hydrate gel: Structure, reactivity, and mechanical properties. *J. Phys. Chem. C* **2015**, *119*, 1346–1358. [CrossRef]
9. Papatzani, S.; Paine, K.; Calabria-Holley, J. A comprehensive review of the models on the nanostructure of calcium silicate hydrates. *Constr. Build. Mater.* **2015**, *74*, 219–234. [CrossRef]
10. Jakobsson, S.; Moore, J.G. Hydrothermal minerals and alteration rates at Surtsey volcano, Iceland. *Geol. Soc. Am. Bull.* **1986**, *97*, 648–659. [CrossRef]
11. Bonaccorsi, E.; Merlino, S. Modular microporous minerals: Cancrinite-davyne group and CSH phases. *Rev. Mineral. Geochem.* **2005**, *57*, 241–290. [CrossRef]
12. Milodowski, A.E.; Rochelle, C.A.; Lacinska, A.; Wagner, D. A natural analogue study of CO_2-cement interaction: Carbonation of calcium silicate hydrate-bearing rocks from Northern Ireland. *Energy Procedia* **2011**, *4*, 5235–5242. [CrossRef]
13. Marincea, S.; Bilal, E.; Verkaeren, J.; Pascal, M.L.; Fonteilles, M. Superposed parageneses in the spurrite-, tilleyite and gehlenite-bearing skarns from Cornet Hill, Apuseni Mountains, Romania. *Can. Mineral.* **2001**, *39*, 1435–1453. [CrossRef]

14. Hoffmann, C.; Armbruster, T. Clinotobermorite, $Ca_5[Si_3O_8(OH)]_2 \cdot 4H_2O$–$Ca_5[Si_6O17] \cdot 5H_2O$, a natural C-S-H(I) type cement mineral: Determination of the substructure. *Z Kristallogr.* **1997**, *212*, 864–873. [CrossRef]
15. Aguirre, L.; Dominguez-Bella, S.; Morata, D.; Wittke, O. An occurrence of tobermorite in tertiary basalts from Patagonia, Chile. *Can. Mineral.* **1998**, *36*, 1149–1155.
16. Maeshima, T.; Noma, H.; Sakiyama, M.; Mitsuda, T. Natural 1.1 and 1.4 nm tobermorites from Fuka, Okayama, Japan: Chemical analysis, cell dimensions, 29Si NMR and thermal behavior. *Cem. Concr. Res.* **2003**, *33*, 1515–1523. [CrossRef]
17. Hu, S.; Xu, Z.; Ma, X.; Huo, Y.; Yang, Y. Preparation of C-S-H seeds from solid waste and its application as Portland cement accelerator. *Constr. Build. Mater.* **2024**, *428*, 136277. [CrossRef]
18. Zampieri, D. Segmentation and linkage of the Lessini Mountains normal Faults, Southern Alps, Italy. *Tectonophys.* **2000**, *319*, 19–31. [CrossRef]
19. De Vecchi, G.P.; Gregnanin, A.; Piccirillo, E.M. Aspetti petrogenetici del vulcanesimo terziario veneto. *Mem. Ist. Geol. Mineral. Dell'universita Padova* **1976**, *30*, 1–32.
20. Barbieri, G.; De Zanche, V.; Medizza, F.; Sedea, R. Considerazioni sul vulcanismo terziario del Veneto occidentale e del Trentino meridionale. *Rend. Soc. Geol. Ital.* **1982**, *4*, 267–270.
21. De Vecchi, G.P.; Sedea, R. The Paleogene basalt ofthe Veneto region (NE Italy). *Mem. Ist. Geol. Mineral. Dell'università Padova* **1995**, *47*, 253–274.
22. Milani, L.; Beccaluva, L.; Coltorti, M. Petrogenesis andevolution of the Euganean magmatic complex, Veneto Region, North-East Italy. *Eur. J. Mineral.* **1999**, *11*, 379–399. [CrossRef]
23. Bonadiman, C.; Coltorti, M.; Milani, L.; Salvini, L.; Siena, F.; Tassinari, R. Metasomatism in the lithospheric mantleand its relationships to magmatism in the Veneto Volcanic Province, Italy. *Period. Di Mineral.* **2001**, *70*, 333–357.
24. Beccaluva, L.; Bianchini, G.; Bonadiman, C.; Coltorti, M.; Milani, L.; Salvini, L.; Siena, F.; Tassinari, R. Intraplate litospheric and sublitospheric components in the Adriatic domain: Nephelinite to tholeiite magma generation in the Paleogene Veneto volcanic province, southern Alps. *Mineral. Soc. Am. Spec. Pap.* **2007**, *418*, 131–152.
25. Mattioli, M.; Cenni, M.; Passaglia, E. Secondary mineral assemblages as indicators of multi stage alteration processes in basaltic lava flows: Evidence from the Lessini Mountains, Veneto Volcanic Province, Northern Italy. *Period. Mineral.* **2016**, *85*, 1–24.
26. Mattioli, M.; Cenni, M. First Occurrence of Willhendersonite in the Lessini Mounts, Northern Italy. *Crystals* **2021**, *11*, 109. [CrossRef]
27. Cametti, G.; Giordani, M. Humidity-and temperature-dependent study of YUG type zeolite. A new dehydrated topology. *Microporous Mesoporous Mater.* **2024**, *363*, 112811. [CrossRef]
28. Giordani, M.; Mattioli, M.; Dogan, M.; Dogan, A.U. Potential carcinogenic erionite from Lessini Mounts, NE Italy: Morphological, mineralogical and chemical characterization. *J. Toxicol. Environ. Health A* **2016**, *79*, 808–824. [CrossRef]
29. Giordani, M.; Mattioli, M.; Ballirano, P.; Pacella, P.; Cenni, M.; Boscardin, M.; Valentini, L. Geological occurrence, mineralogical characterization and risk assessment of potentially carcinogenic erionite in Italy. *J. Toxicol. Environ. Health B* **2017**, *20*, 81–103. [CrossRef]
30. Cangiotti, M.; Battistelli, M.; Salucci, S.; Falcieri, E.; Mattioli, M.; Giordani, M.; Ottaviani, M.F. Electron paramagnetic resonance and transmission electron microscopy study of the interactions between asbestiform zeolite fibers and model membranes. *J. Toxicol. Environ. Health A* **2017**, *80*, 171–187. [CrossRef] [PubMed]
31. Mattioli, M.; Giordani, M.; Arcangeli, P.; Valentini, L.; Boscardin, M.; Pacella, A.; Ballirano, P. Prismatic to asbestiform offretite from Northern Italy: Occurrence, morphology and crystal-chemistry of a new potentially hazardous zeolite. *Minerals* **2018**, *8*, 69. [CrossRef]
32. Giordani, M.; Ballirano, P.; Pacella, A.; Meli, M.A.; Roselli, C.; Di Lorenzo, F.; Fagiolino, I.; Mattioli, M. Fibrous mordenite from Northern Italy: Another potentially hazardous zeolite. *Minerals* **2022**, *121*, 627. [CrossRef]
33. Giordani, M.; Cametti, G.; Di Lorenzo, F.; Churakov, S.V. Real-time observation of fibrous zeolites reactivity in contact with simulated lung fluids (SLFs) obtained by atomic force microscope (AFM). *Minerals* **2019**, *9*, 83. [CrossRef]
34. Goldstein, J.I.; Newbury, D.E.; Echlin, P.; Joy, D.C.; Romig, A.D.; Lyman, C.E.; Fiori, C.; Lifshin, E. *Scanning Electron Microscopy and X-Ray Microanalysis*, 2nd ed.; Plenum Press: New York, NY, USA, 1992.
35. Pacella, A.; Ballirano, P.; Cametti, G. Quantitative chemical analysis of erionite fibres using a micro-analytical SEM-EDX method. *Eur. J. Mineral.* **2016**, *28*, 257–264. [CrossRef]
36. Rammelsberg, C. Über den Xonaltit, ein neues wasserhaltiges Kalksilikat, und den Bustamit aus Mexico. *Z. Deutsch. Geol. Ges.* **1866**, *18*, 33–34.
37. O'brien, J.P.; Rodgers, K.A. Xonotlite and rodingites from Wairere, New Zealand. *Mineral. Mag.* **1973**, *39*, 233–240. [CrossRef]
38. Kalousek, G.L.; Mitsuda, T.; Taylor, H.F.W. Xonotlite: Cell parameters, thermogravimetry and analytical electron microscopy. *Cem. Concr. Res.* **1977**, *7*, 305–312. [CrossRef]
39. Churakov, S.V.; Mandaliev, P. Structure of the hydrogen bonds and silica defects in the tetrahedral double chain of xonotlite. *Cem. Concr. Res.* **2008**, *38*, 300–311. [CrossRef]

40. Hejny, C.; Armbruster, T. Polytypism in xonotlite $Ca_6Si6O_{17}(OH)_2$. *Z. Für Krist. Cryst. Mater.* **2001**, *216*, 396–408. [CrossRef]
41. Bernstein, S.; Fehr, K.T.; Hochleitner, R. Crystal chemistry of Xonotlite Ca6Si6O17(OH)2. Part I: Determination of polytypes using X-Ray Powder Diffraction (XRPD). *N. Jb. Miner. Abh.* **2009**, *186*, 153–162. [CrossRef]
42. Hurlbut, C.S.J.; Baum, J.L. Ettringite from Franklin, New Jersey. *Am. Mineral.* **1960**, *45*, 1137–1143.
43. Knill, D.C. Thaumasite from Co. Down, Northern Ireland. *Mineral. Mag. J. Mineral. Soc.* **1960**, *32*, 416–418. [CrossRef]
44. Edge, R.A.; Taylor, H.F.W. Crystal structure of thaumasite [Ca_3Si (OH) 6.12 H_2O](SO_4)(CO_3). *Acta Crystallogr. Sect. B Struct. Crystallogr. Cryst. Chem.* **1971**, *27*, 594–601. [CrossRef]
45. Duffy, J.A.; Macphee, D.E. The Coordination Number of Silicon in Silicon-Oxygen Compounds: The Special Case of 6-Fold Coordination in Thaumasite. *J. Phys. Chem. B* **2007**, *111*, 8740–8745. [CrossRef]
46. Scholtzová, E.; Kucková, L.; Kožíšek, J.; Pálková, H.; Tunega, D. Experimental and computational study of thaumasite structure. *Cem. Concr. Res.* **2014**, *59*, 66–72. [CrossRef]
47. Crammond, N. The occurrence of thaumasite in modern construction—A review. *Cem. Concr. Compos.* **2002**, *24*, 393–402. [CrossRef]
48. Crammond, N.J. The thaumasite sulfate attack in the UK. *Cem. Concr. Compos.* **2003**, *25*, 809–818. [CrossRef]
49. Tsivilis, S.; Sotiriadis, K.; Skaropoulou, A. Thaumasite sulfate attack (TSA) in limestone cement pastes. *J. Eur. Ceram. Soc.* **2007**, *27*, 1711–1714. [CrossRef]
50. Rahman, M.M.; Bassuoni, M.T. Thaumasite sulfate attack on concrete: Mechanisms, influential factors and mitigation. *Constr. Build. Mater.* **2014**, *73*, 652–662. [CrossRef]
51. Heddle, M.F. Preliminary notice of substances which may prove to be new minerals. *Mineral. Mag.* **1880**, *4*, 117–123. [CrossRef]
52. McConnell, J.D.C. The hydrated calcium silicates riversideite, tobermorite, and plombierite. *Mineral. Mag.* **1954**, *30*, 293–305. [CrossRef]
53. Megaw, H.D.; Kelsey, C.H. Crystal structure of tobermorite. *Nature* **1956**, *177*, 390–391. [CrossRef]
54. Hamid, S.A. The crystal structure of the 11Å natural tobermorite Ca2.25[Si3O7.5(OH)1.5] 1H_2O. *Z. Für Krist. Cryst. Mater.* **1981**, *154*, 189–198. [CrossRef]
55. Merlino, S.; Bonaccorsi, E.; Armbruster, T. Tobermorites: Their real structure and OD character. *Amer. Mineral.* **1999**, *84*, 1613–1621. [CrossRef]
56. Merlino, S.; Bonaccorsi, E.; Armbruster, T. The real structure of tobermorite 11Å: Normal and anomalous forms, OD character and polytypic modifications. *Eur. J. Mineral.* **2001**, *13*, 577–590. [CrossRef]
57. Bonaccorsi, E.; Merlino, S.; Kampf, A.R. The Crystal Structure of Tobermorite 14 Å (Plombierite), a C–S–H Phase. *J. Am. Ceram. Soc.* **2005**, *88*, 505–512. [CrossRef]
58. Biagioni, C.; Merlino, S.; Bonaccorsi, E. The tobermorite supergroup: A new nomenclature. *Mineral. Mag.* **2015**, *79*, 485–495. [CrossRef]
59. Biagioni, C.; Bonaccorsi, E.; Merlino, S.; Bersani, D. New data on the thermal behavior of 14Å tobermorite. *Cem. Concr. Res.* **2013**, *49*, 48–54. [CrossRef]
60. Biagioni, C.; Bonaccorsi, E.; Lezzerini, M.; Merlino, S. Thermal behaviour of Al-rich tobermorite. *Eur. J. Mineral.* **2016**, *28*, 23–32. [CrossRef]
61. Biagioni, C.; Bonaccorsi, E.; Merlino, S.; Bersani, D.; Forte, C. Thermal behaviour of tobermorite from N'Chwaning II mine (Kalahari Manganese Field, Republic of South Africa) II. Crystallographic and spectroscopic study of tobermorite 10 Å. *Eur. J. Mineral.* **2012**, *24*, 991–1004. [CrossRef]
62. Daubrée, M. Mémoire sur le relation des sources thermals de Plombières avec le filons métalliféres er sur la formation contemporaine des zéolithes. *Ann. Min.* **1858**, *13*, 227–256.
63. Constantinescu, E.; Ilinca, G.; Ilinca, A. Contributions to the study of the Oravita–Ciclora skarn occurrence, southwestern Banat. D.S. *Inst. Geol. Geofiz. 2 Zăcăminte* **1988**, *72*, 27–45.
64. Stoppa, F.; Scordari, F.; Mesto, E.; Sharygin, V.; Bortolozzi, G. Calcium-aluminum-silicate-hydrate "cement" phases and rare Ca-zeolite association at Colle Fabbri, Central Italy. *Open Geosci.* **2010**, *2*, 175–187. [CrossRef]
65. Zhang, Y.Q.; Radha, A.V.; Navrotsky, A. Thermochemistry of two calcium silicate carbonate minerals: Scawtite, Ca7(Si_6O_{18})(CO_3)·2H_2O, and spurrite, $Ca_5(SiO_4)_2(CO_3)$. *Geochim. Cosmochim. Acta* **2013**, *115*, 92–99. [CrossRef]
66. Ferrini, V.; De Vito, C.; Mignardi, S. Synthesis of nesquehonite by reaction of gaseous CO_2 with Mg chloride solution: Its potential role in the sequestration of carbon dioxide. *J. Hazard. Mater.* **2009**, *168*, 832–837. [CrossRef] [PubMed]
67. Matter, J.M.; Kelemen, P.B. Permanent storage of carbon dioxide in geological reservoirs by mineral carbonation. *Nat. Geosci.* **2009**, *2*, 837–841. [CrossRef]
68. Orr, F.M.J. CO_2 capture and storage: Are we ready? *Energy Environ. Sci.* **2009**, *2*, 449–458. [CrossRef]
69. Stanton, M.F.; Layard, M.; Tegeris, A.; Miller, E.; May, M.; Morgan, E.; Smith, A. Relation of particle dimension to carcinogenicity in amphibole asbestoses and other fibrous minerals. *J. Natl. Cancer Inst.* **1981**, *67*, 965–975. [PubMed]
70. Fubini, B.; Fenoglio, I. Toxic potential of mineral dusts. *Elements* **2007**, *3*, 407–414. [CrossRef]

71. Aust, A.E.; Cook, P.M.; Dodson, R.D. Morphological and chemical mechanisms of elongated mineral particle toxicities. *J. Toxicol. Environ. Health Part B* **2011**, *14*, 40–75. [CrossRef]
72. Crovella, S.; Bianco, A.M.; Vuch, J.; Zupin, L.; Moura, R.R.; Trevisan, E.; Schneider, M.; Brollo, A.; Nicastro, E.M.; Cosenzi, A.; et al. Iron signature in asbestos-induced malignant pleural mesothelioma: A population-based autopsy study. *J. Toxicol. Environ. Health Part A* **2016**, *79*, 129–141. [CrossRef] [PubMed]
73. Giordani, M.; Mattioli, M.; Cangiotti, M.; Fattori, A.; Ottaviani, M.F.; Betti, M.; Ballirano, P.; Pacella, A.; Di Giuseppe, D.; Scognamiglio, V.; et al. Characterisation of potentially toxic natural fibrous zeolites by means of electron paramagnetic resonance spectroscopy and morphological-mineralogical studies. *Chemosphere* **2022**, *291*, 133067. [CrossRef] [PubMed]
74. Mattioli, M.; Giordani, M.; Dogan, M.; Cangiotti, M.; Avella, G.; Giorgi, R.; Dogan, A.U.; Ottaviani, M.F. Morpho-chemical characterization and surface properties of carcinogenic zeolite fibers. *J. Hazard. Mater.* **2016**, *306*, 140–148. [CrossRef] [PubMed]
75. Mattioli, M.; Ballirano, P.; Pacella, A.; Cangiotti, M.; Di Lorenzo, F.; Valentini, L.; Meli, M.A.; Roselli, C.; Fagiolino, I.; Giordani, M. Fibrous Ferrierite from Northern Italy: Mineralogical Characterization, Surface Properties, and Assessment of Potential Toxicity. *Minerals* **2022**, *12*, 626. [CrossRef]

minerals

MDPI

Article

Chemical and Mineralogical Characterization of Waste from Abandoned Copper and Manganese Mines in the Iberian Pyrite Belt, Portugal: A First Step Towards the Waste-to-Value Recycling Process

Daniel P. S. de Oliveira [1,2,*], Teresa P. Silva [1], Igor Morais [3] and João A. E. Fernandes [1]

[1] Mineral Resources and Geophysics Research Unit, Laboratório Nacional de Energia e Geologia (LNEG), Estrada da Portela, Bairro do Zambujal—Alfragide, Apartado 7586, 2610-999 Amadora, Portugal; teresa.pena@lneg.pt (T.P.S.); joao.fernandes@lneg.pt (J.A.E.F.)

[2] Mineral Resources Expert Group, EuroGeoSurveys, Rue Joseph II, 36–38, 1000 Brussels, Belgium

[3] Mineral Resources and Geophysics Research Unit, Laboratório Nacional de Energia e Geologia (LNEG), Bairro da Vale d'Oca, Apartado 14, 7601-909 Aljustrel, Portugal; igor.morais@lneg.pt

* Correspondence: daniel.oliveira@lneg.pt

Abstract: This study examines the chemical and mineralogical composition of waste materials from abandoned copper and manganese mines in the Iberian Pyrite Belt, Portugal, as a first step toward their potential recycling for critical and strategic raw materials (CRM and SRM). Using portable X-ray fluorescence (pXRF) and other analytical techniques, this research highlights the presence of valuable elements, including copper, manganese, and rare earth elements, in concentrations significantly above their crustal abundance. The findings underscore the dual potential of these wastes: as sources of secondary raw materials and for mitigating environmental hazards such as acid mine drainage (AMD). Recovered materials include chalcopyrite, pyrolusite, and rhodochrosite, with critical elements like cobalt, lithium, and tungsten identified. pXRF proved to be a reliable, cost-effective tool for rapid field and laboratory analyses, demonstrating high precision and good correlation with standard laboratory methods. The study emphasizes the importance of characterizing historical mining waste to support a circular economy, reduce reliance on foreign material imports, and address environmental challenges. This approach aligns with the European Union's Critical Raw Materials Act, promoting sustainable resource use and the recovery of strategic resources from historical mining sites.

Keywords: mineralogical composition; mining waste; historical mines; critical raw materials; strategic raw materials; circular economy

1. Introduction

The EU's concerns about the sourcing and supply of raw materials are more worrisome today than ever. The initial calls for action in this area date back to the Council's 2nd Environment Action Programme, which highlighted the Community's reliance on raw materials sourced from outside its borders [1]. More than two decades later, the European Commission launched the Raw Materials Initiative, the first integrated strategy focused on enhancing access to these resources [2]. This initiative led to the creation of the first list of critical raw materials (CRM).

By definition, a CRM is a material that is both economically important and has a high supply risk and lack of viable substitutes while a strategic raw material (SRM) is a natural resource or material that is crucial for a nation's economic stability, technological development, or national security. Unlike critical raw materials, which are assessed based on economic importance and supply risk, strategic raw materials are defined by their role in fulfilling long-term strategic goals, such as energy independence, military capability, or technological leadership.

Since then, the importance of securing access to raw materials has been reinforced through various high-level EU strategies [3]. However, recent disruptions in supply chains caused by the COVID-19 pandemic and the ongoing conflict in Ukraine have introduced additional complexities to this challenge [4,5]. On 16 March 2023, the European Commission introduced the Critical Raw Materials Act (CRMA) as part of its wider "Green Industrial Plan" and alongside the "Net-Zero Industry Act" (NZIA). According to the European Commission [3], the initiative aims to reduce the EU's reliance on CRM while promoting a sustainable and competitive landscape for CRM value chains within the EU [6]. The search for primary critical and strategic raw materials (CRM and SRM, respectively), and even secondary raw materials, is a major concern nowadays given their high consumption [6–8]. One way to minimise the stress on new mining targets definition and primary mineral extraction is to search for these materials in the tailings of historical mines, which have resulted from the milling of the bulk ore-bearing rock or even from mineral processing (e.g., slag) [9]. Mining waste recycling also helps to prevent acid mine drainage (AMD). Thus, the reprocessing of mining waste provides, beyond the extraction of primary critical and secondary raw materials, the removal of hazardous substances and a new opportunity to manage the remaining waste in historical mining sites.

There are several examples in the literature: (a) several authors claim that copper mine drainage (CMD) generated from copper mining and smelting is a type of AMD, with a high sulphate concentration and large amounts of metal ions such as Fe, Cu, and Mn, e.g., [10]; moreover, these authors proposed a combined process including Fe, Cu, and S pretreatment recovery units through chemical oxidation and the sulphur biocycling of CMD materials; and (b) a copper recuperation plant at the gold mine Pueblo Viejo, in the Dominican Republic, using sulfidogenic microorganisms to remediate metal waste [11].

Equally important in recent years, manganese recovery from waste using the bioleaching of metal using microbes has provided advantages over the traditional method of recovery [12]. The Chvaletice Manganese Project (https://www.mn25.ca/chvaletice-manganese-project, accessed on 10 December 2024) in Czechia is an example of recycling and remediation by reprocessing old tailings from a decommissioned mine to produce battery-grade manganese products for the electric vehicle industry [13].

Mixtures of synthetic tetrahedrite samples and tetrahedrite–tennantite ore samples from abandoned Barrigão mine dumps located in the Portuguese zone of the Iberian Pyrite Belt (IPB) have been produced by solid-state mechanochemical synthesis to explore their use in tellurium-free thermoelectric (TE) materials [14], for application in heavy and maritime industries, and also as primary power sources for off-grid sensors and IoT devices (START project, https://www.start-heproject.com/, accessed on accessed on 10 December 2024).

The aim of this study is to characterize the chemical and mineralogical waste from historical Cu and Mn mine sites in view of the first evaluation of their potential for recycling CRM, SRM, and other valuable elements. The performance of the portable equipment used to carry out the chemical characterization in these complex samples is also assessed on the wastes/tailings/slags from abandoned Cu and Mn exploitations in the Portuguese sector of the IPB.

1.1. Historical Mines Overview

Portugal is one of the European countries with the longest and strongest mining traditions, given the richness of the Iberian Peninsula in mineral resources and the favorable conditions for mineral deposits discovery, e.g., [15].

In the south of the country, there are two large tectonostratigraphic zones that contain a wide variety of mineral deposits. The Ossa Morena Zone (OMZ) to the north is the Iberian Massif geotectonic unit that displays the greatest variety of types of mineralization as well as the largest number of ore deposits and showings (>650). It includes a wide range of commodities such as iron, lead-zinc, copper, gold, silver, antimony, nickel, manganese, tungsten, mercury, barite, variscite, uranium, and coal [16]. The South Portuguese Zone (SPZ) is the southernmost of the zones in the Iberian Massif and includes the Iberian Pyrite Belt (IPB), one of the largest districts of volcanogenic massive sulphide (VMS) deposits in the world, and is a critical source of base metals (Cu, Pb, and Zn) for Europe. Confirmed resources exceed 1700 Mt of massive sulphides with grades of around 1.2% Cu, 1% Pb, and 3% Zn, as well as more than 300 Mt of stockwork-type copper mineralization [17].

For this study, several old mines have been selected—one in the Ossa Morena Zone (Aparis mine) and ten in the South Portuguese Zone (Lousal, Brancanes, Barrigão, Porteirinhos, Algaré, Cova dos Mouros, Ferragudo, Ferrarias e Castelo, Cerro do Serpe, and Balança—representing several types of ore deposits (VMS, copper veins, and manganese lenses) (Figure 1).

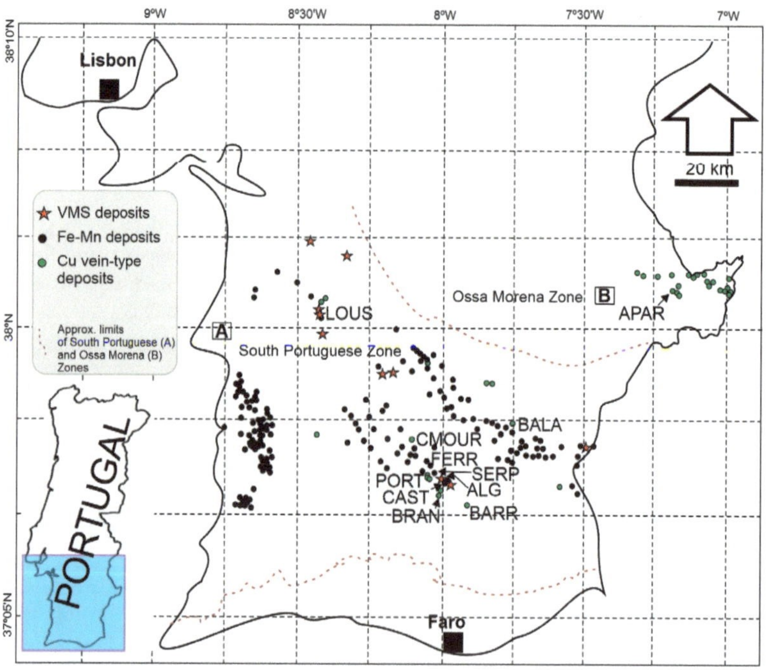

Figure 1. Simplified map of the studied areas with the approximate location of several old mines. (**A**) Iberian Pyrite Belt sector in the South Portuguese Zone; (**B**) Barrancos sector in the Ossa-Morena Zone. Mine name abbreviations: LOUS—Lousal; BALA—Balança; CMOUR—Cova dos Mouros—Castro Verde; FERR—Ferragudo; SERP—Cerro do Serpe; PORT—Porteirinhos; CAST—Ferrarias e Castelo; ALG—Algaré; BRAN—Brancanes; BARR—Barrigão; APAR—Aparis (see references [18,19] for geological information).

1.1.1. VMS-Type Deposits

The Lousal Mine is an old pyrite (FeS_2) mine located in the NW sector of the IPB, approximately 120 SSE of Lisbon (Figure 1A), in a complex structure formed by the volcano–sedimentary complex (VSC) (Famennian to Visean age) and the Phyllite–Quartzite Group (PQG) (Givetian to latest Famennian age—Strunian Biozone), the two lithostratigraphic units of the IPB. Opened in 1900 and closed in 1988, the subvertical deposit was exploited by galleries to a ~500 m depth with an NW direction. It is a VMS hosted in an antiform structure.

1.1.2. Cu Vein-Type Deposits

Late-Variscan hydrothermal copper sulphide veins systems occur in the OMZ and SPZ. The old Aparis Mine located in the Barrancos area (OMZ; Figure 1B) is an ore system associated with N10 E to N20 E subvertical veins (>2000 m extension) and mined up to a 150 m depth. The mineralogy is composed of chalcopyrite + pyrite ± marcasite ± pyrrhotite ± galena ± tetrahedrite ± arsenopyrite + quartz + dolomite + ankerite ± siderite ± calcite ± chlorite [20].

Along the IPB, dozens of Cu-bearing veins associated with fault breccias were exploited in the 19th century for copper, lead, zinc, barium, and antimony. Some of the copper veins are aligned along the Neves-Corvo copper trend indicating a possible remobilization of metals at depth. One such mine, the Barrigão Mine, consists of two converging metric-thick vein structures, extending approximately 1800 m along strike [21,22] SSE of the Neves-Corvo Mine (Figure 1A). The Barrigão copper ore is represented by fault breccias composed of chalcopyrite, tennantite, tetrahedrite, and pyrite, with minor or rare arsenopyrite, löllingite, sphalerite, and native bismuth and an undetermined Cu–Sn–Ge-sulphide in a matrix of quartz, angular fragments of black shale, and carbonate minerals [23].

Another type of such a deposit, the Brancanes Mine, is located close to the Neves-Corvo mine (approximately 3 km to the WSW; Figure 1A). The Brancanes Copper Mining Company initiated mining activity in 1883, exploiting a copper vein with an N45° W, 50° E orientation rich in chalcopyrite and pyrite with quartz and carbonates [24]. The quartz + carbonate + sulphide veins are installed in shales and greywackes of the Mértola Formation. The superficial levels of copper mineralization were largely represented by malachite and more rarely by azurite and covellite.

The Porteirinhos deposit (Figure 1A) resulted in a small old mine composed of a 1.2 m thick Cu-bearing quartz vein with an N70° E orientation in the Mértola Formation. The mineralogy is characterized by chalcopyrite + arsenopyrite + pyrite and azurite, with malachite and tetrahedrite as secondary minerals.

The Algaré Mine is located very close to the Neves-Corvo mine next to the Neves-Corvo thrust (Figure 1A). The structure is characterized by fractures with an orientation of NW-SE and 70° NE and runs sub-parallel to the regional structure. The massive pyrite has an average content of 0.45% of copper and the mineralogy is composed of pyrite + chalcopyrite + arsenopyrite and azurite, stannite, galena, and malachite.

The Cova dos Mouros Mine is located 1 km W of the village of Castro Verde (Figure 1A). The structure corresponds to a gossan (likely a copper vein or VMS deposit) anomalous in base metals. The region was previously investigated by "Serviço de Fomento Mineiro" (currently LNEG).

1.1.3. Manganese Lens-Type Deposits

The manganese deposits in the IPB have been mined throughout history. These are related to chert and jasper horizons and occur at a stratigraphic position roughly similar to that of massive sulphide lenses [25] and Upper Volcano-Sedimentary Complex sequences [26,27]. Their size and metal content make them uneconomic nowadays, although they were important in the past [28].

The most important Fe-Mn in the Portuguese sector of the IPB are the Ferragudo and Balança mines (Figure 1A). The Ferragudo Mine is a stratiform mineralized structure (50° NW, 45° NE) composed of lenses of oxides, carbonates, and silicates of Mn and Fe (several bodies with thicknesses varying from 1 to 12 m over an extension of approximately 150 m). Mineralization is associated with the lithologies of the Volcano-Sedimentary Complex. The main ore was composed of pyrolusite, psilomelane, wad, rhodochrosite, rhodonite, manganite, and hematite, occurring in quartz gangue + carbonates [29–31]. The Balança Mine with a 60° NW, 80° NE direction is very similar to the Ferragudo mine. The paragenetic mineral assemblage is composed of pyrolusite, psilomelane, wad, rhodochrosite, rhodonite, manganite, hematite, and quartz in a quartz and carbonate gangue.

Ferrarias e Castelo and Cerro do Serpe are small manganese mines that had very incipient exploitation at the beginning of the 20th century. Exploitation was essentially carried out in open pits and small shafts and galleries. The mineralization corresponds to small lenticles of manganese oxides (essentially pyrolusite) and is associated with jasper and chert horizons of the Upper Vulcano-Sedimentary Complex.

2. Materials and Methods

2.1. Sampling

A total of 45 samples were collected in the different types of mine wastes and deposit types (see the description of the samples in Table 1). The variations included coarse-, medium-, and fine-grained types and depended on the sample type (Figure 2). The sample type reflects the processing methods that were carried out in each mining area.

At each sampling location, four individual samples were combined to create a composite sample representative of the mine waste at that site.

The total sample collected was between 5 and 6 kg at a depth of 30–40 cm depending on the morphology of the waste.

In the lab, the samples for chemical and mineralogical analyses were initially dried at 35 °C for 3 weeks. Coarse grain samples were previously cut into small portions with a diamond saw.

The samples were initially crushed using a tungsten jaw crusher (Retsch BB250 XL), and a 200 g split was processed using a secondary crusher (Retsch BB50 with tungsten jaws), reducing the particle size to below 500 μm. The material was then sieved through a 75 μm mesh to separate the particles already at the desired grain size for chemical analysis, ensuring proper calibration for further processing in the agate mill (Retsch RS200). After pulverization in the agate mill, the sample was sieved again through a 75 μm mesh to confirm that the particle size was uniformly calibrated.

Table 1. Description of collected samples (sample reference was simplified). Material size: C—Coarse-grained (>0.5 to 1 mm); M—Medium-grained (>0.25 to 0.5 mm); F—Fine-grained (<0.25 mm). Sample type (represent the type of material of mining waste): Processing—Mining waste product of some type of mechanical treatment (example: float plant; manual selection); Dump composite—Mining wastes composed of host rocks with heterogeneous amounts of ore.

Sample Reference	Mine Name	Geotectonic Zone	Mine Type	Material Size	Sample Type	Sample Description
LOUS/001	Lousal	South Portuguese Zone—Iberian Pyrite Belt	VMS	M	Processing	Crushed ore (mainly pyrite) with clasts of quartz. Oxidation cap with 5 cm
LOUS/002				C	Dump composite	Blocks of mineralization host rocks (felsic volcanic rocks, shales, and quartz)
LOUS/003				F	Processing	Tailings inside the acid lagoons. Material very fine, composed of neoformation minerals
LOUS/004				C	Dump composite	Blocks of mineralization host rocks (essentially shales)
LOUS/006				C	Dump composite	Blocks of mineralization host rocks (essentially shales). Occasionally massive pyrite blocks
LOUS/007				C	Dump composite	Blocks of mineralization host rocks (essentially shales and volcanic rocks). Occasionally massive pyrite blocks. Neoformation minerals
LOUS/008				F	Processing	Tailings inside the acid lagoons. Material very fine, composed of neoformation minerals
LOUS/009				M	Processing	Crushed pyrite with blocks of host rocks (volcanic rocks and shales). Neoformation minerals (sulfates)
LOUS/010				C	Processing	Shales with fine pyrite associated. Quartz and neoformation minerals
LOUS/011				C	Dump composite	Shales with pyrite blocks. Neoformation minerals
LOUS/012				C	Dump composite	Pyrite blocks with rare volcanic rocks and quartz. Neoformation minerals
LOUS/013				C	Dump composite	Shales with pyrite blocks
LOUS/014				C	Processing	Crushed pyrite

Table 1. *Cont.*

Sample Reference	Mine Name	Geotectonic Zone	Mine Type	Material Size	Sample Type	Sample Description
PORT/001	Porteirinhos	South Portuguese Zone—Iberian Pyrite Belt	Cu veins	C	Dump composite	Shales and graywackes with blocks of quartz (pyrite, chalcopyrite, and malachite
PORT/001 Ore				C	Ore	Quartz with primary sulfides
BARR/001	Barrigão			C	Dump composite	Shales and graywackes with fine material composed of host rock crushed and cooper neoformation minerals
BARR/002				C	Roasting	Slags probably from 19th century. Blocks of coal are observed
BARR/003				C	Ore	Ore blocks with host rocks associated
ALG/001	Algaré			M	Dump composite	Crushed pyrite with clast of host rocks
Algaré Ore				C	Ore	Massive pyrite
ALG/002				M	Dump composite	Host rocks (shales and volcanic rocks) with neoformation minerals (malachite)
ALG/003				C	Processing	Blocks of volcanic rocks, shales, and quartzites with oxidized sulfides
ALG/004				C	Dump composite	Quartzites and volcanic rocks with iron and manganese oxides
ALG/004 Ore				C	Dump composite	Iron and manganese oxides
BRAN/001	Brancanes			M	Roasting	Slags probably from 19th century. Blocks of coal are observed
BRAN/002				F	Processing	Post flotation tailings
BRAN/003				M	Roasting	Slags probably from 19th century. Blocks of host rocks
Brancanes Ore				C	Ore	Quartz and carbonate veins with sulfides (chalcopyrite, tetrahedrite)
CMOUR/001 Ore	Cova dos Mouros			C	Dump composite	Volcanic rocks, quartz, and cherts with Fe and Mn concretions

Table 1. *Cont.*

Sample Reference	Mine Name	Geotectonic Zone	Mine Type	Material Size	Sample Type	Sample Description
APAR/001	Aparis	Ossa Morena Zone	Cu veins	C	Dump composite	Quartz blocks and host rocks (shales). Copper neoformation minerals
APAR/002				C	Dump composite	Quartz blocks and shales. Sulfides impregnation in the quartz
APAR/003				C	Dump composite	Quartz blocks with copper neoformation minerals
APAR/004				F	Processing	Post flotation tailings
APAR/005				F	Processing	Post flotation tailings
APAR/006				F	Processing	Post flotation tailings
APAR/007				F	Processing	Post flotation tailings
FERR/001	Ferragudo	South Portuguese Zone—Iberian Pyrite Belt	Fe-Mn lens	M	Processing	Crushed material composed of iron and manganese oxides
FERR/002				F	Processing	Black material composed of crushed ore
FERR/003				C	Dump composite	Blocks of host rocks (shales) and cherts with Fe and Mn impregnation
FERR/004				M	Processing	Black material composed of crushed ore
CAST/001	Ferrarias and Castelo			C	Dump composite	Jaspers and cherts with iron and manganese mineralization. Pyrite associated with chert facies
CAST/002				C	Dump composite	Iron and manganese oxides with pyrite
CAST/003				C	Dump composite	Jaspers and cherts with iron and manganese mineralization. Pyrite associated with chert facies
CAST/004				C	Dump composite	Shales with iron and manganese impregnation
SERP/001 Ore	Serpe			C	Dump composite	Jaspers and cherts block with iron and manganese mineralization
BALA/001	Balança			C	Dump composite	Jaspers and cherts block with iron and manganese mineralization

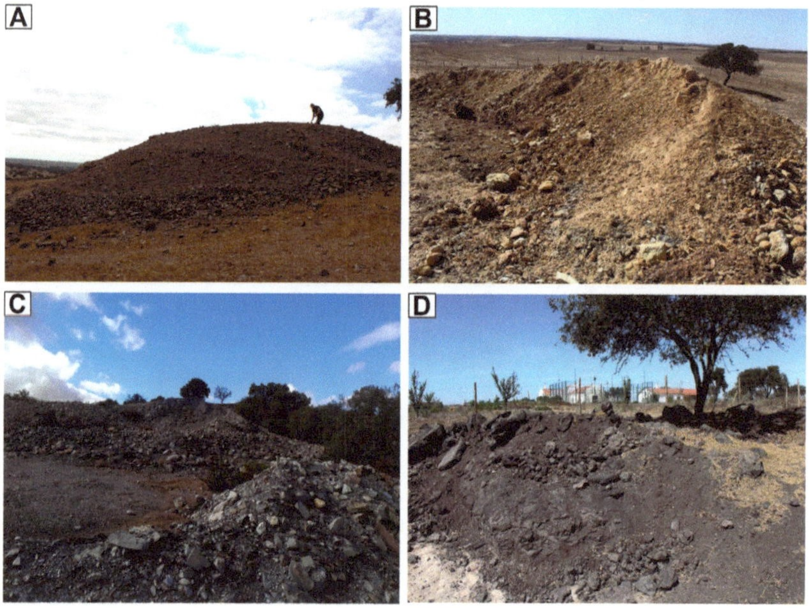

Figure 2. General aspects of mine dump waste in selected mines. (**A**) Mine wastes of Algaré Cu vein deposit; (**B**) Gossan/alteration zone in the Cova dos Mouros—Castro Verde occurrence; (**C**) aspect of mine wastes with high sulphide content (Ge and Sb) in the Barrigão Cu deposit; (**D**) final manganese concentrate (manganese oxides) in the Ferragudo iron-manganese deposit.

2.2. Methodology

Portable X-ray fluorescence (pXRF) equipment, X-MET8000 Expert Geo from HITACHI (Tokyo, Japan), was used for rapid chemical analysis of the powdered samples in the laboratory using a benchtop stand. This apparatus was equipped with a Rh tube (4 W) and a silicon drift detector (SDD). Two analytical programs (calibrations) were used: mining (40 kV, 120 s per sample) and REE (for the rare earth elements, during 180 s, 50 kV).

A multi-analytic approach was used to ensure that the pXRF equipment could be used with confidence in the determination of the elemental composition of waste materials. With this purpose, the results of selected samples were compared with chemical analyses obtained by XRF (wavelength dispersive, WDS) laboratory equipment (Philips PW2404, Andover, MA, USA). It was not possible to obtain fused discs, so the analyses of waste samples were carried out on pressed pellets. The major elements were obtained with a semi-quantitative program and the minor and trace with the "Protrace" program.

Major and trace elements of selected samples were also analysed by Activation Laboratories (ACTLABS, Ancaster, ON, Canada) using analytical package UT-3—inductively coupled plasma mass spectroscopy (ICP-MS) analyses by 4-acid (hydrochloric, nitric, perchloric, and hydrofluoric) digestion and by instrumental neutron activation analysis (INAA). Quality control was ensured by internal protocols at the laboratory (see www.actlabs.com (accessed on 10 December 2024) and Supplementary Materials, Table S1, folder QC-ACTLABS).

Powder X-ray diffraction (XRD) data were collected using a D8 Advance Bruker AXS diffractometer (Bruker AXS GmbH, Karlsruhe, Germany) with Cu Kα radiation, operating at 40 kV and 40 mA. The powder diffractograms (5–70° 2θ) were obtained using a step size of 0.015° and a step time of 0.3 s with a LYNXEYE XE-T detector. The XRD data treatment was performed using DIFFRAC.EVA v5 software for phase identification.

3. Results and Discussion

A raw comparison between methodologies concerning chemical characterisation was performed first. Indeed, pXRF is a useful tool for a quick and inexpensive overview (both in the field and in the laboratory) of the range of elements that a waste sample can contain, e.g., [32]. No significant difference in the element concentration was observed between performing one or an average of three measures collected at different points/areas of the same sample, indicating good precision of the equipment. Some examples can be seen in Figure 3. The complete table and all graphics ordered alphabetically are supplied as Supplementary Materials in Table S1.

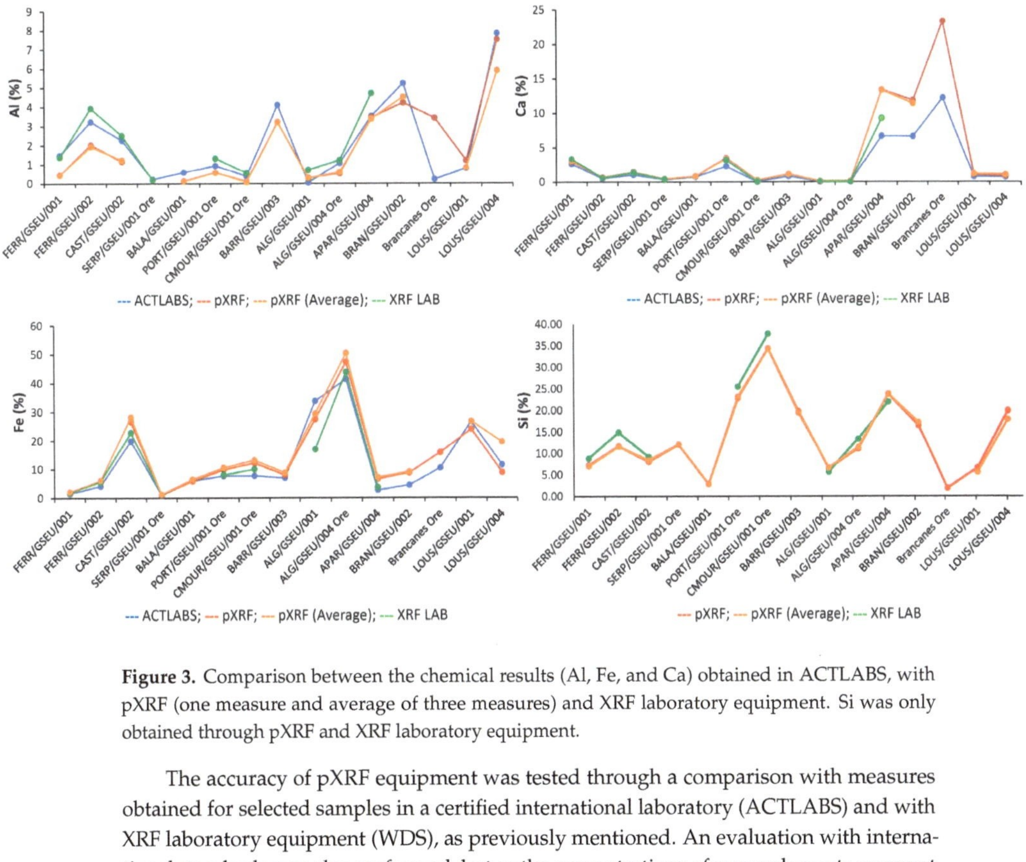

Figure 3. Comparison between the chemical results (Al, Fe, and Ca) obtained in ACTLABS, with pXRF (one measure and average of three measures) and XRF laboratory equipment. Si was only obtained through pXRF and XRF laboratory equipment.

The accuracy of pXRF equipment was tested through a comparison with measures obtained for selected samples in a certified international laboratory (ACTLABS) and with XRF laboratory equipment (WDS), as previously mentioned. An evaluation with international standards was also performed, but as the concentrations of many elements were not known, the methodology described was adopted. The correlation between ACTLABS and pXRF equipment (Table S1) produced good values for Al, As, Ba, Ca, Cu, Fe, Ga, K, Mg, Mn, Mo, Ni, Pb, Rb, S, Sb, Se, Sr, Ti, Zn, and W ($R^2 \geq 0.7$). Despite the strong correlation, the line slopes often deviate from the unity. A slope higher than 1 is assigned to Cu, Rb, Ni, As, K, Mn, Sr, Ca, Pb, Se, W, Mo, S, Zn, and Sb (ordered from largest to smallest). For Ti, the slope is approximately 1, as well as for Si (correlation between pXRF and XRF LAB as ACTLABS did not analyze Si). A slope of less than 1 was observed for Fe, Al, Ga, Ba, and Mg (examples in Figure 4). The lowest reliability measurements ($R^2 < 0.7$) were observed for Ag, Bi, Cd, Co, Cr, Hf, Hg, Nb, P, Sc, Sn, Ta, Th, U, V, Y, and Zr, plus La, Ce, Pr, and Nd (the rare earth analytical program should be revised). The graphic of ACTLABS vs. XRF laboratory shows, for almost all the elements mentioned above with $R^2 \geq 0.7$, slopes closer to 1. However, given the high content of some elements (e.g., Sr, Ba, Sn, Cu, Pb, and As) in some samples, a semi-quantitative analysis program was used in these specific cases. On the other hand, the high content of various elements (e.g., Cu, Pb, S, As, Mn, and Sr) in some samples exceeded the maximum detection limit of the analytical program used in ACTLABS.

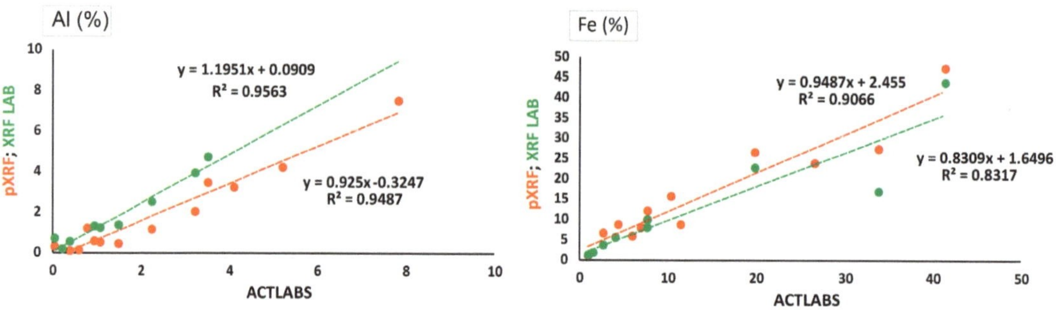

Figure 4. Examples of correlations obtained between chemical analyses: ACTLABS vs. pXRF and ACTLABS vs. XRF laboratory equipment.

The great discrepancy in elemental concentrations due to the complexity of the samples makes it difficult to choose the best analytical method to use. However, for a first step toward a more structured study, the pXRF equipment gave quick and acceptable results for many elements.

The complete table of chemical analyses performed on wastes from copper and manganese abandoned mines is presented in Supplementary Materials, Table S2, where the major contents of each element are assigned, bearing in mind the limitations of the techniques and the correlations/slope obtained (see Table S1). For example, Cu exceeds the maximum detection limit of the analytical program used in ACTLABS; conversely, the semi-quantitative analysis program was used with XRF laboratory equipment instead of the quantitative program; therefore, the pXRF results were considered in this case. Concerning Mo, the correlation between the ACTLABS and pXRF results is high (R^2 = 0.9913) but the slope is slightly higher than 1 (1.3), so the data from the international laboratory were chosen. In general, samples taken from the old copper mines show higher values of Ag (72 ppm), Al (12%), As (8.4%), Au (1110 ppb), Bi (217 ppm), Ca (12%), Co (209 ppm), Cu (27.5%), Fe (47.4%), Ge (9.6 ppm), Hf (9.6 ppm), Hg (106 ppm), In (6 ppm), K (3.3%), Li (83 ppm), Mg (6.6%), Ni (750 ppm), Pb (3%), Rb (132 ppm), S (38%), Sb (665 ppm), Sc (19 ppm), Se (132 ppm), Si (37.8%), Sn (139 ppm), Th (14.9 ppm), Zn (0.2%), Zr (271 ppm), and REE (La—47.6 ppm, Ce—94.7, Pr—11.3, Nd—42.1, Sm—7.4, Eu—1.28, Gd—5.1, Dy—4.1, Tb—0.7, Ho—1.2, Er—4.3, Tm—0.8, Yb—6, and Lu—1 ppm). Conversely, the waste from manganese mines is richer in Ba (1.5%), Be (7 ppm), Cr (111 ppm), Ga (129 ppm), Mn (38.6%), Mo (205 ppm), Sr (0.4%), Tl (94.4 ppm), V (170 ppm), and W (790 ppm). Cd, Cs, Na, Nb, P, Re, Ta, Tb, Te, Ti, U, and Y are random.

Many of these elements are considered critical and/or strategic raw materials [7]: CRM: As, Be, Bi, Co, Ga, Ge, Hf, HREEs (heavy rare earth elements—Dy, Er, Eu, Gd, Ho, Lu, Tb, Tm, Yb, and Y), LREEs (light rare earth elements—Ce, La, Nd, Pr, and Sm), Li, Mg, Mn, Nb, P, Sb, Sc, Sr, Ta, W, and V; SRMs: Bi, Co, Ga, Ge, HREE, LREE, Li, Mg, Mn, Cu, Ni, W, and V. Other elements, e.g., In, Re, Sn, Se, Au, and Ag, are of economic importance but are not considered to be at risk of supply [7].

The concentrations of various elements found in the mining wastes analyzed are several times greater than the abundance in the Earth's crust, as shown in Figure 5. The considerable volume of certain waste associated with the presence of valuable metals raises questions about their sustainable recovery before rehabilitation procedures, e.g., [33,34].

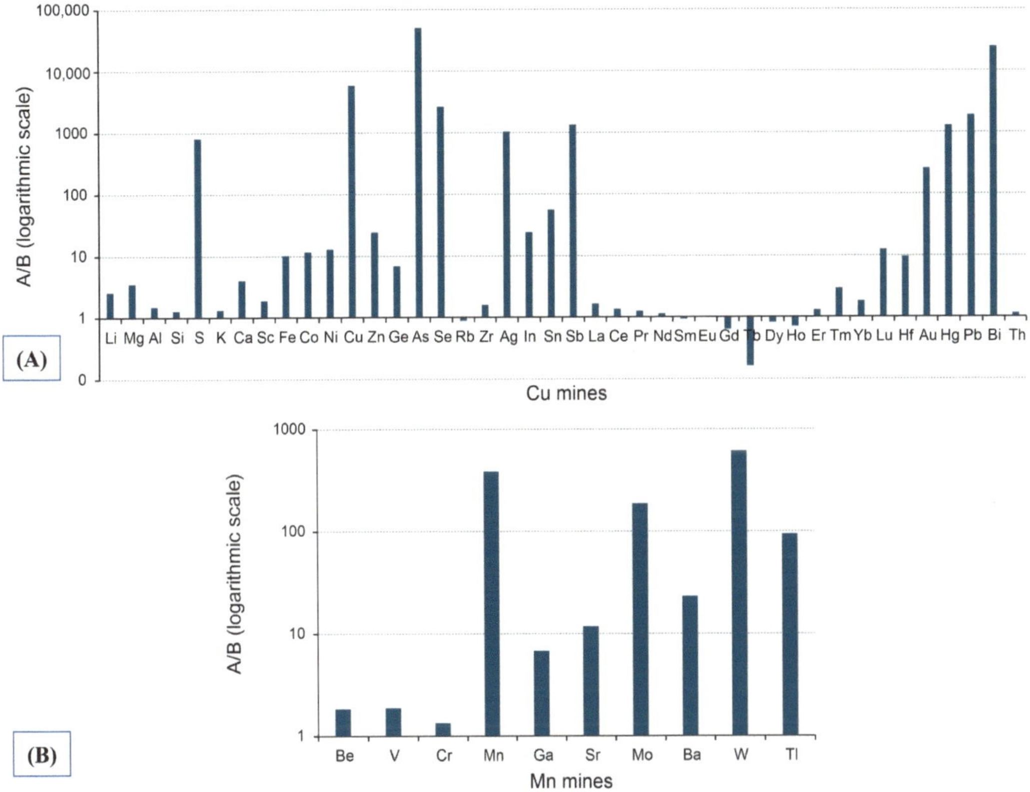

Figure 5. Ratio of the maximum content found in the wastes from Cu and Mn mines (**A**) and the abundance of chemical elements in the Earth's crust (**B**). Elements are ordered by atomic number (adapted from [35]).

With regards to the Cu mines, sample 1 from Porteirinhos is rich in Co, Cu, Mn, and Ni as critical and/or strategic raw materials, while sample 1 ore is concentrated in As, Co, Cu, Li, Mn, Ni, and Sb. This ore also has a high content of Ag, In, and S plus Si and sample 1 has a high content of Al and K plus Si. To link these elements to a mineralogical or carrier phase, XRD analysis was performed (Table 2). Some examples of the XRD spectra referred to in the following tables (Tables 2–4) can be seen in Figure 6. Due to the sample's mineralogical complexity, only a semi-quantitative approach was used to investigate the content of each phase. Phase identification was achieved using the analytical software, bearing in mind the elements previously identified by pXRF (Table S2). The main ore phases are quartz (SiO_2), chalcopyrite ($CuFeS_2$), and dolomite [$CaMg(CO_3)_2$]. A rough calculation gives about 18% chalcopyrite in the ore, considering all the copper at this phase. The same minerals are present in sample 1, but chalcopyrite is in vestigial, which led us to suppose that this waste material was discharged after ore processing; pseudomalachite, $Cu_5(PO_4)_2(OH)_4$, a secondary mineral with low solubility found in oxidized zones of copper deposits [36], is also present.

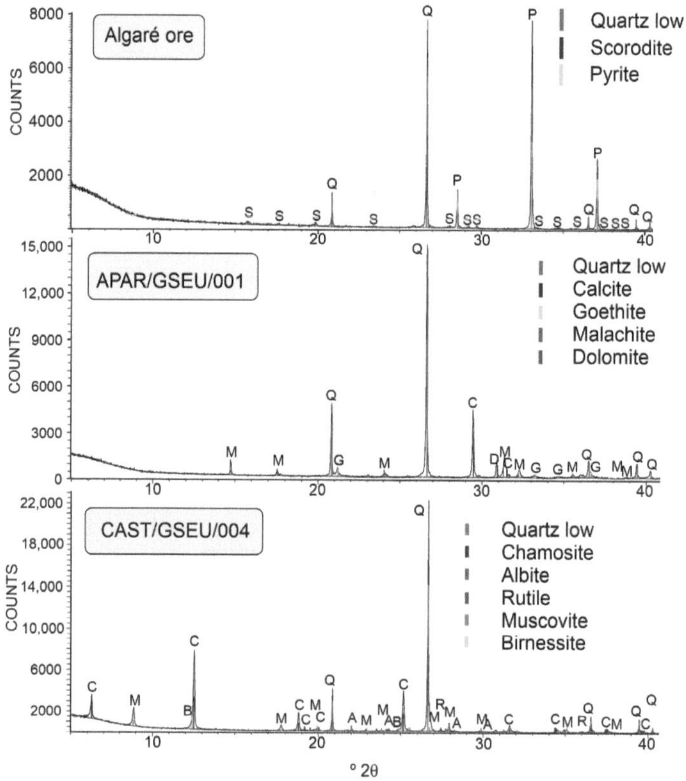

Figure 6. Illustrative examples of XRD spectra, as mentioned in Tables 2–4 (the name of the mineral phases in each spectrum is simplified using the first letter of the identified phase).

The same methodology was used for sample 1 ore from Cova dos Mouros (Table 2), rich in Se, Si, Sn, and W (the latter being both CRM and SRM), where quartz is the main gangue phase present. Tridymite is a high-temperature, low-pressure polymorph of silica. Two Mn phases were also identified, although in vestigial contents: Braunite, $Mn^{2+}Mn^{3+}{}_6(SiO_4)O_8$ and Manganite, $Mn^{3+}O(OH)$.

Samples 1 and 3 from Barrigão present the same mineralogical phases but in markedly different concentrations (Table 2). Copper phases—chalcopyrite, $CuFeS_2$, and tennantite—(Fe), $Cu_6(Cu_4Fe^{2+}{}_2)As_4S_{12}S$, are very well represented in the last sample, which is rich in Ag, Al, As, Bi, Cu, Ga, Ge, Li, S, Si, Sn, Zn, and REE (La, Ce, Nd, Sm, and Gd), making it the richest sample in Bi content of all wastes analyzed. From these elements, As, Bi, Cu, Ga, Ge, Li, and REE are considered critical or strategic. Sample 1 is only rich in Sb (CRM) and Si. A slag sample (BARR/GSEU/002) showed a high content of Al, K, Mn (CRM and SRM), Rb, Si, and Sr (CRM). It should be noted that the direct use of tennantite–tetrahedrite from the Barrigão mine dump has been studied for its application in thermoelectric (TE) material processing [37].

The XRD results of Algaré samples (Table 2) show the presence of scorodite, $Fe^{3+}AsO_4{\cdot}2H_2O$ (sample 1 and Algaré ore). This secondary mineral is a member of the variscite group and results from the oxidation of arsenopyrite or other mineral species containing arsenic and is a very stable phase [38]. Quartz and pyrite (FeS_2) are also present. High concentrations of Ag, As (CRM), Au, Co (both a CRM and SRM), Fe, and In plus S were detected in sample 1 and As, Co, and Fe plus S in Algaré ore. Samples 2 and 3 are also mineralogically similar (mainly quartz and chamosite—$(Fe^{2+})_5Al(Si,Al)_4O_{10}(OH,O)_8$). Sample 2 also presents malachite, $Cu_2(CO_3)(OH)_2$ in a vestigial content (secondary mineral). Al, As, Co, Cu, Fe, Mn (CRM and SRM), Pb, Si, and Zn are concentrated in sample 2. Sample 3 has high contents of Al, As, Fe, and Si. Sample 4 is mainly constituted by quartz, hematite—Fe_2O_3, and kaolinite—$Al_2(Si_2O_5)(OH)_4$, having Al, Fe, Nb, Si, Sn, Y, and Zr in good concentrations. Sample 4 ore was collected nearby and was considered ore, presenting similar phases (quartz, hematite, and goethite—α-$Fe^{3+}O(OH)$). The chemical elements with high values are Fe, Ga, Hf, In, Nb, Sc, Si, Th, Zr, and REE (La, Ce, Nd, Gd, Tb, Ho, Er, Tm, Yb, Lu, and Y), with Ga, Hf, Nb, Sc, and REE considered critical and/or strategic raw materials.

Table 2. Mineralogical phases from Porteirinhos (PORT), Cova dos Mouros (CMOUR), Barrigão (BARR), and Algaré (ALG) samples identified by XRD (in alphabetical order for better visualization; symbols from IMA-CNMNC [39]). The main phases in each sample were roughly estimated as more (+++) or less (+) represented by the intensity of the principal lines. "+" signs indicate relative abundance of mineral. Legend: Ab—Albite, $Na(AlSi_3O_8)$; Alu—Alunite, $KAl_3(SO_4)_2(OH)_6$; Ank—Ankerite, $Ca(Fe^{2+},Mg)(CO_3)_2$; Ath—Anthophyllite, $\square\{Mg_2\}\{Mg_5\}(Si_8O_{22})(OH)_2$; Bir—Birnessite, $(Na,Ca)_{0.5}(Mn^{4+},Mn^{3+})_2O_4\cdot 1.5H_2O$; Bnt—Braunite, $Mn^{2+}Mn^{3+}{}_6(SiO_4)O_8$; Btl—Butlerite, $Fe^{3+}(SO_4)(OH)\cdot 2H_2O$; Ccp—Chalcopyrite, $CuFeS_2$; Chm—Chamosite, $(Fe^{2+})_5Al(Si,Al)_4O_{10}(OH,O)_8$; Cld—Chloritoid, $(Fe^{2+},Mg,Mn^{2+})Al_2(SiO_4)O(OH)_2$; Cst—Cassiterite, SnO_2; Dol—Dolomite, $CaMg(CO_3)_2$; En—Enstatite, $Mg_2Si_2O_6$; Fau-Ca—Faujasite-Ca, $(Ca,Na_2,Mg)_{3.5}[Al_7Si_{17}O_{48}]\cdot 32H_2O$; Gp—Gypsum, $CaSO_4\cdot 2H_2O$; Gth—Goethite, α-$Fe^{3+}O(OH)$; Hem—Hematite, Fe_2O_3; Ilm—Ilmenite, $Fe^{2+}TiO_3$; Kln—Kaolinite, $Al_2(Si_2O_5)(OH)_4$; Mlc—Malachite, $Cu_2(CO_3)(OH)_2$; Mnn—Manganite, $Mn^{3+}O(OH)$; Ms/Bt—Muscovite/Biotite, $Al_2(AlSi_3O_{10})(OH)_2/K(Mg,Fe)_3AlSi_3O_{10}(OH)_2$; Mul—Mullite, $Al_{4+2x}Si_{2-2x}O_{10-x}$ (x~0.4); Ncr—Nacrite, $Al_2(Si_2O_5)(OH)_4$; Phl—Phlogopite, $KMg_3(AlSi_3O_{10})(OH)_2$; Pmlc—Pseudomalachite, $Cu_5(PO_4)_2(OH)_4$; Prl—Pyrophyllite, $Al_2Si_4O_{10}(OH)_2$; Py—Pyrite, FeS_2; Qz—Quartz, SiO_2; Rt—Rutile, TiO_2; Scd—Scorodite, $Fe^{3+}AsO_4\cdot 2H_2O$; Tmgh—Titanomaghemite, $(Ti^{4+}{}_{0.5}\square_{0.5})Fe^{3+}{}_2O_4$; Tnt-Fe—Tennantite-(Fe), $Cu_6(Cu_4Fe^{2+}{}_2)As_4S_{12}S$; Trd—Tridymite, SiO_2; Var—Variscite, $AlPO_4\cdot 2H_2O$; Zrn—Zircon, $Zr(SiO_4)$. Vtg: vestigial content; ?: dubious identification.

Sample Reference	Phase Identification	+++	++	+
PORT/GSEU/001	Ab + Bir + Ccp (vtg) + Chm + Dol + Ms/Bt + Pmlc (vtg) + Qz + Rt (vtg)	Quartz	Muscovite/Biotite, Chamosite, Albite	Birnessite, Dolomite
PORT/GSEU/001 Ore	Ank (vtg) + Bir (vtg) + Ccp + Chm (vtg) + Dol + En? (vtg) + Ms/Bt (vtg) + Qz + Var (vtg)	Quartz		Chalcopyrite, Dolomite
CMOUR/GSEU/001 Ore	Bnt (vtg) + Cst + Hem + Mnn (vtg) + Qz + Trd + Zrn (vtg)	Quartz		Cassiterite, Hematite, Tridymite
BARR/GSEU/001	Ccp (vtg) + Chm + Dol + Gp (vtg) + Kln (vtg) + Ms/Bt + Qz + Tnt-Fe (vtg)	Quartz		Dolomite, Chamosite, Musc./Biotite
BARR/GSEU/002	Ab + Mul + Qz +Tmgh (vtg)	Quartz	Mullite	Albite
BARR/GSEU/003	Ccp + Chm + Dol + Kln + Ms/Bt + Ncr + Qz + Tnt-Fe	Quartz	Tennantite-(Fe), Chalcopyrite, Kaolinite	Chamosite, Musc./Biotite, Dolomite

Table 2. *Cont.*

Sample Reference	Phase Identification	+++	++	+
ALG/GSEU/001	Alu (vtg) + Ath (vtg) + Btl (vtg) + Phl (vtg) + Prl (vtg) + Py + Qz + Scd	Quartz	Pyrite	Scorodite
ALG/GSEU/002	Chm + Cld (vtg) + Fau-Ca? (vtg) + Gth (vtg) + Hem (vtg) + Kln (vtg) + Mlc (vtg) + Ms/Bt + Qz	Quartz	Chamosite	Musc./Biotite
ALG/GSEU/003	Chm + Fau-Ca? (vtg) + Gth (vtg) + Kln (vtg) + Ms/Bt (vtg) + Qz	Quartz	Chamosite	
ALG/GSEU/004	Cld (vtg) + Gth (vtg) + Hem + Ilm (vtg) + Kln + Ms/Bt (vtg) + Qz	Quartz		Hematite, Kaolinite
ALG/GSEU/004 Ore	Gth + Hem + Kln (vtg) + Ms/Bt (vtg) + Qz	Quartz	Hematite	Goethite
Algaré Ore	Py + Qz + Scd	Quartz, Pyrite		Scorodite

Table 3. Semi-quantitative approach to the mineralogical content of Aparis (APAR) and Brancanes (BRAN) samples. The main phases in each sample were roughly estimated as more (+++) or less (+) represented by the intensity of the principal lines. "+" signs indicate relative abundance of mineral. Legend: Ab—Albite, $Na(AlSi_3O_8)$; An—Anorthite, $Ca(Al_2Si_2O_8)$; Ank—Ankerite, $Ca(Fe^{2+},Mg)(CO_3)_2$; Bct—Brochantite, $Cu_4(SO_4)(OH)_6$; Bir—Birnessite, $(Na,Ca)_{0.5}(Mn^{4+},Mn^{3+})_2O_4{\cdot}1.5H_2O$; Cal—Calcite, $CaCO_3$; Ccp—Chalcopyrite, $CuFeS_2$; Cer—Cerussite, $PbCO_3$; Chm—Chamosite, $(Fe^{2+})_5Al(Si,Al)_4O_{10}(OH,O)_8$; Cpr—Cuprite, Cu_2O; Dol—Dolomite, $CaMg(CO_3)_2$; Fa—Fayalite, $Fe^{2+}{}_2SiO_4$; Gp—Gypsum, $CaSO_4{\cdot}2H_2O$; Gth—Goethite, α-$Fe^{3+}O(OH)$; Hem—Hematite, Fe_2O_3; Kln—Kaolinite, $Al_2(Si_2O_5)(OH)_4$; Mag—Magnetite, $Fe^{2+}Fe^{3+}{}_2O_4$; Mlc—Malachite, $Cu_2(CO_3)(OH)_2$; Ms/Bt—Muscovite/Biotite, $Al_2(AlSi_3O_{10})(OH)_2/K(Mg,Fe)_3AlSi_3O_{10}(OH)_2$; Mul—Mullite, $Al_{4+2x}Si_{2-2x}O_{10-x}$ (x~0.4); Qz—Quartz, SiO_2; Spl—Spinel, $MgAl_2O_4$; Tns—Ternesite, $Ca_5(SiO_4)_2(SO_4)$; Tnt-Fe—Tennantite-(Fe), $Cu_6(Cu_4Fe^{2+}{}_2)As_4S_{12}S$; Wwf—Wroewolfeite, $Cu_4(SO_4)(OH)_6{\cdot}2H_2O$. Vtg: vestigial content; ?: dubious identification.

Sample Reference	Phase Identification	+++	++	+
APAR/GSEU/001	Cal + Dol + Gth + Mlc + Qz	Quartz	Calcite	Goethite, Malachite, Dolomite
APAR/GSEU/002	Bir (vtg) + Ccp + Chm (vtg) + Dol + Ms/Bt (vtg) + Qz	Quartz, Dolomite		Chalcopyrite
APAR/GSEU/003	Bct (vtg) + Ccp + Chm (vtg) + Dol + Mlc (vtg) + Ms/Bt (vtg) + Qz	Quartz	Dolomite	Chalcopyrite
APAR/GSEU/004	Ank (vtg) + Cal (vtg) + Cer (vtg) + Chm + Cpr (vtg) + Dol + Ms/Bt + Qz	Quartz	Dolomite	Chamosite, Muscovite/Biotite
APAR/GSEU/005	Chm + Dol + Ms/Bt + Qz	Quartz	Dolomite	Chamosite, Muscovite/Biotite
APAR/GSEU/006	Chm + Dol + Ms/Bt + Qz	Quartz	Dolomite	Chamosite, Muscovite/Biotite

Table 3. *Cont.*

Sample Reference	Phase Identification	+++	++	+
APAR/GSEU/007	Chm + Dol + Ms/Bt + Qz	Quartz		Dolomite, Chamosite, Muscovite/Biotite
BRAN/GSEU/001	Ab + An + Fa + Mag + Mul + Qz + Spl(1) + Spl(2) + Tns	Spinel(1), Anorthite, Albite	Magnetite, Spinel(2), Fayalite	Quartz,Mullite, Ternesite
BRAN/GSEU/002	Ank + Dol + Gp + Kln + Mlc + Ms/Bt + Qz	Quartz	Dolomite, Gypsum, Ankerite	Muscovite/Biotite, Malachite, Kaolinite
BRAN/GSEU/003	Ab (vtg) + Hem (vtg) + Mag + Mul + Qz	Quartz		Mullite, Magnetite
Brancanes Ore	Ccp + Dol + Gp (vtg) + Qz + Tnt? (vtg) + Wwf (vtg)	Dolomite		Chalcopyrite, Quartz

Table 4. Semi-quantitative approach to the mineralogical content of Ferragudo (FERR), Ferrarias e Castelo (CAST), Cêrro do Serpe (SERP), and Balança (BALA) samples. The main phases in each sample were roughly estimated as more (+++) or less (+) represented by the intensity of the principal lines. "+" signs indicate relative abundance of mineral. Legend: Ab—Albite, $Na(AlSi_3O_8)$; Alm—Almandine, $Fe^{2+}{}_3Al_2(SiO_4)_3$; Ank—Ankerite, $Ca(Fe^{2+},Mg)(CO_3)_2$; Bir—Birnessite, $(Na,Ca)_{0.5}(Mn^{4+},Mn^{3+})_2O_4\cdot1.5H_2O$; Bnt—Braunite, $Mn^{2+}Mn^{3+}{}_6(SiO_4)O_8$; Bob—Bobierrite, $Mg_3(PO_4)_2\cdot8H_2O$; Cbz-Ca—Chabazite-Ca, $(Ca,K_2,Na_2)_2[Al_2Si_4O_{12}]_2\cdot12H_2O$; Chm—Chamosite, $(Fe^{2+})_5Al(Si,Al)_4O_{10}(OH,O)_8$; Clb-Fe—Columbite-(Fe), $Fe^{2+}Nb_2O_6$; Cpr-Cuprite, Cu_2O; Dpt—Dioptase, $CuSiO_3\cdot H_2O$; Fau-Na—Faujasite-Na, $(Na_2,Ca,Mg)_{3.5}[Al_7Si_{17}O_{48}]\cdot32H_2O$; Gh—Gehlenite, $Ca_2Al[AlSiO_7]$; Gnp—Ganophyllite, $(K,Na,Ca)_2Mn_8(Si,Al)_{12}(O,OH)_{32}\cdot8H_2O$; Gp—Gypsum, $CaSO_4\cdot2H_2O$; Mnn—Manganite, $Mn^{3+}O(OH)$; Mog—Mogánite, SiO_2; Hem—Hematite, Fe_2O_3; Hsm—Hausmannite, $Mn^{2+}Mn^{3+}{}_2O_4$; Mnt—Montmorillonite, $(Na,Ca)_{0.33}(Al,Mg)_2(Si_4O_{10})(OH)_2\cdot nH_2O$; Ms/Bt—Muscovite/Biotite, $Al_2(AlSi_3O_{10})(OH)_2/K(Mg,Fe)_3AlSi_3O_{10}(OH)_2$; Ncr—Nacrite, $Al_2(Si_2O_5)(OH)_4$; Phl—Phlogopite, $KMg_3(AlSi_3O_{10})(OH)_2$; Pyc—Pyrochroite, $Mn(OH)_2$; Pyl—Pyrolusite, $Mn^{4+}O_2$; Qz—Quartz, SiO_2; Rds—Rhodochrosite, $MnCO_3$; Rt—Rutile, TiO_2; Sd—Siderite, $FeCO_3$; Sid—Siderophyllite, $KFe^{2+}{}_2Al(Al_2Si_2O_{10})(OH)_2$; Sps—Spessartine, $Mn^{2+}{}_3Al_2(SiO_4)_3$, Whm—Whitmoreite, $Fe^{2+}Fe^{3+}{}_2(PO_4)_2(OH)_2\cdot4H_2O$. Vtg: vestigial content; ?: dubious identification.

Sample Reference	Phase Identification	+++	++	+
FERR/GSEU/001	Ab + Alm calcian + Cbz-Ca (vtg) + Chm + Dpt + Gnp (vtg) + Gp (vtg) + Hsm + Mnt (vtg) + Phl (vtg) + Qz + Rds	Rhodochrosite, Quartz	Albite	Dioptase, Hausmannite, Almandine calcian, Chamosite
FERR/GSEU/002	Ab (vtg) + Alm calcian + Bir (vtg) + Fau-Na (vtg) + Ms/Bt (vtg) + Pyl + Qz + Sps	Quartz		Pyrolusite, Almandine calcian, Spessartine
FERR/GSEU/003	Ab (vtg) + Bir (vtg) + Chm (vtg) + Mnt (vtg) + Ms/Bt + Qz + Rt (vtg)	Quartz		Muscovite/Biotite
FERR/GSEU/004	Ab (vtg) + Alm calcian + Bir + Fau-Na + Mnt (vtg) + Ms/Bt (vtg) + Qz + Sps	Quartz	Faujasite-Na, Almandine calcian, Spessartine	Birnessite

Table 4. *Cont.*

Sample Reference	Phase Identification	+++	++	+
CAST/GSEU/001	Ab (vtg) + Bir (vtg) + Chm (vtg) + Mnn + Qz + Rds + Sid (vtg)	Rhodochrosite		Quartz, Manganite
CAST/GSEU/002	Ab (vtg) + Bir (vtg) + Chm + Clb-Fe? (vtg) + Cpr (vtg) + Hem (vtg) + Ms/Bt (vtg) + Py (vtg) + Pyc (vtg) + Qz + Rds + Rt (vtg) + Sd	Quartz	Chamosite	Rhodochrosite, Siderite
CAST/GSEU/003	Ab (vtg) + Ank (vtg) + Bir (vtg) + Chm + Gh? (vtg) + Hem (vtg) + Qz + Rds + Sd + Whm? (vtg)	Chamosite	Quartz	Rhodochrosite, Siderite
CAST/GSEU/004	Ab (vtg) + Bir (vtg) + Chm + Ms/Bt + Qz + Rt (vtg)	Quartz	Chamosite	Muscovite/Biotite
SERP/GSEU/001 Ore	Ab + Bob? (vtg) + Mog (vtg) + Pyl + Qz	Quartz		Albite, Pyrolusite
BALA/GSEU/001	Ab + Bnt + Hem + Ncr? (vtg) + Pyl + Qz (vtg) + Rds	Rhodochrosite, Braunite	Pyrolusite	Hematite, Albite

Quartz and dolomite [$CaMg(CO_3)_2$] are present in all samples from Aparis (Table 3). Several Cu phases were also identified, and among those were malachite (APAR/GSEU/001) and chalcopyrite (samples 2 and 3). Similar critical and strategic raw materials were found (As, Cu, and Mn). Samples 5–7 are mineralogically identical and do not have significant CRMs or SRMs. Although sample 4 is also identical, the presence of Li, Mg, and REE (Ce, Nd, and Sm) should be highlighted.

Two slag samples from Brancanes (1 and 3) contain minerals such as fayalite, mullite, and several spinels (Table 3). High contents of Al, Ba, Co, Fe, Mn, Ni, and Si plus Sr were found in sample 1, while Al, As, Ca, Co, Cs, Cu, K, Mg, Mn, Ni, Rb, Sb, Si, Th, and REE (La, Ce, Pr, Nd, Sm, Eu, and Gd) were found in sample 2: CRMs (As, Co, Mg, Mn, Sb, Sr, and REE), SRMs (Co, Cu, Mg, Mn, Ni, and REE). Sample 2 is rich in Al, Ba, Fe, and Mn plus Sr and Brancanes ore is rich in As, Ca, Cu, In, Mg, Mn, Ni, S, and Sb plus REE (Gd, Dy). This last sample has the highest Ni content of all wastes studied (750 ppm).

The mineralogy of the Lousal samples studied previously [33] mainly showed the presence of pyrite, quartz, mica (muscovite or biotite), feldspar (albite), chlorite (chamosite), and numerous neogenic sulphates with variable degrees of hydration, of which gypsum ($CaSO_4{\cdot}2H_2O$), rhomboclase [$(H_5O_2)Fe^{3+}(SO_4)_2{\cdot}2H_2O$], ferricopiapite [$Fe^{3+}_{0.67}Fe^{3+}_4(SO_4)_6(OH)_2{\cdot}20H_2O$], coquimbite [$AlFe_3(SO_4)_6(H_2O)_{12}{\cdot}6H_2O$], and jarosite [$KFe_3{}^{+3}(SO_4)_2(OH)_6$] stood out. New chemical analyses of the two samples were performed in ACTLABS for comparison purposes (Table S2). For instance, interesting values of Au (sample 1) and Ag (samples 1 and 14) were obtained. A higher content of Sb and Pb is present in samples 1, 9, 10, and 14, while As is concentrated in samples 10, 11, 12, and 14. Sample 4 is rich in several CRMs and SRMs, Hf, Ga, Nb, Sc, V, and REE (La, Ce, Pr, Nd, Sm, Eu, Gd, Dy, Tb, and Er plus Yb), sample 1 is rich in Co and Ce, samples 3 and 8 are rich in Mn, and sample 14 is rich in Sr.

In the Mn mines, several phases with Mn were identified in the Ferragudo samples (Table 4), namely hausmannite ($Mn^{2+}Mn^{3+}{}_2O_4$), pyrolusite ($Mn^{4+}O_2$), birnessite

$[(Na,Ca)_{0.5}(Mn^{4+},Mn^{3+})_2O_4{\cdot}1.5H_2O]$, ganophyllite $[(K,Na,Ca)_2Mn_8(Si,Al)_{12}(O,OH)_{32}{\cdot}8H_2O]$, rhodochrosite ($MnCO_3$), and spessartine $[Mn^{2+}{}_3Al_2(SiO_4)_3]$ with Mn in various oxidation states. This critical and raw material is concentrated in all samples (Table S2). Sample 1 is rich in Ba, Be, Ce, Co, Cs, Mo, V, and W, while sample 2 has a high content of Ba, Be, Ce, Co, Cs, Mo, Nb, Rb, Si, Sr, Th, U, V, W, Zn, and REE (Y, La, Nd, Eu, Gd, Dy, Tb, Er, and Yb). These two samples have the highest Mo content of all wastes analyzed. Sample 3 is less rich with good values only for Al, K, and Rb plus Si, while sample 4 has high levels of Co, V, and W. The presence of spessartine is an indicator of temperature conditions above those regularly quoted for massive primary Mn(-Fe) oxide ores in the IPB (>350 °C) [40].

The mineralogy of the Ferrarias and Castelo samples (Table 4) revealed two new Mn-phases: manganite $[Mn^{3+}O(OH)]$ and pyrochroite $[Mn(OH)_2]$. Mn is concentrated in all samples. Sample 1 is rich in Ba, Fe, and Sr, while sample 2 has high concentrations of Ba, Co, Cr, Fe, Nb, V, and REE (Nd). Sample 3 displays high contents of Ba and Cr plus Sr; however, sample 4 is rich in Cr, Nb, and Si.

The sample ore from Cêrro do Serpe displays quartz, albite, and pyrolusite as the main phases (Table 4) and high contents of Ba, Be, Co, Ga, Mn, Si, and Sr plus W. It also shows remarkable concentrations of Ba (1.5%) and Ga (129 ppm), with the latter being both a CRM and SRM.

Several Mn phases are present in the Balança sample (Table 4): rhodochrosite, braunite $[Mn^{2+}Mn^{3+}{}_6(SiO_4)O_8]$, and pyrolusite. The chemical elements with high contents are Ba, Be, Mn, Sr, and Tl (with Mn content reaching about 48%). Although not considered critical or strategic, Tl is very well represented in this waste sample (94 ppm).

Table 5 depicts a summary of where to search for CRM and/or SRM on the samples analyzed, according to the chemical contents obtained. The waste volume was not considered in this first approach. Heavy rare earth elements (HREEs) include Dy, Er, Eu, Gd, Ho, Lu, Tb, Tm, Yb, and Y, while light rare earth elements represent Ce, La, Nd, Pr, and Sm [7].

Table 5. Distribution of relevant CRM and/or SRM contents in the samples analyzed according to the 2023 list published by the EU Commission [7].

Sample Reference	Sb	As	Bi	Co	Cu	Ga	Ge	Hf	HREE	Li	LREE	Mg	Mn	Ni	Nb	Sc	Sr	W	V
PORT/GSEU/001					X														
PORT/GSEU/001 Ore	X	X		X	X					X			X	X					
CMOUR/GSEU/001 Ore																		X	
BARR/GSEU/001	X	X			X							X							
BARR/GSEU/002																	X		
BARR/GSEU/003		X	X		X	X	X		X	X	X								
ALG/GSEU/001		X		X															
ALG/GSEU/002		X			X														
ALG/GSEU/003		X																	
ALG/GSEU/004															X				
ALG/GSEU/004 Ore		X				X		X	X		X					X			
Algaré Ore																			
APAR/GSEU/001					X														
APAR/GSEU/002					X														
APAR/GSEU/003					X														
APAR/GSEU/004										X	X	X							
APAR/GSEU/005																			
APAR/GSEU/007																			
BRAN/GSEU/001																	X		

Table 5. *Cont.*

Sample Reference	Sb	As	Bi	Co	Cu	Ga	Ge	Hf	HREE	Li	LREE	Mg	Mn	Ni	Nb	Sc	Sr	W	V
BRAN/GSEU/002	X	X		X	X							X	X	X					
BRAN/GSEU/003									X		X						X		
Brancanes Ore	X	X			X							X	X	X					
LOUS/GSEU/001	X			X					X										
LOUS/GSEU/002											X								
LOUS/GSEU/003																			
LOUS/GSEU/004						X		X	X		X				X	X			X
LOUS/GSEU/006																			
LOUS/GSEU/007																			
LOUS/GSEU/008																			
LOUS/GSEU/009																			
LOUS/GSEU/010																			
LOUS/GSEU/011																			
LOUS/GSEU/012																			
LOUS/GSEU/013																			
LOUS/GSEU/014																	X		
FERR/GSEU/001				X							X		X					X	X
FERR/GSEU/002				X					X		X		X		X		X	X	X
FERR/GSEU/003													X						
FERR/GSEU/004													X						
CAST/GSEU/001													X				X		
CAST/GSEU/002				X							X		X		X				X
CAST/GSEU/003													X				X		
CAST/GSEU/004												X	X						
SERP/GSEU/001 Ore				X		X							X				X	X	
BALA/GSEU/001PORT1													X				X		

4. Conclusions

The performance of the XRF portable equipment in mineralogically complex mine waste samples was evaluated. No significant differences in the element concentrations were observed between performing one or an average of three measures collected in different points/areas of the same sample, which suggests good precision of the equipment. The accuracy of the pXRF equipment, tested through a comparison with element concentrations obtained for selected samples in a certified international laboratory (ACTLABS) and with XRF laboratory equipment (WDS), showed good correlations for Al, As, Ba, Ca, Cu, Fe, Ga, K, Mg, Mn, Mo, Ni, Pb, Rb, S, Sb, Se, Si, Sr, Ti, Zn, and W. Therefore, the use of pXRF equipment has proven to be useful for a quick and inexpensive overview (both in the field and in the laboratory) of the many elements that a waste sample can contain.

The high concentrations of Fe (47.4%), Cu (27.5%), As (8.4%), Pb (3%), and Sn (139 ppm) found in wastes from old Cu mines are consistent with the various minerals identified, namely, chalcopyrite [$CuFeS_2$], pyrite [FeS_2], tennantite-(Fe) [$Cu_6(Cu_4Fe^{2+}{}_2)As_4S_{12}S$], cuprite [$Cu_2O$], malachite [$Cu_2(CO_3)(OH)_2$], pseudomalachite [$Cu_5(PO_4)_2(OH)_4$], brochantite [$Cu_4(SO_4)(OH)_6$], wroewolfeite [$Cu_4(SO_4)(OH)_6{\cdot}2H_2O$], jarosite [$KFe_3{}^{+3}(SO_4)_2(OH)_6$], butlerite [$Fe^{3+}(SO_4)(OH){\cdot}2H_2O$], rhomboclase [$(H_5O_2)Fe^{3+}(SO_4)_2{\cdot}2H_2O$], ferricopiapite [$Fe^{3+}{}_{0.67}Fe^{3+}{}_4(SO_4)_6(OH)_2{\cdot}20H_2O$], coquimbite [$AlFe_3(SO_4)_6(H_2O)_{12}{\cdot}6H_2O$], ankerite [$Ca(Fe^{2+},Mg)(CO_3)_2$], fayalite [$Fe^{2+}{}_2SiO_4$], chamosite [$(Fe^{2+})_5Al(Si,Al)_4O_{10}(OH,O)_8$], chloritoid [$(Fe^{2+},Mg,Mn^{2+})Al_2(SiO_4)O(OH)_2$], hematite [$Fe_2O_3$], magnetite [$Fe^{2+}Fe^{3+}{}_2O_4$], goethite [$\alpha\text{-}Fe^{3+}O(OH)$], ilmenite [$Fe^{2+}TiO_3$], titanomaghemite [$(Ti^{4+}{}_{0.5}\square_{0.5})Fe^{3+}{}_2O_4$], scorodite [$Fe^{3+}AsO_4{\cdot}2H_2O$], cerussite [$PbCO_3$], and cassiterite [$SnO_2$]. Minor and vestigial contents of Mg (6.6%), Ni (750 ppm), Sb (665 ppm), Zr (271 ppm), Bi (217 ppm), Co (209 ppm), Rb (132 ppm), Se (132 ppm), Li (83 ppm), Ag (72 ppm), Sc (19 ppm), Th (14.9 ppm), Hf (9.6 ppm), Ge (9.6 ppm), In (6 ppm), Au (1110 ppb), and REE (La—47.6 ppm,

Ce—94.7, Pr—11.3, Nd—42.1, Sm—7.4, Eu—1.28, Gd—5.1, Dy—4.1, Tb—0.7, Ho—1.2, Er—4.3, Tm—0.8, Yb—6, and Lu—1 ppm) were also found.

Conversely, the Mn minerals identified in mine waste dumps from the studied Mn mines were hausmannite [$Mn^{2+}Mn^{3+}{}_2O_4$], pyrolusite [$Mn^{4+}O_2$], birnessite [$(Na,Ca)_{0.5}(Mn^{4+},Mn^{3+})_2O_4 \cdot 1.5H_2O$], ganophyllite [$(K,Na,Ca)_2Mn_8(Si,Al)_{12}(O,OH)_{32} \cdot 8H_2O$], manganite [$Mn^{3+}O(OH)$], pyrochroite [$Mn(OH)_2$], rhodochrosite [$MnCO_3$], spessartine [$Mn^{2+}{}_3Al_2(SiO_4)_3$], and braunite [$Mn^{2+}Mn^{3+}{}_6(SiO_4)O_8$]. The Mn concentration reached 38.6% in waste materials, where Ba (1.5%), Sr (0.4%), W (790 ppm), V (170 ppm), Mo (205 ppm), Ga (129 ppm), Cr (111 ppm), Tl (94.4 ppm), and Be (7 ppm) were also observed.

Studying mine waste is a crucial step toward sustainable resource management and environmental stewardship, ensuring economic and ecological resilience with obvious environmental and economic gains, ensuring global competitiveness, and fostering the circular economy. Furthermore, the study of mine waste is critical in addressing resource supply challenges and mitigating environmental impacts (e.g., acid mine drainage, AMD).

Although we are lacking total volume calculation for these mine wastes, the study reveals the importance of knowing the chemical and mineralogical potential of mining wastes of abandoned mines in order to potentially increase CRM and SRM sources, use them in the waste-to-value recycling process, and reduce potential environmental impacts.

Supplementary Materials: The following supporting information can be downloaded at: https://www.mdpi.com/article/10.3390/min15010058/s1, Table S1: Chemical characterization of waste materials: comparison of results obtained through ACTLABS, pXRF, and XRF laboratory equipment; Table S2: Chemical analyses performed on wastes from copper and manganese abandoned mines.

Author Contributions: T.P.S.; methodology, D.P.S.d.O., T.P.S. and I.M.; software, T.P.S.; validation, D.P.S.d.O., T.P.S. and I.M.; investigation, D.P.S.d.O., T.P.S., I.M. and J.A.E.F.; data curation, T.P.S. and J.A.E.F.; writing—original draft preparation, D.P.S.d.O., T.P.S. and I.M.; writing—review and editing, D.P.S.d.O., T.P.S., I.M. and J.A.E.F.; visualization, D.P.S.d.O., T.P.S. and I.M.; supervision, D.P.S.d.O., T.P.S., and I.M. All authors have read and agreed to the published version of the manuscript.

Funding: This research was partly funded by the Geological Service for Europe project (GSEU), Grant Agreement number 101075609—GSEU—HORIZON-CL5-2021-D3-02, and through the funds of LNEG.

Data Availability Statement: All new research data acquired and used in this study are published herein.

Acknowledgments: The authors would like to thank the revision and comments from four anonymous reviewers who have improved the manuscript.

Conflicts of Interest: The authors declare no conflicts of interest.

References

1. Council of the European Communities. Resolution of the Council of the European Communities and of the Representatives of the Governments of the Member States Meeting Within the Council of May 1977 on the continuation and implementation of a European Community policy and action programme on the environment. *Off. J. Eur. Commun.* **1977**, *139*, 1–46. Available online: https://eur-lex.europa.eu/legal-content/EN/TXT/?uri=CELEX:41977X0613 (accessed on 5 December 2024).
2. European Commission. The Raw Materials Initiative—Meeting Our Critical Needs for Growth and Jobs in Europe, Communication from the Commission to the European Parliament and the Council, COM (2008) 699 Final. Available online: https://eur-lex.europa.eu/legal-content/EN/ALL/?uri=CELEX:52008DC0699 (accessed on 21 October 2024).
3. European Commission. A Secure and Sustainable Supply of Critical Raw Materials in Support of the Twin Transition, Communication from the Commission to the European Parliament, the Council, the European Economic and Social Committee and the Committee of the Regions, COM (2023) 165 Final. Available online: https://eur-lex.europa.eu/legal-content/EN/TXT/?uri=COM:2023:165:FIN (accessed on 21 October 2024).

4. Rizos, V.; Righetti, E. Low-carbon technologies and Russian imports: How far can recycling reduce the EU's raw materia dependency? *CEPS Policy Insight* **2022**, *2022–17*, 36180, Centre for European Policy Studies..
5. Righetti, E.; Rizos, V. The EU's Quest for Strategic Raw Materials: What Role for Mining and Recycling? *Intereconomics* **2023**, *5*ε 69–73. Available online: https://www.intereconomics.eu/pdf-download/year/2023/number/2/article/the-eu-s-quest-fo strategic-raw-materials-what-role-for-mining-and-recycling.html (accessed on 4 December 2024). [CrossRef]
6. Hool, A.; Helbig, C.; Wierink, G. Challenges and opportunities of the European Critical Raw Materials Act. *Miner. Econ.* **2024**, *3*, 661–668. [CrossRef]
7. Grohol, M.; Veeh, C.; European Commission. Study on the Critical Raw Materials for the EU: Final Report, Publications Office o the European Union. 2023. Available online: https://data.europa.eu/doi/10.2873/725585 (accessed on 10 September 2024).
8. European Commission. Proposal for a Regulation of the European Parliament and of the Council Establishing a Framework fo Ensuring a Secure and Sustainable Supply of Critical Raw Materials and Amending Regulations (EU) 168/2013, (EU) 2018/85 2018/1724 and (EU) 2019/1020, COM/2023/160 Final. 2023. Available online: https://eur-lex.europa.eu/legal-content/EN TXT/?uri=CELEX:52023PC0160 (accessed on 4 October 2024).
9. Carvalho, J.; Diamantino, C.; Rosa, C.; Carvalho, E. Potential recovery of mineral resources from mining tailings of abandone mines in Portugal. In Proceedings of the 3rd International Symposium on Enhanced Landfill Mining, Lisbon, Portugal, 8–1 February 2016; Pereira, M.J., Carvalho, M.T., Neves, P.F., Eds.; Instituto Superior Técnico: Lisbon, Portugal, 2016; pp. 501–51 ISBN 978-989-98342-4-8.
10. Hu, X.; Yang, H.; Wu, F.; Fang, X.; Tan, K. Recovery of copper-dominated resources from copper mine drainage by chemica oxidation and sulfur biocycling: A pilot-scale study. *J. Clean. Prod.* **2022**, *378*, 134525. [CrossRef]
11. Sánchez-Andrea, I.; Stams, A.J.M.; Weijma, J.; Contreras, P.G.; Dijkman, H.; Rozendal, R.A.; Johnson, D.B. A case in suppor of implementing innovative bio-processes in the metal mining industry. *FEMS Microbiol. Lett.* **2016**, *363*, fnw106. [CrossRef [PubMed]
12. Mohanty, S.; Ghosh, S.; Bal, B.; Das, A.P. A review of biotechnology processes applied for manganese recovery from wastes. *Rev Environ. Sci. Biotechnol.* **2018**, *17*, 791–811. [CrossRef]
13. Euro Manganese Inc. Chvaletice Manganese Project. 2024. Available online: https://www.mn25.ca/chvaletice-manganese project (accessed on 23 September 2024).
14. Neves, F.; Esperto, L.; Figueira, I.; Mascarenhas, J.; Salgueiro, R.; Silva, T.P.; Correia, J.B.; Carvalho, P.A.; de Oliveira, D Mechanochemical synthesis of tetrahedrite materials using mixtures of synthetic and ore samples collected in the Portugues zone of the Iberian Pyrite Belt. *Miner. Eng.* **2021**, *164*, 106833. [CrossRef]
15. de Oliveira, D.P.S.; Filipe, A.; Gonçalves, P.; Santos, S.; Albardeiro, L. Critical Raw Materials Deposits Map of Mainland Portugal New Mineral Intelligence in Cartographic Form. *Cartog. J.* **2021**, *58*, 222–232. [CrossRef]
16. Tornos, F.; Inverno, C.M.C.; Casquet, C.; Mateus, A.; Ortiz, G.; Oliveira, O. The metallogenic evolution of the Ossa-Morena Zone *J. Iber. Geol.* **2004**, *30*, 143–181.
17. Sáez, R.; González, F.; Donaire, T.; Toscano, M.; Yesares, L.; de Almodóvar, G.R.; Moreno, C. Updating Geological Information about the Metallogenesis of the Iberian Pyrite Belt. *Minerals* **2024**, *14*, 860. [CrossRef]
18. Díez-Montes, A.; Matos, J.X.; Dias, R.; Carmona, J.J.H.; Albardeiro, L.; Oliveira, J.T.; Morais, I.; Fernandes, P.; Inverno, C. Machado, S.; et al. Geological Map of the South Portuguese Zone, Mapa Geológico de la Zona Surportuguesa/Carta Geológica da Zona Sul Portuguesa, Escala 1/400 000. Proj. Geo-FPI/Interreg POCTEP; Instituto Geológico y Minero de España/LNEG/Junta de Andalucía-SGIEM/CM Aljustrel. 2020. Available online: https://info.igme.es/geofpi/docs/mapas/GEOLOGICO_400K_ ZSP_2020.pdf (accessed on 10 December 2024).
19. Piçarra, J.M. Estudo Estratigráfico do Sector de Estremoz—Barrancos, Zona de Ossa Morena, Portugal. Ph.D. Thesis, University of Évora, Évora, Portugal, 2000; Volume I and II, 268p.
20. Gaspar, O.C. O Jazigo de Cobre de Aparis. *Est. Not. Trab. SFM* **1968**, *18*, 253–290.
21. Matos, J.X.; Rosa, C. Diagnóstico Preliminar de Minas Abandonadas—Área Sul. In *Internal IGM Report*; IGM Eds.: Lisbon, Portugal, 2001; 276p.
22. Matos, J.X.; Martins, L.; Rosa, C. Parque Mineiro da Cova dos Mouros—IGM: Contribution for the sustainable development of the mining park. *IGME Pub. Mus. Geom.* **2003**, *2*, 487–494.
23. Reiser, F.K.M.; Rosa, D.R.N.; Pinto, Á.M.M.; Carvalho, J.R.S.; Matos, J.X.; Guimarães, F.M.G.; Alves, L.C.; de Oliveira, D.P.S Mineralogy and geochemistry of tin- and germanium-bearing copper ore, Barrigão re-mobilized vein deposit, Iberian Pyrite Belt, Portugal. *Int. Geol. Rev.* **2011**, *53*, 1212–1238. [CrossRef]
24. Matos, J.X. Recursos geológicos—Minérios metálicos. In *Notícia Explicativa da Folha 46D Mértola*; Oliveira, J.T., Silva, J.B., Eds.; Dep. Geologia INETI: Lisbon, Portugal, 2007; pp. 28–34.
25. Sáez, R.; Pascual, E.; Toscano, M.; Almodóvar, G. The Iberian type of volcano-sedimentary massive sulphide deposits. *Mineral. Depos.* **1999**, *34*, 549–570. [CrossRef]

26. Matos, J.X.; Martins, L.P.; Oliveira, J.T.; Pereira, Z.; Batista, M.J.; Quental, L. Rota da pirite no sector português da Faixa Piritosa Ibérica, desafios para um desenvolvimento sustentado do turismo geológico e mineiro. In *Rutas Minerales en Iberoamérica*; Carrion, P., Ed.; Esc. Sup. Politécnica del Litoral: Guayaquil, Equador, 2008; pp. 136–155.
27. Oliveira, J.T.; Relvas, J.; Pereira, Z.; Matos, J.X.; Rosa, C.; Rosa, D.; Munhá, J.M.; Fernandes, P.; Jorge, R.; Pinto, A. Geologia da Zona Sul Portuguesa, com ênfase na estratigrafia e na vulcanologia física, geoquímica e mineralizações da Faixa Piritosa. In *Geologia de Portugal Vol. I—Geologia Pré-Mesozóica de Portugal*; Dias, R., Araújo, A., Terrinha, P., Kullberg, J., Eds.; Escolar Editora: Lisbon, Portugal, 2013; pp. 673–767.
28. Pinedo Vara, I. *Piritas de Huelva. Su Historia, Minería y Aprovechamiento*; Edit. Summa: Madrid, Spain, 1963; 1003p.
29. Fernandes, C. Jazigos de Mn do Alentejo. Breve estudo das minas de Mn do concelho de Castro Verde. In *Internal IGM Report*; IGM Eds.: Lisbon, Portugal, 1947.
30. Goinhas, F. *Minas de Manganês do Baixo Alentejo. Internal Report from DGGM*; DGGM Eds.: Lisbon, Portugal, 1986.
31. Silva, F. Géologie et génese des gisements de Manganese du Baixo Alentejo Portugal. *Est. Not. Trab. SFM* **1956**, *XI*, 28–66.
32. Lohmeier, S.; Gallhofer, D.; Lottermoser, B.G. Field-portable X-ray fluorescence analyzer for chemical characterization of carbonate-bearing base metal tailings: Case study from Namib Pb-Zn Mine, Namibia. *J. South. Afr. Inst. Min. Metall.* **2024**, *124*, 421–436. [CrossRef]
33. Figueiredo, M.O.; Silva, T.P.; Veiga, J.P.; de Oliveira, D.; Batista, M.J. Towards the recovery of by-product metals from mine wastes: An X-ray absorption spectroscopy study on the binding state of rhenium in debris from a centennial Iberian Pyrite Belt mine. *J. Miner. Mater. Charact. Eng.* **2014**, *2*, 135–143. [CrossRef]
34. de Oliveira, D.; Gonçalves, P.; Morais, I.; Silva, T.P.; Matos, J.X.; Albardeiro, L.; Filipe, A.; Batista, M.J.; Santos, S.; Fernandes, J. Unlocking the secondary critical raw material potential of historical mine sites, Lousal Mine, southern Portugal. *Minerals* **2024**, *14*, 127. [CrossRef]
35. Yaroshevsky, A.A. Abundances of Chemical Elements in the Earth's Crust. *Geochem. Int.* **2006**, *44*, 48–55. [CrossRef]
36. Abreu, M.M.; Matias, M.J.; Magalhães, M.C.F.; Basto, M.J. Impacts on water, soil and plants from the abandoned Miguel Vacas copper mine, Portugal. *J. Geochem. Explor.* **2008**, *96*, 161–170. [CrossRef]
37. De Oliveira, D.; Salgueiro, R.; Silva, T.P.; Reiser, F.; Guimarães, F.; Neves, F. The Barrigão copper deposit: Tennantite-tetrahedrite for thermoelectric and high-technology applications. In *Extended Abstracts Boock of the XII Congresso Ibérico de Geoquímica and XX Semana da Geoquímica, Évora, Portugal, 22–26 September 2019*; Nogueira, P., Moreira, N., Roseiro, J., Maia, M., Eds.; University of Évora: Évora, Portugal, 2019; pp. 255–258. ISBN 978-972-778-121-8.
38. Tabelin, C.B.; Corpuz, R.D.; Igarashi, T.; Villacorte-Tabelin, M.; Ito, M.; Hiroyoshi, N. Hematite-catalysed scorodite formation as a novel arsenic immobilisation strategy under ambient conditions. *Chemosphere* **2019**, *233*, 946–953. [CrossRef] [PubMed]
39. Warr, L.N. IMA–CNMNC approved mineral symbols. *Mineral. Mag.* **2021**, *85*, 291–320. [CrossRef]
40. Moreira, B.; Figueiras, J.; Mateus, A.; Rodrigues, P.; Jorge, R.; Gonçalves, L. A new manganese mineralisation type in the Iberian Pyrite Belt? In *Abstract Book of the X Congresso Ibérico de Geoquímica—XVIII Semana de Geoquímica, Lisbon, Portugal, 19–23 October 2015*; LNEG, Eds.: Lisbon, Portugal, 2015; pp. 133–136, ISBN 978-989-675-039-8. Available online: https://www.researchgate.net/publication/320945506 (accessed on 10 December 2024).

MDPI AG
Grosspeteranlage 5
4052 Basel
Switzerland
Tel.: +41 61 683 77 34

Minerals Editorial Office
E-mail: minerals@mdpi.com
www.mdpi.com/journal/minerals

www.ingramcontent.com/pod-product-compliance
Lightning Source LLC
Chambersburg PA
CBHW041354210926
51850CB00013B/3082

9783725841554